An excellent introductory guide to the complexities of political geography. Combining discussion of events, theories and the discipline's self understanding, it manages to be clear enough for undergraduates while giving their lecturers much to think about.

Professor Stuart Elden, University of Warwick, UK and Monash University, Australia

This richly illustrated text on political geography will be invaluable for both geographers and other social scientists. It usefully demonstrates the complex associations between various forms of power, geography and socio-spatial relations. The authors bring together the traditions of political geography with more contemporary approaches and themes. A major strength of this book compared with many other textbooks is that it provides the reader with strong links to social and political theory. The book is very clearly written and benefits from numerous useful examples and case studies.

Professor Anssi Paasi, University of Oulu, Finland

Political Geography has undergone a renaissance during the past two decades. This important introductory text for students brings them up to date with the key themes of the discipline, linking the global with the local and the territorial with the political. The authors successfully cross the disciplines between Geography, Political Science and International Relations. Students desiring a deeper understanding of the contemporary geopolitical change should be reading this.

Professor David Newman, Ben-Gurion University of the Negev, Israel

This is a lively, well-written and up-to-date introduction to political geography. The authors are all leadings scholars in the field and their accessible approach provides an excellent overview of this rapidly developing subject. The treatment is thorough and thoughtful, and the book is packed with ideas and fascinating case studies.

Professor Joe Painter, University of Durham, UK

AN INTRODUCTION TO POLITICAL GEOGRAPHY: SPACE, PLACE AND POLITICS
Second Edition

An Introduction to Political Geography continues to provide a broad-based introduction to contemporary political geography for students following undergraduate degree courses in geography and related subjects.

The text explores the full breadth of contemporary political geography, covering not only traditional concerns such as the state, geopolitics, electoral geography and nationalism; but also increasing important areas at the cutting-edge of political geography research including globalization, the geographies of regulation and governance, geographies of policy formulation and delivery, and themes at the intersection of political and cultural geography, including the politics of place consumption, landscapes of power, citizenship, identity politics and geographies of mobilization and resistance.

This second edition builds on the strengths of the first. The main changes and enhancements are:

- four new chapters on: political geographies of globalization, geographies of empire, political geography and the environment and geopolitics and critical geopolitics
- significant updating and revision of the existing chapters to discuss key developments, drawing on recent academic contributions and political events
- new case studies, drawing on an increasing number of global examples
- additional boxes for key concepts and an enlarged glossary.

As with the first edition, extensive use is made of case study examples, illustrations, explanatory boxes, guides to further reading and a glossary of key terms to present the material in an easily accessible manner. Through employment of these techniques this book introduces students to contributions from a range of social and political theories in the context of empirical case study examples. By providing a basic introduction to such concepts and pointing to pathways into more specialist material, this book serves, both as a core text for first and second year courses in political geography, and as a resource alongside supplementary textbooks for more specialist third year courses.

Martin Jones, Department of Geography, University of Sheffield, UK.

Rhys Jones, Mark Whitehead, and Michael Woods, all at the Department of Geography and Earth Sciences at Aberystwyth University, UK.

Deborah Dixon, School of Geographical and Earth Sciences, University of Glasgow, UK.

Matthew Hannah, Fakultät II (Bio-Chem-Geo), University of Bayreuth, Bayreuth, Germany.

AN INTRODUCTION TO POLITICAL GEOGRAPHY

Space, Place and Politics

Second Edition

Martin Jones
Department of Geography
The University of Sheffield, Sheffield, UK.

Rhys Jones, Mark Whitehead, Michael Woods
Department of Geography and Earth Sciences,
Aberystwyth University, Aberystwyth, UK.

Deborah Dixon
School of Geographical and Earth Sciences
University of Glasgow, Glasgow, UK.

Matthew Hannah
Fakultät II (Bio-Chem-Geo)
University of Bayreuth, Bayreuth, Germany.

 Routledge
Taylor & Francis Group

LONDON AND NEW YORK

First published 2015
by Routledge
2 Park Square, Milton Park, Abingdon, Oxon OX14 4RN

and by Routledge
711 Third Avenue, New York, NY 10017

Routledge is an imprint of the Taylor & Francis Group, an informa business

First Edition © 2004 Martin Jones, Rhys Jones and Michael Woods

This Edition © 2015 Martin Jones, Rhys Jones, Michael Woods, Mark Whitehead, Deborah Dixon and Matthew Hannah

The right of Martin Jones, Rhys Jones, Michael Woods, Mark Whitehead, Deborah Dixon and Matthew Hannah to be identified as authors of this work has been asserted by them in accordance with sections 77 and 78 of the Copyright, Designs and Patents Act 1988.

British Library Cataloguing in Publication Data
A catalogue record for this book is available from the British Library

Library of Congress Cataloging in Publication Data
Jones, Martin
An introduction to political geography : space, place and politics / Martin Jones, Rhys Jones, Michael Woods, Mark Whitehead, Deborah Dixon, Matthew Hannah.—Second Edition.
pages cm
Includes bibliographical references and index.
JC319.J66 2015
320.1'2—dc23
2014025139

ISBN: 978-0-415-45796-5 (hbk)
ISBN: 978-0-415-45797-2 (pbk)
ISBN: 978-0-203-09216-3 (ebk)

Typeset in Garamond
by Keystroke, Station Road, Codsall, Wolverhampton

Printed and bound by CPI Group (UK) Ltd, Croydon, CR0 4YY

Contents

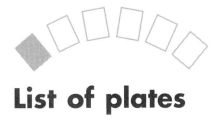

List of plates

List of figures

List of tables

Acknowledgements

Like many projects, this book has had a long gestation period. Between us, it is the result of nearly fifty aggregate years' curiosity with the broad field of political geography. This book is developed from a number of undergraduate and postgraduate courses that we have taught at Aberystwyth University since 1995 and we would like to thank our many students for their perseverance and enthusiasm.

An *Introduction to Political Geography* would not have been completed without the assistance of a number of individuals. We owe a huge debt to Andrew Mould, who commissioned the first edition of this book way back in August 2000. Andrew has been an enthusiastic editor, mixing a number of well-needed on-the-account meals and drinks – progress meetings in disguise – with emails asking 'exactly when are you going to deliver?' The answer is now (at last) for the second edition. We are also grateful to the anonymous reviewers, whose comments have proved invaluable in reworking the manuscript.

We would like to thank (the late) Ian Gulley and Anthony Smith at Aberystwyth University for redrawing some of the figures and maps that appear in the book. The authors and publishers would like to thank the following for granting permission to reproduce images: Elsevier Ltd for Figure 3.1, Ivan Turok for Figure 3.3, and Elsevier Ltd for Table 3.2. Every effort has been made to contact copyright holders for their permission to reprint material in this book. The publishers would be grateful to hear from any copyright holder who is not acknowledged here and will undertake to rectify any errors or omissions in future editions of this book.

We would also like to acknowledge our various teachers, tutors, supervisors, mentors and colleagues for putting us on the right path over the years and thanks also go to our friends and family for support during this project.

Aberystwyth, October 2013

Introduction: power, space and 'political geography'

A journey into political geography

Westminster Bridge in London is an interesting vantage point from which to start our journey into political geography. It will be a location familiar to many readers, but you can also find it by putting the coordinates 51° 30' 6" N 0° 7' 24" W into Google Earth. To the south-west is the Palace of Westminster, the seat of the British parliament and a symbol of the electoral politics that many people will first think of when 'political geography' is mentioned. Indeed, Westminster Bridge crosses the boundary between two parliamentary constituencies, reminding us that parliamentarians tend to be elected to represent particular geographical territories, and that the way in which these territories are drawn up and the geographical spread of votes across them can shape the outcome of elections. Across the River Thames is St Thomas's Hospital, representing the function of the state in providing public services such as healthcare and education – objects of political debate in parliament, but also the places that we most commonly encounter the state on an everyday basis in our local communities.

To the north-east of Westminster Bridge is the old County Hall, formerly the headquarters of the Greater London Council until it was abolished in 1982 in an ideological struggle with the UK central government – represented by the offices of Whitehall directly across the river. Together with the new City Hall down-river, built when London was given an elected mayor in 2000, these are sites in the shifting internal political geography of the British state – reminding us that the subdivision of a nation-state into local

territories and the distribution of power between different scales of governance are fluid and contested.

Following Whitehall to the north we come to Trafalgar Square, whose name and monumental centrepiece – Nelson's Column – commemorate a British battle victory over France in the Napoleonic wars (as does the name of Waterloo railway station across the river). Rich in iconography, Trafalgar Square symbolises a discourse of British national identity that is nostalgic for imperial glory and encodes into the landscape the geopolitics of the nineteenth century. Yet, all around are reminders of the volatile geopolitics of the twenty-first century, not least the red London buses that were the target for a murderous al-Qaeda terrorist attack on 7 July 2005.

Round the river bend, the shimmering glass towers of the City of London form a different landscape of power, representing the dominance of global capitalism. The way in which the corporate offices dwarf the buildings of regulatory institutions such as the Bank of England resonates with the limited capacity of nation-states to control the global economy, and the political-geographical challenge of creating new forms of transnational governance. But places like the City of London also provide spaces for dissent. Near to the river the cruciform of St Paul's Cathedral stands out, whose grounds were colonised by anti-capitalist 'Occupy' protestors in 2011, demonstrating the power of controlling and occupying space.

To the east of the City of London, the river passes the gentrified districts of Whitechapel and Stepney, the redeveloped docklands at Canary Wharf, and, a little inland to the north, the 2012 Olympic Park at Stratford. Here we encounter a more localised political geography – conflicts over the physical and social displacement of working class residents and local businesses for new elite developments and mega-events, and debates over the impact on community coherence. Further east, suburbs such as Plaistow and Barking host populations rich in ethnic diversity, with residents engaged in the everyday negotiation of the politics of identity and citizenship in a multicultural society. The streets and industrial estates of East London tell other stories of everyday political geography too: labour disputes over pay and conditions; struggles over the use of public spaces; community mobilisation and voluntary action to represent tenants' interests, fight crime, and fill gaps in local services; and the gendered politics of the home and the workplace.

Returning to the river, we come finally to the Thames Flood Barrier at Woolwich Reach. Completed in 1982 to protect London from tidal flooding, it is a reminder of the threat of sea level rises with climate change and the challenge of environmental politics. Understanding political responses to climate change, the growth of transnational environmental campaigns and the negotiation of international agreements on the environment are increasing concerns for political geographers, as are the politics of managing our dwindling natural resources.

Our journey along the River Thames through London has presented a vivid and varied panorama of political geography, but it is not exceptional. Political geography can be encountered on a stroll through any city, town or village. James Sidaway (2009) illustrates this with his account of a walk through Plymouth, in south-west England, making connections to stories of imperialism and colonialism, global trade, twentieth-century geopolitics, urban development and gentrification, neoliberal housing policy and the politics of land access. In our own small college town of Aberystwyth in west Wales, a waterfront walk would pass a bridge blockaded in a pivotal protest for Welsh nationalism, the former offices of a regional development agency before territorial re-organisation, a diminished fishing fleet that prompts thoughts about the politics of resource management and the up-scaling of policy to the European Union, the ruins of a castle

that would have dominated the mediaeval landscape of power as a symbol of English control, and an ornate war memorial reflecting the dark side of geopolitics and the pride and ambition of a provincial town.

Put simply, political geography is everywhere. From the 'big P' Politics of elections and international relations, to the 'small p' politics of social relations and community life, politics not only shapes and infiltrates our everyday lives, but is everywhere embedded in space, place and territory. It is this intersection of 'politics' and 'geography' that we understand as 'political geography' and which we seek to examine in its many diverse forms in this book.

Defining political geography

Political geographers have taken a number of different approaches to defining the field of political geography. To some, political geography has been about the study of political territorial units, borders and administrative subdivisions (Alexander 1963; Goblet 1955). For others, political geography is the study of political processes, differing from political science only in emphasis given to geographical influences and outcomes and in the application of spatial analysis techniques (Burnett and Taylor 1981; Kasperson and Minghi 1969). Both of these definitions reflected the influence of wider theoretical approaches within geography as a whole – regional geography and spatial science, respectively – at particular moments in the historical evolution of political geography and have generally been superseded as the discipline has moved on. Still current, however, is a third approach which holds that political geography should be defined in terms of its key concepts, which the proponents of this approach generally identify as territory and the state (e.g. Cox 2002). This approach shares with the earlier two approaches a desire to identify the 'essence' of political geography such that a definitive classification can be made of what is and what isn't 'political geography'. Yet, political geography as it is actually researched and taught is much messier than these essentialist definitions suggest. Think, for example, about the word 'politics'. Essentialist definitions of political geography have tended to conceive of politics in very formal terms, as being about the state, elections and international relations. But, 'politics' also occurs in all kinds of other, less formal, everyday situations, many of which have a strong geographical dimension – issues about the use of public space by young people for skateboarding, for example, or about the symbolic significance of a landscape threatened with development. Whilst essentialist definitions of political geography would exclude most of these topics, they have become an increasingly important focus for geographical research.

As such, a fourth approach has been taken by writers who have sought to define political geography in a much more open and inclusive manner. John Agnew, for example, defines political geography as simply 'the study of how politics is informed by geography' (2002: 1; see also Agnew *et al.* 2003), whilst Painter and Jeffrey (2009) describe political geography as a 'discourse', or a body of knowledge that produces particular understandings about the world, characterised by internal debate, the evolutionary adoption of new ideas, and dynamic boundaries. As indicated above, the way in which political geography is conceived of in this book fits broadly within this last approach.

We define political geography as a cluster of work within the social sciences that seeks to engage with the multiple intersections of 'politics' and 'geography', where these two terms are imagined as triangular configurations (Figure 1.1). On one side is the triangle of power, politics and policy. Here power is the

commodity that sustains the other two – as Bob Jessop puts it, 'if money makes the economic world go round, power is the medium of politics' (Jessop 1990a: 322) (see Box 1.1). Politics is the whole set of processes that are involved in achieving, exercising and resisting power – from the functions of the state to elections to warfare to office gossip. Policy is the intended outcome, the things that power allows one to achieve and that politics is about being in a position to do.

The interaction between these three entities is the concern of political science. Political geography is about the interaction between these entities and a second triangle of space, place and territory. In this triangle, space (or spatial patterns or spatial relations) is the core commodity of geography. Place is a particular point in space, whilst territory represents a more formal attempt to define and delimit a portion of space, inscribed with a particular identity and characteristics. Political geography recognises that these six entities – power, politics and policy, space, place and territory – are intrinsically linked, but a piece of political geographical research does not need to explicitly address them all. Spatial variations in policy implementation are a concern of political geography, as is the influence of territorial identity on voting behaviour, to pick two random examples. Political geography, therefore, embraces an innumerable multitude of interactions, some of which may have a cultural dimension that makes them also of interest to cultural geographers, some of which may have an economic dimension also of interest to economic geographers, some of which occurred in the past and are also studied by historical geographers. To employ a metaphor that we will explain in Chapter 2, political geography has frontier zones, not borders.

In this book we explore these various themes and topics by drawing on and discussing contemporary research in political geography. The case studies and examples that we refer to are taken predominantly from books and journal articles published in the last twenty years, the most recent of which may be regarded as sitting at the 'cutting edge' of political geography research. However, current and recent work in political geography of this kind does not exist in a historical vacuum. It builds on the foundations of earlier research

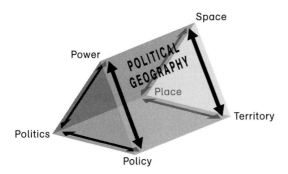

Figure 1.1 Political geography as the interaction of 'politics' and 'geography'

BOX 1.1 POWER

Put simply, power is an ability to get things done, yet there are many different theories about what precisely power is and how it works. In broad terms there are two main approaches to conceptualising power. The first defines power as a property that can be possessed, building on an intellectual tradition that stems from Thomas Hobbes and Max Weber. Some writers in this tradition suggest that power is relational and involves conscious decision-making, as Robert Dahl describes: 'A has power over B to the extent that he can get B to do something that B would not otherwise do' (Lukes 1974: 11–12). Others have argued that power can be possessed without being exercised, or that the exercise of power does not need conscious decision-making but that ensuring that certain courses of action are never even considered is also an exercise of power. The second approach contends that power is not something that can be possessed, as Bruno Latour remarks: 'When you simply have power – *in potentia* – nothing happens and you are powerless; when you exert power – *in actu* – others are performing the action and not you. . . . History is full of people who, because they believed social scientists and deemed power to be something you can possess and capitalise, gave orders no one obeyed!' (Latour 1986: 264–5). Instead, power is conceived of as a 'capacity to act', which exists only when it is exercised and which requires the pooling together of the resources of a number of different entities.

Key readings: Allen (2003), Allen (2011).

---

and writing, advancing an argument through critique and debate and through the exploration of new empirical studies that allow new ideas to be proposed. Knowing something about this genealogy of political geography helps us to understand the nature, approach and key concerns of contemporary political geography. To provide this background, the remainder of this chapter outlines a brief history of political geography, from the emergence of the sub-discipline in the nineteenth century through to current debates about its future direction.

A brief history of political geography

The history of political geography as an academic sub-discipline can be roughly divided into three eras: an era of ascendency from the late nineteenth century to the Second World War; an era of marginalisation from the 1940s to the 1970s; and an era of revival from the late 1970s onwards. However, the trajectory of

political geographic writing and thinking can be traced back long before even the earliest of these dates. Aristotle, writing some 2,300 years ago in ancient Greece, produced a study of the state in which he adopted an environmental deterministic approach to considering the requirements for boundaries, the capital city, and the ratio between territory size and population; whilst, the Greco-Roman geographer, Strabo, examined how the Roman Empire was able to overcome the difficulties caused by its great size to function effectively. Interest in the factors shaping the form of political territories was revived in the European 'Age of Enlightenment' from the sixteenth through to the eighteenth centuries, as writers combined their new enthusiasm for science and philosophy with the practical concerns generated by a period of political reform and instability. Most notable was Sir William Petty, an English scientist and economist who in 1672 published *The Political Anatomy of Ireland* in which he explored the territorial and demographic bases of the power of the British state in Ireland. Petty developed these ideas further in his second book, *Essays in Political Arithmetick*, begun in 1671 and published posthumously, which outlined theories on, among other things, a state's sphere of influence, the role of capital cities, and importance of distance in limiting the reach of human activity. In this way Petty foreshadowed the concerns of many later political geographers, but his books were, like other geographical writing of the time and the classical texts of Strabo and Aristotle, popular works of individual scholarship by polymaths, which did not stand as part of a coherent field of 'political geography'. To find the real beginnings of 'political geography' as an academic discipline we need to look to nineteenth-century Germany.

The era of ascendency

The significance of Germany as the cradle of political geography lies in its relatively recent formation. Modern Germany had come into being as a unified state only in 1871 and under ambitious Prussian leadership sought in the closing decades of the nineteenth century to establish itself as a 'great power' on a par with Britain, France, Austria-Hungary and Russia. However, Germany was constrained by its largely landlocked, central European location, which restricted its potential for territorial expansion. In these circumstances, ideas about the relationship between territory and state power became key concerns for Germany's new intellectual class and in particular for Friedrich Ratzel, sometimes referred to as 'the father of political geography'.

Much of Ratzel's work was driven by a desire to intellectually justify the territorial expansion of Germany and in writing such as *Politische Geographie* (1897) he embarked on a 'scientific' study of the state (see Bassin 1987). Ratzel drew on earlier political geographical work, notably that of Carl Ritter, but his innovation was to borrow concepts from the evolutionary theories of Darwin and his followers. In particular, Ratzel was influenced by a variation on Social Darwinism known as neo-Lamarckism, which held that evolution occurred through species being directly modified by their environments rather than by chance. Translating these ideas to the political

sphere, Ratzel argued that the state could be conceived of as a 'living organism' and that like every living organism the state 'required a specific amount of territory from which to draw sustenance [Ratzel] labelled this territory the respective *Lebensraum* or living space of the particular organism' (Bassin 1987: 477).

Extending the metaphor, Ratzel contended that states followed the same laws of development as biological units and that when a state's *Lebensraum* became insufficient – for example because of population growth – the state needed to annex new territory to establish a new, larger, *Lebensraum*. As such he posited seven laws for the spatial growth of states, which held that a state must expand by annexing smaller territories, that in expanding a state strives to gain politically valuable positions, and that territorial expansion is contagious, spreading from state to state and intensifying, such that escalation towards warfare becomes inevitable. In this way, Ratzel not only provided an 'intellectual justification' for German expansionism, but suggested that it was an entirely natural and necessary process. Ratzel himself argued that the only way Germany could acquire additional *Lebensraum* was through colonial expansion in Africa – a policy he actively promoted – but his theories were seen by some more militant nationalists as justifying the more aggressive and more dangerous strategy of expanding German territory in the crowded space of continental Europe itself.

Ratzel's ideas were developed further by Rudolf Kjellén, a Swedish conservative whose own political motives were fired by opposition to Norwegian independence. Kjellén's intellectual project was to develop a classification of states based on the Linnaean system. By adapting Ratzel's theories, he attempted to identify the 'world powers' and predicted a future dominated by large continental imperialist states. Although he received some support in Germany, Kjellén's work would probably have been long forgotten had he not in an 1899 article coined the term *geopolitisk*, which – translated into German as *geopolitik* and by 1924 into English as *geopolitics* – came to describe that part of political geography that is essentially concerned with the external relations, strategy and politics of the state, and which seeks to employ such knowledge to political ends (see Chapter 11).

Whilst Ratzel and Kjellén were wrestling with the dynamics of state power and territoriality, a second strand of political geography was being developed in Britain by Sir Halford Mackinder. Like Ratzel, Mackinder is regarded as a founding father of modern geography, having popularised the subject in a series of public lectures in the 1880s and 1890s leading to his appointment as Oxford University's first Professor of Geography. Also like Ratzel, Mackinder saw the benefits of proving the political usefulness of his infant discipline. As Ó'Tuathail (1996: 25) has commented,

> to an ambitious intellectual like Mackinder, the governmentalizing of geographical discourse so that it addressed the imperialist dilemmas faced by Britain in a post-scramble world order was a splendid way of demonstrating the relevance of his 'new geography' to the ruling elites of the state.

However, unlike Ratzel, Mackinder was primarily concerned with issues of global strategy and the balance of power between states – topics that better suited the interests of British foreign policy. He was not the first to consider such matters. In the United States a retired naval officer, Alfred Mahan, had established himself as a newspaper pundit by arguing that global military power was dependent on sea power, and expounding on the geographical factors that enabled the development of a state as a sea power. Mackinder, though, disagreed with Mahan's thesis, suggesting that as the age of exploration came to end, so the balance of power was shifting (Kearns 2010). In 1904 Mackinder published a paper entitled 'The geographical pivot of history' in the *Geographical Journal*, in which he divided history into three eras – a pre-Columbian era in which land power had been all important, a Columbian era in which sea power had become predominant, and an emergent post-Columbian era. In this new era, Mackinder argued, the end of the imperialist scramble had demoted the importance of sea power whilst new technologies,

which enabled long distances on land to be more easily overcome – such as the railways – would help to swing the balance of power back to continental states. Applying this hypothesis, Mackinder ordered the world map into three political regions – an 'outer crescent' across the Americas, Africa and the oceans; an 'inner crescent' across Europe and southern Asia; and the 'pivot area', located at the heart of the Eurasian land mass. Whoever controlled the pivot area, Mackinder argued, would be a major world power.

The First World War put the theories produced by the new political geography to the test, and Mackinder clearly felt that his ideas were vindicated. Writing in his 1919 book, *Democratic Ideals and Reality*, he dismissed Ratzel's models as misguided and outdated:

> Last century, under the spell of the Darwinian theory, men came to think that those forms of organisation should survive which adapted themselves best to their natural environment. To-day we realise, as we emerge from our fiery trial, that human victory consists in our rising superior to such mere fatalism.
>
> (Mackinder 1919: 3)

In *Democratic Ideals and Reality*, Mackinder expanded on his thesis of the shift from sea-power to land-power and recast his map of the world's seats of power to suit the new post-war order. He renamed the 'pivot area' the 'heartland', but left it centred on the Eurasian land-mass, which he labelled the 'world island'. Significantly, he proposed that control of Eastern Europe was crucial to control of the heartland – and hence to global domination. To maintain peace, therefore, Mackinder argued that Western Europe had to form a counterweight to Russia, which occupied the heartland, and that the key priority of the West's strategy had to be to prevent Germany and Russia forming an alliance that would dominate Eastern Europe.

Mackinder participated as a British delegate in the Versailles Peace Conference in 1919 and whilst his direct influence in deliberations is debatable, the outcomes of the conference strongly resonate with his world-view, most notably in the creation of 'buffer states' in Eastern Europe, separating Germany and Russia. His legacy can be found in US strategy on containing the Soviet Union during the Cold War (Ó'Tuathail 1992) and in analyses of the twenty-first century 'new world order' by both political leaders and popular writers such as Robert Kaplan (Kaplan 2012) – despite repeated critiques by political geographers (see Kearns 2009, 2010; Megoran 2010) (see also Chapter 11).

Ironically, Mackinder's thesis was also consumed with interest in the country that suffered most from its practical application at Versailles – Germany. For German nationalists enraged by the way in which Germany had had its territory reduced and its military dismantled after the First World War, the geopolitical ideas of Ratzel and Mackinder offered a blueprint for revival (Paterson 1987). Most prominent in this movement was Karl Haushofer, a former military officer and geographer who became an early member of the Nazi party. Haushofer sought to build public support for a new expansionist policy by popularising interest in geopolitics. In 1924 he founded the *Zeitschrift für Geopolitik* (*Journal of Geopolitics*) and the following year was involved in establishing the German Academy, aimed at 'nourishing all spiritual expressions of Germandom', of which he later became president. Haushofer's 'pseudo-science' of *geopolitik* took from Ratzel the concept of *Lebensraum* and twisted it, arguing that densely populated Germany needed to annex additional territory from more sparsely populated countries such as Poland and Czechoslovakia. From Mackinder it took the idea that control of Eastern Europe and the heartland would lead to global dominance, arguing for the construction of a continental bloc comprising of Germany, Russia and Japan, which would control the heartland and form a counterweight to the British Empire (see Ó'Tuathail 1996).

Geopolitik provided the intellectual justification for Nazi Germany's annexation of Czechoslovakia and Poland, for the Hitler–Stalin pact, and, later, for Germany's ill-fated invasion of the Soviet Union (see Chapter 11 for further discussion). However, the extent of Haushofer's influence on the Nazi leadership is questionable (see Heske 1987). More significant was the contribution of *Geopolitik* in shaping public opinion, most effectively achieved through the promotion of a new form of cartography in which highly subjective maps were used to emphasise the mismatch between Germany's post-1919 borders and its 'cultural sphere', to justify the annexation of territory, and to suggest that it was vulnerable to aggression by its Slavic neighbours (see Herb 1997 for examples).

The misadventures of *Geopolitik* inextricably associated political geography with the brutality and racism of the Nazi regime and led to its discrediting as a serious academic pursuit.

The era of marginalisation

The excesses of German *Geopolitik* cast a pall over all of political geography. Writing in 1954, the leading American geographer Richard Harteshorne mournfully remarked of political geography that, 'in perhaps no other branch of geography has the attempt to teach others gone so far ahead of the pursuit of learning by the teachers' (Harteshorne 1954: 178). In an attempt to 'depoliticise' political geography and to put it on what he regarded as a more scientific footing, Harteshorne (1950) promoted a 'functional approach' to political geography. In this he argued that political geography should not be concerned with shaping political strategy, but rather with describing and analysing the internal dynamics and external functions of the state. Included in the former were the centrifugal forces that placed pressures on the cohesion of states (such as communication problems and ethnic differences), the centripetal forces that held states together (such as the state-idea and the concept of a 'nation'), and the internal organisational mechanisms through which a state governed its territory. The external functions, meanwhile, included the territorial, economic, diplomatic and strategic relations of a state with other states.

The functional approach led political geographers to become concerned with questions such as the distribution of different ethnic populations in a state, the match between a state's boundaries and physical geographical features, and the structure of a state's local government areas, as well as with mapping patterns of communication networks within states and of trade routes between states (some examples of this type of work include: Cole 1959; East and Moodie 1956; Moodie 1949; Soja 1968; and Weigert 1949). However, whilst the functional approach was popularised after the Second World War, it was pioneered in Britain and North America between the wars and arguably can be traced back to the work of American geographer Isaiah Bowman in the early 1920s.

Like Mackinder, Bowman straddled both academic geography and politics, participating in the US delegation at Versailles and, as Smith (2004) demonstrates, playing a key role in shaping the United States' geopolitical vision during its emergence as a world superpower. Influenced by Mackinder, Bowman supported the creation of buffer states in Eastern Europe, and influenced by Ratzel he advocated the need for the USA to create 'economic *lebensraum*' through trade relations. However, he detested the German *geopolitik* and American attempts to copy it, and was aghast at being described as a geopolitician (Smith 2004). Bowman tried instead to create a new 'political geography', distinct from 'geopolitics', based on scientific analysis not political strategising. During the First World War he had produced detailed maps and analyses of different countries that had informed America's position in the Versailles talks, yet he regarded the new world map that emerged to be extremely unstable. His pessimism stemmed from a concern not with strategic models, but with social and economic factors such as access to natural resources and the distribution of population, which he considered to be the real sources of political instability. Bowman set out these concerns in *The New World: Problems in Political Geography* (1921), in which he identified the 'major problems' facing the new world order as national debts and reparations, control over the production and distribution of raw materials, population movement and the distribution of land, the status of mandates and colonies, trade barriers and control over communications and transit links, the limitation of armaments, the status of minority populations and disputed boundaries between states. As such, Bowman changed the scale at which political geography was focused and set the foundations for a new, arguably more scientific and more objective, form of analysis.

This new style of political geography was more explicitly outlined by East (1937) in a paper that Johnston (1981) identifies as laying down the principles of the functional approach later championed by Harteshorne. East argued that 'the proper function of political geography is the study of the geographical results of political differentiation' and 'that the visible landscape is modified by the results of state and inter-state activities is a matter of common observation and experience' (p. 263). As such, East continued,

> political geography is distinguishable from other branches of geography only in its subject matter and specific objectives. . . . Whereas the regional geographer has for his objective the discovery and description of the distinct components of a physical and human landscape . . . the political geographer analyses geographically the human and physical texture of political territories.
>
> (East 1937: 267)

Political geography as practised in the immediate post Second World War period therefore had little by way of a distinctive identity separate to mainstream regional geography, and became largely fixated with the territorial state as its object of analysis. Moreover, fear of the sub-disciplines' past made political geographers wary of modelling and theorising, such that research remained essentially descriptive and empirically driven. The consequences of this self-restraint were two-fold. First, political geography largely missed out on theoretical developments taking place elsewhere in geography, notably the 'quantitative revolution' of the late 1960s. Second, (and relatedly) political geography became marginalised within geography and began to disappear as a university subject. Berry (1969: 450) famously described it as 'that moribund backwater' (see also Antonsich *et al.* 2009), and by the mid 1970s, Muir (1976) found that political geography was taught in only half of Britain's university geography departments, with over two-thirds of heads of geography departments considering that the development of political geography literature was unsatisfactory compared with other branches of geography.

However, Muir's article, which was provocatively titled 'Political geography: dead duck or phoenix?', found grounds for optimism. He noted that over half of respondents to his survey had felt that political geography was 'an underdeveloped branch of geography that *should* increase in importance' (Muir 1976: 196), and pointed to theoretical innovations that were beginning to take place on the fringes of the sub-discipline. As such he concluded, 'the contemporary climate of geographical opinion augers well for the future of political geography, and a promising trickle of progressive contributions suggests stimulating times to come' (p. 200).

The era of revival

The revival of political geography that Muir detected in the 1970s was driven by two parallel processes – a re-introduction of theory into political geography, and a 'political turn' in geography more broadly. Significantly, neither resulted from developments in the established mainstream of political geography, but rather reflected innovation at the fringes of political geography, producing research clusters that eventually came to eclipse the old-style 'functional approach'. One illustration of this is the rise of quantitative electoral geography from the late 1960s onwards. Although the quantitative revolution tended to pass political geographers by, some quantitative geographers realised that the spatially structured nature of elections, combined with the large amount of easily available electoral data, made them an ideal focus for the application of quantitative geographical analysis. Elections had not traditionally been a concern of mainstream political geographers and the new electoral geographers did not therefore have to challenge any orthodoxies as they employed quantitative techniques to develop models and test hypotheses across their tripartite interests of geographies of voting, geographical influences on voting, and geographical analyses of electoral districts (Busteed 1975; McPhail 1971; Taylor and Johnston 1979). The lure of technical and theoretical innovation made electoral geography the fashionable 'cutting edge' of political geography in the 1970s, such that by 1981 Muir was moved to comment

that its output had become 'disproportionate in relation to the general needs of political geography' (Muir 1981: 204) (we discuss electoral geography in fuller detail in Chapter 7).

The growth of electoral geography was the most prominent aspect of the belated introduction of a systems approach to political geography, drawing on the broader development of systems theory in geography as part of a focus on processes not places (Cohen and Rosenthal 1971; Dikshit 1977). Electoral geographers viewed the electoral process as a system – comprising various interacting parts, following certain rules and having particular spatial outcomes – but they also realised that other parts of the political world could also be conceived of and analysed as systems, including the state, local government, policy making and public spending (see Johnston 1979). Significantly, the mechanical principles underlying systems theory meant that adopting the approach rendered complex political entities suitable to mathematical analysis and modelling. However, the extent to which a full-bodied systems analysis was adopted in political geography varied. At the most basic level, 'systematic political geography' implied no more than re-ordering the way in which political geography was taught and researched to start from themes or concepts rather than regions (see de Bilj 1967). Whilst this allowed for generalisation in a way that the regionally focused approach did not, it did not necessarily lead to in-depth theorising. Yet even the most conscientious attempts to produce models and theories through quantitative analysis were constrained by their positivist epistemology – that is, the belief that the world might be understood through the construction and testing of laws based on empirical observation. As critics pointed out, positivism is problematic because it creates a false sense of objectivity, filters out social and ethical questions, oversimplifies the relationship between observed events and theoretical languages, and fails to engage with the part played by both human agency and social, economic and political structures in shaping the human world. Thus, because of these epistemological short-comings, positivist political geography continued to be strangely apolitical (Johnston 1980). Moreover, the 'time-lag' that inflicted the introduction of concepts into political geography meant that positivism was being championed in political geography at a time when these criticisms were already widely accepted elsewhere (Walsh 1979).

Ironically, the challenge to positivism was led by theoretical approaches that were intrinsically political, not least the development of Marxist political economy within geography (see Box 3.1 in Chapter 3 for more on models of political economy). David Harvey in *Social Justice and the City* (1973), for example, proposed a new analysis of urban systems as embedded in capitalism, which described an urban geography saturated by class, corporate and state power and forged through political conflict. However, the infusion of these ideas into political geography was slow. Despite the calls of commentators such as Walsh that 'what political geography needs most urgently . . . is a comprehensive analysis of the state as a political-economic entity' (Walsh 1979: 92), political-economic research within political geography remained the exception not the rule, and the task of studying urban conflicts, the geography of the state and the political-geographic expressions of capitalism was taken up primarily by urban and economic geographers, political scientists and sociologists. It was not until the 1980s that mainstream political geography really started to take the political-economy approach seriously, with a blossoming of work on the state, localities and urban politics (see Johnston 1989) (we discuss more recent applications of the political-economy approach in Chapters 3 and 9).

One of the relatively few attempts to link the traditional concerns of political geography with theoretical insight from Marxist political economy was Peter Taylor's introduction of world systems analysis. The world systems approach had been developed by a political sociologist, Immanuel Wallerstein, who was himself influenced by the materialist school of historical analysis associated with Fernand Braudel and Karl Polányi and by neo-Marxist development studies (see Wallerstein 1979, 1991). As Box 1.2 details, Wallerstein rejected the idea that societal change could be studied on a country-by-country basis and argued instead that change at any scale can only be understood in the context of a 'world-system'. The modern world-system, Wallerstein argues, is global in scope, but he

recognises that it is only the latest of a series of historical systems and proposes that it is the changes within and between historical systems that are the key to understanding contemporary society, economy and politics. For Taylor, the world systems approach was particularly attractive to political geography not only because spatial pattern was core to its analysis (Taylor 1988), but also because it offered the potential to develop a comprehensive, unifying theory of political geography that could include traditional areas such as geopolitics and electoral geography and accommodate political-economic analysis of the state, urban politics and so on. However, despite its superficial attractiveness, world systems analysis is open to a number of criticisms (Box 1.2), and although it has formed the framework for Taylor's series of textbooks (see Taylor 1985 and Flint and Taylor 2011 as the first and most recent editions), the world systems approach has not been widely adopted by political geographers.

Far more influential has been the legacy of the 'cultural turn' in human geography during the late 1980s, and the related engagement with theoretical and methodological perspectives from post-structuralism and feminism. These developments impacted on political geography in three key ways.

First, the cultural turn expanded the scope for political geography research by enlarging the understanding of 'politics'. Conventionally, political geographers had followed political scientists in defining politics around the state – its form, function, apparatus, external relations (geopolitics), internal structures, policies and democratic processes (electoral geography). New perspectives drawn from cultural studies, anthropology and sociology located politics in a wider range of social relations apart from the state, in communities, workplaces and families. Accordingly, cultural geographers began to study the politics of identity – especially for marginalised groups – the politics of consumption, the politics of representation, the politics of public and private space, and so on (Jackson 1989; Mitchell 2000). Feminist geographers championed the notion that 'the personal is political', taking political geography into the gendered politics of the home and positioning the body as a political object (Kofman and Peake 1990; England 2003), as

BOX 1.2   PETER TAYLOR, IMMANUEL WALLERSTEIN AND WORLD SYSTEMS ANALYSIS

World systems analysis forms the basis of the best known attempt to construct a comprehensive theoretical framework for political geography, undertaken by Peter Taylor. It was initially developed by Immanuel Wallerstein as a critique of analyses of social change that focused on one country and considered only a short-term perspective. In contrast, two of the fundamental principles of world systems analysis are that social change at any scale can only be understood in the context of a wider world system, and that change needs to be approached through a long-term historical perspective (the latter principle is derived from economic historians such as Fernand Braudel and Karl Polányi).

Wallerstein holds that a single modern world system is now globally dominant, but that it has been preceded by numerous historical systems. These systems can be categorised as one of three types of 'entity', characterised by their mode of production. In the most basic, the 'mini-system', production is based on hunting, gathering or rudimentary agriculture where there is limited specialisation of tasks, and exchange is reciprocal between producers. In the second type, the 'world empire', agricultural production creates a surplus that can support the expansion of non-agricultural production and the establishment of a military-bureaucratic elite. The third type, the 'world-economy', is based on the capitalist mode of production where the aim of production is to create profit. From the sixteenth century onwards, Wallerstein argues, the European 'world-economy' system expanded to subjugate all other systems and monopolise the globe. Transformation from one system to another can occur as a result of either internal or external factors, but changes can also occur within systems (termed 'continuities') – for example, in cycles of economic growth and stagnation. In the modern world-economy these cycles are mapped by the Kondratieff waves, which describe fifty-year cycles of growth and stagnation in the global economy since 1780/90.

Wallerstein further described the modern world-economy as being defined by three basic elements. First, there is a single world market, which is capitalist, and in which competition results in uneven economic development across the world. Second, there is a multiple state system. The existence of different states is seen as a necessary condition for economic competition, but also results in political competition between states creating a variety of 'balances of power' over time. Third, the world-economy always operates in a three-tier format. As Flint and Taylor (2011) explain: 'in any situation of inequality three tiers of interaction are more stable than two tiers of confrontation. Those at the top will always manoeuvre for the "creation" of a three-tier structure, whereas those at the bottom will emphasise the two tiers of "them and us". The continuing existence of the world-economy is therefore due in part to the success of the ruling groups in sustaining three-tier patterns throughout various fields of conflict' (p. 17). Examples cited by Flint and Taylor include 'centre' parties in democratic political systems and the 'middle class', but also, crucially, a geographical ordering of the world into 'core', 'periphery' and 'semi-periphery'. For Wallerstein, core areas are associated with complex production regimes, and the periphery with more rudimentary structures. But there is also a 'semi-periphery' in which elements of both core and peripheral processes can be found, and which forms a dynamic zone where opportunities for political and economic change exist.

By drawing on these different components of world systems theory, Taylor identified 'space-time matrix' for political geography, structured by the Kondratieff cycle and spatial position (core, periphery or semi-periphery), which formed a context for the analysis of all types of political interaction from the global scale down to the household, hence providing a unifying framework for political geography.

However, the world systems approach can be criticised on a number of grounds. First, it is economically reductionist – it sees the driving processes of change as purely economic; it positions political action as secondary; and it reduces sexism and racism to reflections of the economy. Second, it is totalising in that it incorporates everything under one big umbrella and fails to fully acknowledge the heterogeneity of political or cultural relations. Third, it is functionalist, not recognising that what causes something to exist may have nothing to do with the effects it produces. For example, the factors behind the creation of a nation-state may not be related to subsequent nationalistic actions.

Key readings: For more on world systems analysis see Flint and Taylor (2011), especially chapter 1, and Wallerstein (1991). For more on the critique of world systems analysis see Giddens (1985).

well as critiquing the masculine bias of conventional political geography (Drake and Horton 1983).

Second, new theoretical perspectives introduced new ways of thinking about power and its interaction with space. Particularly influential has been the work of Michel Foucault and his conceptualisation of power as diffuse and only existing in its practice. Foucault proposed that 'space is fundamental in any exercise of power' (Rabinow 1984: 252), both through the physical control and occupation of territory (informing political geography research such as Herbert 1996 and Ogborn 1992), and through the translation of space into maps, statistics and other discursive representations in technologies of governmentality (that is, ways of making society governable) (investigated through work such as Hannah 2000). Moreover, Foucault argued that power is also countered by resistance, promoting geographers to explore not only geographies of power, but also the inseparable geographies of resistance (Pile and Keith 1997; Sharp *et al.* 2000). Ideas were also drawn from post-colonial theory, including writers such as Homi Bhabha, Edward Said and Gayatri Chakravorty Spivak who challenged Western-centric representations of the world, inspiring geographers to critique the discursive geographies through which the West has exerted global domination and to give voice to non-Western perspectives (see Gregory 2004; Laurie and Calla 2004).

Third, the period also saw the introduction of new methodological techniques to political geography. Most important of these has been discourse analysis, a technique for deconstructing the ways in which a

world-view is constructed, encoded and communicated through various 'texts', which include not only written documents, but also films, cartoons, murals, performances, landscapes and so on. Discourse analysis was adopted as a key method in the emerging field of 'critical geopolitics', which combined influences from post-structuralism and critical international relations theory to interrogate and critique the discourses and practices of conventional geopolitical 'statecraft' (Dalby 2010; Ó'Tuathail 1996). This included analysis of policy documents (Dodds 1994; Ó'Tuathail and Agnew 1992), news reports and photography (Campbell 2007; McFarlane and Hay 2003), cartoons and comics (Dittmer 2005; Dodds 1996, 2007), film (Dodds 2003; Power and Crampton 2005) and monumental landscapes (Atkinson and Cosgrove 1998).

These developments helped to further reinvigorate political geography research and bring it back towards the mainstream of human geography, but they also presented a challenge to political geography as a sub-discipline by contesting the 'discourse' of what political geography is. The danger identified by some commentators was that if 'politics' are defined so broadly such that almost every social interaction can be seen as 'political', and if cultural geographers, social geographers, historical geographers and others are all engaged in studying politics, then the distinctive of 'political geography' and its coherence as a community of scholars gets lost.

These anxieties were articulated in a debate at the conference of the Association of American Geographers in Los Angeles in 2002 and a related theme issue of the journal *Political Geography* published in 2003. The perceived problem was expressed in the contribution by Flint (2003) who pointed to the 'paradox' that whilst political geography (at least in the United States) was in good institutional health, it appeared to lack coherence and faced uncertainty about its direction. Flint identified the uncertainty with the dilemma of whether political geography should concentrate on politics with a big 'P' or a little 'p':

Identity politics, the environment, post-colonialism, and feminist perspectives are all relatively 'new' politics, placed on the agenda by the political upheavals of the 1960s . . . and can be classified as politics with a small 'p'. They stand in contrast to the old politics of the state and its geopolitical relations, statemanship or politics with a large 'P'.

(Flint 2003: 618)

Flint argued that knowledge of both Politics and politics is required to understand the contemporary world, and that a coherence could be maintained for 'political geography' by focusing on 'the way that different spatial structures are the product of politics and the terrain that mediates those actions' (p. 619), and by showing the relevance of spatiality for all types of power. Yet, he also noted that much work on the 'new', small 'p' political geography is undertaken by individuals who are not 'card-carrying' political geographers, thus raising concerns about disciplinary boundaries that were echoed by Cox (2003). For Cox, the answer was to reassert the centrality of territory to political geography, whilst Low (2003) similarly argued for political geography to be defined by a focus on the state.

However, other contributors to the debate saw less cause for concern. Agnew (2003), for example, emphasised the historical fluidity of political geography, and argued that 'much of what is of interest to me in contemporary political geography is exciting precisely because there is more limited agreement than was once the case' (p. 603). Kofman (2003) similarly contended that 'there isn't necessarily a contradiction between a heightened interest in political questions in human geography and the existence of something called

political geography' (p. 621), whilst Marston (2003) observed that 'the migration of the political to other areas of the discipline seem to me to be compelling evidence that we have failed to attend to a large portion of what is legitimately and centrally the purview of political geography' (p. 635). These contributors and others celebrated the emergence of a 'post-disciplinary political geography' (Painter 2003), and argued that there was scope for political geography to be further enriched by engagement with allied subjects such as peace and conflict studies (Flint 2003), socio-legal studies (Kofman 2003), political ecology (Robbins 2003), feminist geography (England 2003) and political theory (Painter 2003), as well as with political geographies produced from outside the insular environment of Anglo-American geography (Mamadouh 2003; Robinson 2003).

Twenty-first century political geography

In spite of the concerns expressed in the above debate, the actual development of political geography since the start of the new century has suggested that political geography will both remain open to new theoretical and methodological influences and empirical topics, and continue to accommodate work on traditional themes. Indeed, in the last decade (and particularly in the period since the publication of the first edition of this book), there has been a notable resurgence of interest in the areas of geopolitics, electoral geography, and political-economy analysis, as well as a growth in newer areas of political geography research. In updating this book for the second edition, we have sought to reflect and engage with these key trends.

The continuing revival of interest in geopolitics – following on from pioneering work in critical geopolitics in the 1990s – has been prompted both by global current affairs and by new theoretical insights. The terrorist attacks on New York and Washington, DC, on 11 September 2001 were a pivotal moment, ushering in a new era of geopolitical strategising as the United States and its allies launched interventions in Afghanistan and Iraq. Critical geopolitics accordingly refocused from an initial retrospective concern with Cold War statecraft, to analysis of the new geopolitical discourses of the unfolding world order, with research interrogating narratives of 9/11 and its impacts (Dalby 2004; Dittmer 2005; L. Jones 2010), the framing of 'the war on terror' (Coleman 2004; Dodds 2007; Hannah 2006) and the justification and conduct of the interventions in Afghanistan and Iraq (Dalby 2009; Gregory 2004; Mercille 2010; Ó'Tuathail 2004).

As Dalby (2010) has observed, critical geopolitics has also broadened its scope to critique discourses of militarism and securitisation (Cowen and Gilbert 2008; Flint 2005; Ingram and Dodds 2009; Kobyashi 2009; Woodward 2004), and to examine geopolitical framings of climate change and energy resources (Busby 2008; Ciuță and Klinke 2010), and the extension of geopolitics into new settings of the city and outer space (Graham 2004, 2010; MacDonald 2007). In these endeavours, critical geopolitics has continued to draw insights from post-structuralist theory, including, for example, Giorgo Agamben's (2005) concept of 'states of exception', which has been employed to investigate the extra-legal treatment of detainees in the 'war on terror' (Gregory 2006; Minca 2006). Especially notable, though, has been the influence of feminist theory in shaping a new 'feminist geopolitics' that 'is about putting together the quiet, even silenced, narratives of violence and loss that do the work of taking apart dominant geopolitical scripts of "us" and "them"' (Hyndman 2007: 39). Feminist geopolitics has criticised both conventional and critical geopolitics for fixing on male-dominated domains of statecraft, military and the media, and marginalising the predominantly female victims of war and state violence. As Hyndman argues in her analyses of 'body counts' in Afghanistan and Iraq, 'the visibility, or lack thereof, of civil deaths contributes to a gendered geopolitics that values (masculinized) U.S. lives over (feminized) Afghan ones' (Hyndman 2007: 39).

In contrast, feminist geopolitics has rescaled and refocused critical enquiry, producing narratives of 'geopolitics from below' (Fluri 2009) and emphasising the embodied nature of geopolitics, with the body positioned as an object of geopolitical violence not only in war and state violence, but also in the regulation of irregular migrants and refugees, and in struggles over religion and culture (Dowler and Sharp 2001; Hyndman and Mountz 2008; Mountz and Hyndman 2006; Secor 2001). Feminist perspectives have also informed analysis of the 'emotional geopolitics of fear' as a tool of contemporary statecraft (Pain 2009). These new directions in geopolitics are discussed further in Chapter 11.

Less predictably, perhaps, there has also been revitalisation of research in electoral geography. The closeness of the 2000 US presidential election, in which geographical factors were critical in determining the outcome (as demonstrated in the first edition of this book), and the apparent entrenchment of a polarised electoral geography in subsequent American elections, prompted a slew of new research and analyses, as discussed in Chapter 7. Electoral geography has also benefited from new technology – just as the quantitative revolution spurred advances in electoral geography in the 1960s, so GIS, GPS and georeferencing innovations have opened up new possibilities for representing and analysing electoral data. Furthermore, as Shin (2009) observes, the popular use of GIS technologies by the news media and political websites has generated new public interest in electoral geography, such that

> these new modes and means of production and consumption of political and geographic data and information, and the individual or collective responses to such visualizations, represent a renaissance of sorts for electoral geography that is changing the way people view and think about maps, geography, elections, politics, and even democracy itself.
>
> (Shin 2009: 152)

Indeed, democracy itself has become an increasingly significant focus of research by political geographers, who have studied the embedded nature of democracy and the role of place in promoting anti-democratic politics (Barnett and Low 2004; Staeheli 2010), as well as examining the spatialities of the public sphere and of social movements that represent an alternative form of democratic engagement (see Chapter 7).

A renewed interest in political-economy approaches has similarly been witnessed, supported by engagement with new theoretical perspectives that have helped to overcome some of the difficulties that political-economy analysis had run into at the end of the 1980s, with its over-emphasis on structural processes and reification of territorial units of the state. First, insights from 'cultural political economy' developed in sociology and, informed by elements of post-structuralist theory, have focused attention on the discursive practices and structuring principles through which objects of economic regulation and governance are constituted and institutions of governance consolidated (see Chapter 3). Second, new relational thinking on place in human geography has unlocked political-economy research by showing that places are not bounded territories but dynamic entanglements of wider social, economic and political relations. Here work by Massey (2005), Amin (2004) and Allen (2011) has pointed to the need to explore both the politics of connections between places, and the 'politics of propinquity' (Amin 2004) that arises from the juxtaposition of competing interests within place. These new perspectives have empowered a political-economy critique of neoliberalism as the ascendant ideology of the early twenty-first century, both in terms of the neoliberal restructuring of the state, its function and spatial form (which we discuss further in Chapter 3), and with respect to the reproduction of neoliberal globalisation and its challenge to settled political geographies (which we discuss further in Chapter 9).

At the same time, political geography has continued to expand into new areas of enquiry, including, perhaps most significantly, a growth of interest in the politics of the environment. This includes not only the geopolitics of climate change and resource struggles, but also investigation of the ways in which the environment is constructed as an object of governance, and the structures and techniques of governmentality that are employed; of the mobilisation of environmental movements, and the rescaling of environmental politics; and of the role of place in conflicts over environmental management, conservation and the development of renewable energy sources. This research has also extended the scope of political geography in considering non-human entities as both objects of governance and as potential political actants, including animals (Hobson 2007) and even carbon molecules (Whitehead 2009). Much of this work has drawn on theoretical perspectives from political ecology, an interdisciplinary approach that emphasises the embedding of environmental processes and issues within structures of power and politics, which had previously tended to co-exist with, rather than interact with, political geography (Robbins 2003). The political geographies of the environment are discussed further in Chapter 10.

Finally, there has also been a rise in a more activist political geography, which seeks not just to analyse the political world, but also to put geographical knowledge to use in challenging and shaping political actions and structures. In some ways this echoes the framing of early political geography and the ambitions of political geographers such as Mackinder, and as such it radically departs from attempts to de-politicise political geography in the post-war period. However, in stark contrast to Mackinder and Haushofer, contemporary activist political geographers aim not to assist the state, but rather to aid subordinate and marginalised populations in challenging power (see also Chapter 12 for more on the broader engagement of political geography in public policy). This has been a core agenda for feminist geopolitics, with Koopman (2011), for example, proposing an 'alter-geopolitics' that proactively engages with grassroots movements resisting state violence and constructing political alternatives on the ground. Similarly, the growth of political geography research on social movements is entwined with the active involvement of political geographers in environmental, global justice and peace movements – generating auto-ethnographic accounts from the 'inside' (Clough 2012; Routledge 2012). As such, activist political geographies reflect back on the sub-discipline as a whole, asking questions about what we study, how and for whom? For instance, in one notable intervention, Megoran (2011) sets out an argument for peace research and practice in political geography that can be part of the solution in areas of conflict, not part of the problem. As he observes, 'we are better at researching war than peace. For our discipline to play a serious role in addressing the problems wracking twenty-first century humanity, it is imperative that this imbalance be redressed' (Megoran 2011: 188).

Our challenge to you, therefore, is to read this book? reflexively: to not only (we hope) develop an understanding of how politics interacts with geography to shape the world around us, but also to think critically about your own thoughts and opinions on political issues, and your everyday practice of political geography.

Further reading

Agnew's *Making Political Geography* (2002) provides a more detailed history of political geography than that outlined here, albeit one that emphasises the traditional concerns of the sub-discipline more than recent innovations. Regular 'progress reports' in political geography published in the journal *Progress in Human Geography* are also a good way of following developments in the field.

The debate about the scope and direction of political geography at the start of the twenty-first century is published in *Political Geography*, 22, 6 (2003). Megoran's paper on geography and peace is published in *Political Geography*, 30, 178–189 (2011), and Sidaway's account of his geopolitical walk through Plymouth can be found in *Environment and Planning D: Society and Space 27* (2009), 1091–1116.

Many of the classic texts in political geography can still be found in university libraries, but it is often more informative to read more contemporary commentaries on these books and articles rather than the originals themselves. For more on Ratzel's theories, Bassin's paper on 'Imperialism and the nation state in Friedrich Ratzel's political geography' in *Progress in Human Geography* 11 (1987), 473–495, is a good overview. Ó'Tuathail's paper 'Putting Mackinder in his place' in *Political Geography* 11 (1992), 100–118, is a similarly good source on Halford Mackinder, as is Kearns's more recent book, *Geopolitics and Empire: The Legacy of Halford Mackinder* (2009). Smith's book, *American Empire: Roosevelt's Geographer and the Prelude to Globalization* (2004), details Isaiah Bowman's geopolitical work, whilst Herb's *Under the Map of Germany* (1997) is an interesting exploration of the perversion of cartography by German *Geopolitik*.

States and territories

Introduction

Rather like the air we breathe, states are organisations that surround us as individuals, influencing, and in many ways, offering sustenance to the lives we lead. Similar to the air we breathe, states are also organisations that often lie beyond the limits of our critical reflection. We may question the priorities of political parties; we may also disagree with the policies implemented by various governments. We do not often question, however, the character of the organisation that political parties, while in government, seek to govern. In other words, we rarely think about what states actually are, how they are constituted, how they come into being and how they change over time. These are some of the questions concerning the form of the state that we will ask, and ultimately seek to answer, in this chapter.

The first and most fundamental issue we need to deal with, of course, is what exactly is a state? Fortunately, a number of eminent social scientists have sought to answer this question. Max Weber, for instance, has argued that a state is a 'human community that (successfully) claims the monopoly of the legitimate use of physical force within a given territory' (Gerth and Mills 1970: 78). Michael Mann (1984) has built on this definition by arguing that any definition of states should incorporate a number of different elements:

- first, a set of institutions and their related personnel
- second, a degree of centrality with political decisions emanating from this centre point
- third, a defined boundary that demarcates the territorial limits of the state

- fourth, a monopoly of coercive power and law-making ability.

An extended definition such as this makes us think about a number of important aspects of the state. Significantly, it encourages us to think about the state in a far more abstract sense. In addition to the various paraphernalia associated with states in individual countries, there are, or at least there should be, certain underlying constants in states throughout the world, ones that are highlighted in the above definition. Rather than viewing the state in purely personal terms, therefore – as a supplier of public utilities, or something that is embodied in a senate or parliament, for instance – it makes us think of the underlying processes and institutions that (usually) help to constitute state bureaucracies.

So what do geographers and, more specifically, political geographers, have to offer in any study of the character of states? In other words, what can we gain from studying the state from a geographical perspective? We argue that there are three main reasons for doing so. First, geographers can help to illuminate the fact that states, when considered at a global scale, vary from region to region. One geographer whose work can be said to demonstrate this is Peter Taylor (see Box 1.2 in Chapter 1). Taylor has deployed the ideas of Immanuel Wallerstein (1974; 1979; 1980; 1989) regarding the existence of a capitalist world-economy – one in which the northern states of the first world thrive through their exploitation of southern states and people of the third world – in order to structure his understanding of the economic disparities that exist from one region of the world to another (see

Chapter 1). Crucially, this process of exploitation leads to significant differences in the political, economic and social viability of southern states in the 'periphery'. To put it bluntly, they do not have the money either to support their state institutions or their citizens. To states such as these, the badges of statehood – described by Mann – are ones to be aspired to, rather than being ones that reflect political reality. This is a theme that we will return to later in the chapter.

Second, geographers can help to highlight the unequal effect of particular policies on different areas within the territory of the state. These may range from policies that are explicitly targeted towards certain areas, for instance, policies of urban renewal in Europe and North America, to more general or 'national' policies. Even though these latter set of policies may be directed towards the state's territory and population as a whole, they may in turn have different impacts in different areas of the state, due to pre-existing cultural, social, economic and political geographies. So, for instance, a policy that seeks to encourage the use of public transport amongst the general public through the raising of taxes on private car ownership may work successfully in urban areas, where opportunities for the use of public transport proliferate, but may further disadvantage the inhabitants of rural areas, where levels of car dependency are far higher. Obviously, geographers have a key role here in studying, and attempting to alleviate, these problems. This is a theme that we will discuss at length in Chapter 12.

Third, and most fundamentally, geographers can contribute to our understanding of the state because of the state's effort to govern a demarcated territory. Indeed, this is the main justification for studying the state from a geographical perspective and it is a point worth emphasising. Broadly speaking, states use the notion of territoriality in two main ways. In the first place, territoriality is important in a material or physical sense. This is a relatively straightforward idea and relates to the fact that states always try to demarcate the physical limits of their power. A good illustration of the growing power of the state over the modern period (since approximately 1500) has been its efforts to demarcate its physical boundaries in a more precise manner. In this way, diffuse and ill-defined frontiers

became precisely delineated boundaries (Newman and Paasi 1998). One of the best examples of this physical territoriality lies in the context of the shifting boundary between the US federal state and Mexico (see Prescott 1965: 77–87; Donnan and Wilson 1999: 34–39). For much of the nineteenth century, a conflict existed between the two countries regarding their territorial extent. Much of the wrangling revolved around the precise location of the boundary between the two states. Critical here were natural features used to designate the boundary, such as the Rio Grande river. A large amount of the conflict concerned the changing course of the river over time and the implications this had for the territorial extent of the two states.

A more recent set of examples revolve around the geographical limits of the Chinese state. Hong Kong, for a century an integral part of the British Empire, was reinstated into the People's Republic of China in 1997, and there have been long-term conflicts regarding the independent status of Taiwan – which refers to itself as the Republic of China, but which is claimed by the mainland People's Republic of China to be part of its sovereign space. The ambiguous status of Taiwan can be traced back to the period immediately following the Second World War. In 1949 the Communist Party of China defeated the nationalist forces of Jiang Kaishek and gained control of the whole of the country, apart from the remote area of Tibet and the islands of Hong Kong and Taiwan. Since this period, the Chinese state has tried to gain control of these enclaves. Chinese subjugation of Tibet began in 1950 and Hong Kong was ceded from the British Empire to China in 1997. This has left Taiwan as the only remaining blot on the territorial integrity of China and the main focus of the Chinese state's enmity. Indeed, Taiwan has been a thorn in China's side for many reasons (Calvocoressi 1991). First, Taiwan was the place of refuge for the Nationalist forces of Jiang Kaishek after the revolution of 1949. Second, Taiwan's status as an independent state has been consistently been supported by the US. The main reason for US interest in Taiwan was its perceived need to maintain a series of territorial footholds throughout the Asian continent. American troops were stationed on the island until 1978 and contributed to the frosty relationship that existed between the US and China.

With the US withdrawal from Taiwan, China has begun to reaffirm its belief that Taiwan is an island that it can legitimately lay claim to. Much recent military posturing between the two countries – including missile testing and a series of military manoeuvres by the Chinese armed forces – has helped to reinforce the territorial significance of Taiwan. In this example, the state's key role in taking and maintaining physical control over a defined territory is clearly illustrated.

But, of course, the state's physical territoriality does not merely exist in the context of its external boundaries; all modern states are internally subdivided into a series of spatial units. Some of these spatial units may be the basis for the collection of taxes, as in the case of local taxes in the United States. Others may act as a territorial basis of law and order, such as prefectures that exist in France. Others still are key to the state's effort to collect information about its citizens. There are no better examples of this latter line of reasoning that the various territorial units used as a basis of collecting information for national censuses. As Hannah (2000: 115) has argued, 'state power requires at least the mapping and bounding of territory, the identification and registration of people living within it, and an inventory of resources', and the census acts as a crucial way of furthering this aim. Censuses in every country are dependent upon the existence of a series of territorial subdivisions, on the basis of which different kinds of data are collected, before being aggregated upwards into generic yet detailed statistical commentaries on a state's population (see Box 2.1).

BOX 2.1 STATE AND TERRITORY: THE US CENSUS IN THE NINETEENTH CENTURY

Matthew Hannah (2000) charts the emergence of a more systematic and dependable way of collecting information about the distribution and characteristics of the population of America during the late nineteenth century. As Hannah (ibid.: 118) notes, such a census of population is predicated upon the existence of the state as a territorial – and territorially subdivided – entity: 'it is impossible to undertake an accurate census unless there is some geographical framework on which to define and precisely locate enumeration districts which exhaust the territory without any overlap'. Such ideas were especially crucial during the late nineteenth century, when the US Superintendent of Censuses, Francis A. Walker, was considering ways of making the American census more efficient. The key issue for Walker was the need to have a territorial framework of subdivisions within the US, which could act as the basis of the collection and collation of information. Luckily for Walker, the vast majority of the US had already been subdivided into a series of territorial units – based on reference grids – and these acted as the territorial templates for the designation of enumeration districts. And yet, the use of reference grids as a way of framing the designation of enumeration districts also presented its own difficulties. In certain regions, enumeration districts were too large, being the same as counties. In other more populous regions, enumeration districts could contain over 20,000 people. In both these cases, the collection of data could be arduous. Despite these difficulties, the existence of such territorial units was – and has been – a crucial facilitator of the collection of information for the US state. Hannah's work is important for two reasons. First, he demonstrates the way in which the emergence of a more efficient kind of census in America derived in large part from the leadership of Francis A. Walker and, most especially, the lessons that he had learned as a serving officer in the Union Army during the American Civil War. Important military concerns regarding the mapping of space, resources and people were translated by Walker into the civilian sphere in the form of the census. In more general terms, Hannah's work shows how a territorial logic was central to the emergence of an effective census in America and, by extension, of state power in general.

Key readings: Hannah (2000).

In the second place, a state's territory is of key significance in an ideological context. What we mean by this is that the notion of territoriality is used by the state as a way of explaining its way of governing and ruling its population. In a sense, it is possible to argue that states do not govern people as such. Rather, they govern a defined territory and it is by doing so that they subsequently govern the people living within it. As Robert Sack (1983; 1986) has argued, this territorial method of control is in many ways a far easier way of governing than one that emphasises the direct control of people. Within this system, anyone living or working in, or even passing through, a state's territory is subject to the laws and policies of that state, regardless of their social, ethnic or cultural background. In this context, therefore, a state's internal and external boundaries are not only lines drawn on a map, ditches dug, or stones laid on a barren moor. They are this, but they are also far more: they represent the ideological basis of state power.

Our aim in this chapter is to explore the changing nature of the state over time. Adopting this historical approach enables us to demonstrate the development of some of the key features of state power. In order to accomplish this, we will focus first on the process of state 'consolidation' (Tilly 1990) that occurred from approximately the sixteenth century onwards. This gradual process led to the formation of the all-powerful states of the twentieth century. One useful way of exploring these changes is to examine the changing territoriality of the state, something we address in the next section. Here, we also briefly discuss the arguments that suggest that states are being systematically undermined by the processes of globalisation operating in the contemporary world. The final theme we discuss in this chapter is the process of 'exporting' the state from northern to southern states. This process has not been unproblematic, and we discuss the issues facing southern states in detail. Taken together, the themes discussed in the chapter make us appreciate the importance of territory to the state, as well as the need to explore the temporal and spatial variations in state forms.

The consolidation of the state

In this section we discuss the development of the state over the past four hundred years. We do not argue in this context that the state has only existed for this relatively short period of time. This is patently not the case. Ancient states existed as far back as 3,000 BC in Mesopotamia, the region occupied by modern-day Iraq (Mann 1986). We focus our attention here on the state from approximately 1500 onwards, since it is this relatively recent process of state formation that directly informs the political geography of contemporary states.

The sociologist Charles Tilly (1975, 1990) has argued that fundamental changes affected the state from approximately 1500 onwards, most clearly in Europe. During this so-called modern period, a series of far-reaching developments occurred to the nature of states, as they gradually became 'consolidated' into their present forms. By 'consolidation', Tilly means the way in which the states of this period – mainly in Europe – became territorially defined, centralised and possessed a monopoly of coercive power within their boundaries. Of course, this process of consolidation, which echoes some of the themes raised in Mann's definition of states, speaks of an earlier period when European states did not adhere to this organisational formula. The state of the earlier medieval period was a haphazard affair and included a plethora of different individuals and organisations claiming power over territory and space. These included kings, lay and ecclesiastical lords, religious organisations and free townspeople (see Anderson 1996). The process of consolidation that affected state forms during the modern period entailed the gradual abandonment of this medieval legacy and the development of states that possessed clearly defined territories, and which were capable of exploiting their land, people and resources in an efficient manner.

We can put some empirical meat on these bare bones by discussing specific studies that have sought to examine the consolidation of state power. Miles Ogborn (1998), for instance, has explored some of the key changes to have affected the English and Welsh state during the seventeenth century. Crucial to the consolidation of state power during this period was its ability to collect revenue from its people, resources and land. The main method employed by the leaders of the English and Welsh state was the excise duties – or in other words, the taxes raised – on the consumption of beer. Furthermore, a series of new mechanisms were employed in order to ensure the efficiency and consistency of this process. At one level, this meant that a number of different officers were paid to survey the process of brewing beer. This involved much travel throughout the country to ensure consistency in the way in which beer was produced and sold. At a more fundamental level, efforts to raise excise duties on beer production led to sustained attempts to comprehend the internal geometry of barrels and casks. According to Ogborn (1998), by 'mapping' the internal 'geographies' of barrels, the state could ensure that excise duties were raised in an efficient and consistent manner. Efficiency was important, of course, since it enabled the state to raise as much revenue as it possibly could. The consistency and equality of collecting excise duties was just as significant since it helped the state to legitimise the whole process to its citizens.

This example helps to draw our attention to an important feature of the consolidation of state power. States during the modern period were not solely concerned with collecting monetary resources from their people and territory, though this factor was without doubt crucial. Ogborn's study of the development of excise duties in early modern England and Wales is significant for another reason, for it emphasises the key role played by the collection of information and surveillance to the changing power of the state. In this case, the state had to know exactly how much beer was being produced. This general point has been well made by Anthony Giddens (1985). He has explored the crucial role of surveillance in the consolidation of state power. It is the act of collecting information, of recording it within the state bureaucracy, and of using it in order to govern a population, that is so critical to the consolidation of state power. Giddens (1985: 179) argues succinctly: 'as good a single index as any of the movement from the absolutist state to the nation-state is the initiation of the systematic collection of "official statistics".'

Key here is the notion of the infrastructural power of the state, or in other words, the power of the state to affect the life of its citizens in a routine manner (Mann 1984). The promotion of surveillance is both the product of, and the precursor to, the growth of the infrastructural power of the state. States require information about their population, their resources and their land in order to develop higher levels of bureaucratic control. In the same vein, the formation of state bureaucracies enables the development of more sophisticated means of monitoring the state's population and territory. Significantly for geographers, a key facilitator of this process of collecting information was the development of ever-more sophisticated ways of mapping a state's territory. We have already mentioned the significance of this issue in the introduction to this chapter but it is important to emphasise its significance. During the seventeenth and eighteenth centuries, in particular, new mapping techniques and procedures were developed in Europe in order to facilitate the consolidation of state power. While some efforts had been made by landowners prior to this period to map their own property, the significance of the seventeenth century was that it witnessed the beginning of state-driven efforts to produce property or so-called cadastral maps (Whitehead et al. 2007: 92). A new cadre of surveyors and mathematicians was employed and trained with a variety of states in order to produce systematic maps of the state's territory. As is shown in Box 2.2, it was only through this attempt to know a state's territory in detailed fashion that the process of state consolidation could proceed throughout the modern period.

As well as collecting information about the state's population and territory, another important aspect of the growth of state power during this period relates to the production of certain knowledges. Michel Foucault (1977; 1979) has consistently argued that a key feature of the growth of state power during the modern period lay in the state's ability to produce knowledges concerning its population. So, for instance, before the state could impose more restrictive rules and regulations on the practices of its population, it first had to develop a series of knowledges concerning the difference between acceptable and unacceptable

---

BOX 2.2   MAPPING TERRITORY IN SWEDEN DURING THE 'AGE OF GREATNESS'

The state-sponsored mapping of Sweden began in 1628 during the reign of Gustav II Adolf. Gustav II Adolf is best remembered for his military successes against Denmark, Norway and Poland and it is for this reason that his reign is remembered as an 'age of greatness' within Swedish history. As well as being a successful military leader, Gustav II Adolf also sought to promote far-reaching changes to the socio-political order within Sweden. The tax Swedish tax system and the Parliament (*Riksdag*) were reformed and similar developments took place in more cultural settings; a national education system was developed and there was a great emphasis on promoting the value of a single Swedish language within the state's borders. Authors such as Baigent (2003: 33) have argued that the far-reaching developments that took place during Gustav II Adolf's reign can be traced back to the king's own educational history; he was an individual who valued science and logic and, thus, was keen to use more systematic ways of governing his state. The implementation of a state-sponsored mapping exercise of the Swedish territory certainly formed a significant part of this general trend. State-sponsored cadastral mapping began in Sweden in 1628, when Gustav II Adolf formed the *Lantmäteriet* (the National Land Survey Board). The statutes that formed the *Lantmäteriet* stressed 'the first aim of mapping to be the defence of the realm [and] the rigorous recoding of the nation's resources' (quoted in ibid.: 51). A mapmaker general (*generalmatematiker*) was appointed as a way of furthering this aim and his first act was to replace the map surveyors, who had previously produced cadastral maps for private landowners, with state-trained cartographers known as *Lantmätare*. These officials of the state then began the process of producing systematic maps of the Swedish territory: *Skifte* maps, which showed property boundaries, and *Konceptkarta*, which were more detailed base maps showing the qualities of the land being depicted on the map (ibid.: 52). The value of these two kinds of map was that they were produced in a systematic and regular manner and, thus, for the first time, state administrators could actually compare the geographies of property boundaries and land quality from one region of Sweden with another. In broader terms, it can be argued that the creation of more systematic cadastral maps of Sweden was at once a product and producer of the consolidation of the Swedish state at this time. The development of a more sophisticated state apparatus was necessary for the emergence of more systematic ways of mapping Sweden's territory. Conversely, the production of more systematic maps of the Swedish territory enabled the state, in turn, to be more systematic in its methods of governing.

Further reading: Whitehead *et al.* (2007).

---

behaviour. In this way, considerable efforts were made to define deviancy, various forms of sanity and insanity and of morality and immorality (ibid.). Only by developing these sets of very specific knowledges could the state ensure that it was able to classify its population in the correct manner. In developing these knowledges and in classifying its population, the state was then able to take action against them, for instance by building prisons, madhouses, workhouses, where criminals, the insane and the immoral could be taken.

In focusing on these issues, Foucault seeks to draw our attention to notions of governmentality or, in other words, the rationality involved in government. By this, Foucault means the development of a 'way or system of thinking about the nature of the practice of government (who can govern; what governing is; what or who is governed), capable of making some form of that activity thinkable and practicable both to its practitioners and to those upon whom it was practised' (Gordon 1991: 1). This process of developing knowledges and rational forms of government, therefore, further fuelled the growth of the infrastructural power of the state.

One useful way in which we can think of these changes to the nature of state power is through reference to James Scott's (1998) ideas regarding states' efforts to make the society and territory that they govern more 'legible'. States during the modern period faced significant problems in their efforts to govern in an effective manner. At least part of these problems lie in the complexity of the societies and territories that states sought to govern. For instance, Scott (1998: 27–9) notes that methods of measuring varied greatly throughout Europe during the early modern period (approximately between 1600 and 1800). Since there were so many different means of measuring, Scott (1998: 28) argues that there was considerable potential for geographical variation in the amount of taxes and excise paid by various communities, so much so that we can refer to a politics of measurement:

> Even when the unit of measurement – say the bushel – was apparently agreed upon by all, the fun had just begun. Virtually everywhere in early modern Europe were endless micropolitics about how baskets might be adjusted through wear, bulging, tricks of weaving, moisture, the thickness of the rim, and so on . . . . How the grain was to be poured (from shoulder height, which packed it

somewhat, or from waist height?), how damp it could be, whether the container could be shaken down, and, finally, if and how it was to be leveled off when full were subjects of long and bitter controversy (see also Kula 1986).
>
> (Scott 1998: 28)

Key to the consolidation of state power, according to Scott, was their efforts to make the society and territory that they governed more 'legible'. What Scott means by this is the state's attempt to simplify the society that it governed and to make it more 'understandable' to the state's institutions. Instead of the great variety in methods of measurement, therefore, the state sought to impose its own, standardised means of measuring. Perhaps the best example of this process were the efforts made by the French state during the eighteenth and nineteenth centuries to replace the various local and regional means of measuring distance and to impose a rational and standardised system of measurement based on the metre and kilometre. Box 2.3 discusses another instance of states attempting to make society and territory more legible in the context of the growth of scientific forestry in Prussia (modern-day Germany). In all these projects, linked to attempts to standardise methods of naming, measuring, growing and even brewing, we see the state's determined efforts to create a more legible and, therefore, more governable and exploitable population and territory.

BOX 2.3 MAKING NATURE LEGIBLE: SCIENTIFIC FORESTRY IN NINETEENTH-CENTURY PRUSSIA (LOCATED IN CONTEMPORARY GERMANY)

A recurring theme with regard to the consolidation of state power has been the attempts made by the state to exploit its territory and people in a more efficient manner. One particularly critical resource for the state during the modern period was its forests. These enabled them to support indigenous industry and also acted as the material for the construction of buildings and ships. Forests, though, were originally a problematic resource in that they were very disorderly in the way in which they were organised. In mixed forests in particular, valuable timber was interspersed with trash and other varieties of trees and shrubs. For Scott, this resource was especially 'illegible' in nature. Starting in Prussia during the nineteenth century, however, a concerted effort was made to make this resource more legible, and therefore more manageable, for state foresters. New sampling and mapping techniques were developed that enabled foresters to quantify the amount of useful timber within a given area of woodland. Building on this, state foresters began to impose their own rationality on the forests of Prussia, so that they could create new forests, ones that were easier to 'count, manipulate, measure and assess' (Scott 1998: 15). Here, therefore, we see the beginnings of the regimented and uniform stands of trees, by today so familiar in many states. What is crucial here, however, is that this attempt to impose a certain rationality on a complex natural resource was linked to the efforts of states to consolidate their power. One key way of achieving this aim was to create a more rational, simplified, standardised, and as Scott puts it, more legible society and territory for the state to govern.

Key reading: Scott (1998) *Seeing Like a State: How Certain Schemes to Improve the Human Condition Have Failed.* New Haven: Yale University Press.

Here, therefore, we see the efforts of the modern state to consolidate its power over the long term. This is the process that led to the shift from the so-called puny states of the medieval period, ones that were 'marginal to the lives of most Europeans', into the all-powerful organisations that are 'of decisive importance in structuring the world we live in today' (Mann 1984: 209). The main question that arises in this context is what motivated these massive changes to the nature of the state? What was the driving force behind the consolidation of state power?

Michael Mann has convincingly argued that the main reason for the consolidation of state power during the modern period was the need to conduct wars. Using empirical evidence from the English state, Mann (1988) has mapped out the changing patterns of the state's finances over the very long term. There has been a general tendency towards an ever-increasing economic exploitation of the people and land of the English state. Critically, the main periods of growth correspond almost exactly to wars that the English state was involved in. Even during the limited periods of stability and peace that followed each conflict, state expenditure remained at a higher level than it had been during the pre-war period. This so-called 'ratchet effect' relates to the need to pay back the loans that were taken out by the state in order to sustain its war effort. Mann's work demonstrates clearly that the main impetus for the growing consolidation of state power over the long term was the need to promote its ability to wage war against other states, and the

equally crucial need to defend itself against aggressive neighbours.

An important theme, in this respect, is the close relationship between the external or international relations of states, and their internal or domestic political geographies. This point was made over fifty years ago by one of the fathers of political geography, Richard Hartshorne (1950), but it is an idea that is also apparent in Ogborn's work (see Chapter 3). Ogborn has argued that it was the English state's attention to the domestic political and economic geographies of excise duty collection that enabled it to build external colonies that extended throughout much of the known world. The same relationship between the external and internal political geographies of the state can be seen in the context of the interaction between the state's need to conduct wars and its efforts to tax its citizens in a more effective way (see Figure 2.1). The need for tax revenue had two main consequences. At one level, resistance to taxation could lead to increased surveillance of the population and the creation of a more repressive political regime. One could argue, for instance, that it is this resistance of war-induced taxation that – at its root – explains the formation of the police forces that exist in all states, as well as the development of ever-more detailed systems of law and order. Paradoxically, the need to sustain a war effort could lead to the political and civic emancipation of the citizens of a particular state. Painter (1995: 45) has argued that the state's need to raise taxes in order to support its armed forces in many ways explains the creation of more representative democracies within the state. In other words, state leaders were forced by different factions within the state to pay a price for the financial support they received during times of war. As Tilly has succinctly put it with regard to the English state of this period:

> Kings of England did not *want* a Parliament to assume ever-greater power; they conceded to barons, and then to clergy, gentry, and bourgeois, in the course of persuading them to raise the money for warfare.
>
> (Tilly 1990: 64, original emphasis)

Warfare *between* states could, therefore, have both repressive and emancipatory consequences *within* each individual state (see Figure 2.1). The exact nature of these consequences depended in large part on the balance of power that existed within each state (see Rokkan 1980 for a discussion of these varying alliances and conflicts within European states). Moreover these consequences were socio-economically and geographically uneven in nature, depending on the ability of each social, economic, political and cultural factions to resist state repression and, furthermore, to mobilise the state machinery for its own ends (in a general context see Jessop 1990a).

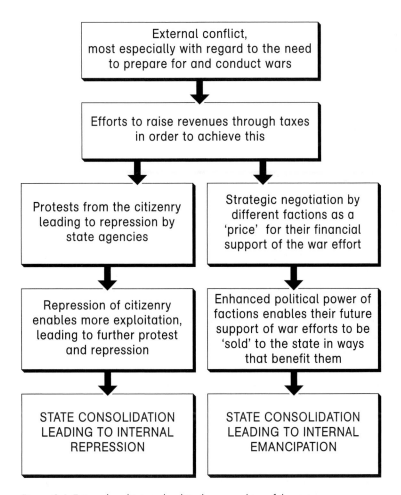

Figure 2.1 External and internal political geographies of the state

The other key point that comes from Mann's work on the financing of the English and Welsh state over the long term relates to the function of the state as an organisation (see Clark and Dear 1984). Contrary to the work of some theorists, who view the state as an organisation solely concerned with furthering the process of capitalist accumulation, Mann's work demonstrates that states – for a considerable period in their history – have been concerned with fostering and sustaining the means to wage war on their competitors. Admittedly, states broadened their range of functions to deal with the regulation of capital accumulation and related welfare issues during the nineteenth and twentieth centuries. This, however, is a theme that we discuss at greater length in Chapter 3.

In this section, we have discussed the consolidation of state power and have explored different theoretical contributions that have sought to explain this change. In the following section, we want to proceed to explore some of these themes in greater detail by focusing explicitly on the growing territorialisation of the state. In addition, we will explore those processes – most particularly revolving around the forces of globalisation – that are allegedly undermining the territorial integrity of the state.

Building state territoriality

One of the main consequences – and driving forces – of the consolidation of the state during the modern period was the increasing emphasis placed by state rulers on governing defined territories. As discussed in the introduction to this chapter, territoriality plays a key role in the material and ideological constitution of every modern state. In other words, states have to govern territories in order to secure their physical form. Similarly, states derive much of their legitimacy from the fact that they govern territories and not people. As Robert Sack (1986) has argued, the territorialisation of power can often cloud the repressive and exploitative nature of power relationships. This is part of the reason for the political success of states as they seek to govern groups of people.

A focus on the territoriality of the state helps us to answer a number of questions regarding state power. First, it enables us to think about one common aspect of the physicality of state forms, namely the extent to which most states during the modern period were of a medium size. Here it is useful to turn to the work of Hendrik Spruyt (1994) who has examined the changing territoriality of state forms in Europe during the modern period. Spruyt's key argument is that the European state has tended to gravitate towards a medium size and scale. Importantly, this was not, in any way, predestined to happen. Indeed, three different territorial formats were open to state rulers on the eve of the modern period: the city-state, such as Florence and Genoa; the extensive empires, such as the Holy Roman Empire and the Hanseatic League; and the medium-sized state, such as England and France. Crucially, according to Spruyt (see also Tilly 1990), the medium-sized state possessed a distinct advantage compared with the other two possible territorial formats, for the reason that it offered the most appropriate combination of economic and military power. Both were needed in order to sustain successful war efforts. Generally speaking, cities are sites of the production and consumption of capital. They are, therefore, well-suited to producing the economic resources needed to sustain war efforts. The economic power of artisans and guilds, however, made it difficult to coerce a large proportion of the inhabitants of cities into the state's armies. As such, city-states were not able to cope with the specific functions required of a state during the modern period. Large empires of the early modern period, however, were characterised by considerable coercive capabilities. They were filled with a warrior nobility at the head of large populations. Unfortunately, these extensive empires did not possess the economic might to support the large-scale military activity needed of a modern state. In effect, Spruyt argues that a successful warring state would require a combination of these two factors – cities in order to produce capital and a warrior class leading large armies. Medium-sized states possessed this balance between cities, as sites of capitalist production, and large agricultural hinterlands, acting as the territorial basis for its warrior leaders.

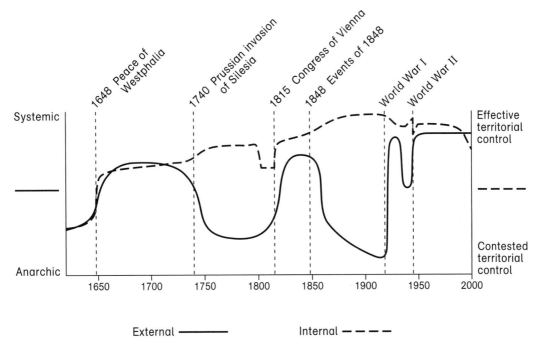

Figure 2.2 Changing 'external' and 'internal' state territorialities (after Murphy 1996)

Spruyt's work helps to draw our attention, therefore, to the key role played by a particular form of territoriality in promoting and sustaining the power of given states. States were far more likely to be successful in wars, and were far more likely therefore to survive as coherent political units, if they were of a particular size and scale. Of necessity, this point demonstrates the key role of geography in the constitution of states during the modern period.

Another key point made in the introductory section of this chapter regarding the territoriality of the state was the crucial role it played in sustaining the ideological integrity of the state as an organisation. Importantly, this territoriality has not been static in any sense. Alexander Murphy (1996) has explored these themes at length and he has suggested that we should distinguish between two interrelated aspects of state territoriality (see Figure 2.2). First, we need to think about territoriality as something that governs the relations between states: this, in other words, is the degree to which states within an international

state's system live by a series of rules and regulations, responsibilities and obligations. This aspect of territoriality largely relates to the international or external activities of states such as in the context of war and diplomacy. Second, state territoriality also exists at a far more fundamental level in the way it relates to the 'relationship between territory and power in a sovereign state system; its central focus is the degree to which the map of individual states is also a map of effective authority' (Murphy 1996: 87). Murphy refers here to the internal territoriality of a state, or in other words, the extent to which a state possesses a practical and effective control over all its territory. As can be noted from the schematic representation in Figure 2.2, there have been variations in the nature of state territoriality over time. Territoriality in an external or international sense has fluctuated wildly over time, ranging from predominantly stable or 'systemic' periods, when states, on the whole, respected each other's territorial integrity – for instance, during the period after the Peace of Westphalia in 1648, or after the Treaty of Versailles in 1918 – to more unstable or 'anarchic' periods, or in other words, times of war and territorial encroachment. We can think here of the international anarchy characteristic of the First and Second World Wars. Focusing our attention on the internal territoriality of states, the diagram portrays a far more stable pattern. Generally speaking, states have increased their ability to promote a more territorialised form of power over time. This relates back to Mann's (1984) ideas concerning the growing infrastructural power of the state over time. One of the main ways that states have ensured that they can reach out and govern their citizens in an effective manner during the modern period has been through proclaiming that all people living, and all the land lying, within the boundaries of the state are subject to the state's laws and coercive powers. In effect, the majority of European states for much of the modern period have sought to create a homogeneous and isomorphic state territory. The aim for many states has been to reach a situation in which distance from the centre of the state has little bearing on state's ability to govern and rule.

The aim of creating a homogeneous state territory was a difficult one, for the simple reason that it meant changing age-old traditions and customs in the various localities of the state. It was often easier to achieve this goal in new lands, less structured by community traditions. It is no surprise, therefore, that some of the more successful attempts to impose a rational territoriality on state space have been achieved beyond the boundaries of Europe. In Figure 2.3, showing an early map of North America, we see a clear effort to create what Scott (1998) has referred to as a 'legible' society and territory. The map shows the precise partitioning of lands that took place to the north-west in the late eighteenth century as territory was subdivided into townships and smaller parcels of land.

The efforts to create a territorial rationality for state space often involved a degree of conflict. In this regard, various options were open to state leaders who wanted to enforce their territorial order on state space. First, states could try to sponsor a process of centralisation whereby the political, economic, social and cultural norms of the centre were imposed on the state's periphery. Graham Smith (1989) has argued that this

is a significant feature of socialist states during much of the twentieth century. Socialist states' position as the prime directors of economic, political, social and cultural processes within their boundaries left little room for other actors – for instance, the forces of capital – to distort their vision of a wholly uniform state territoriality. For instance, the Soviet Union's monopoly of power enabled it to 'shape the spatial structure of society' in 'accordance with its own political preferences', based on notions of communism (ibid.: 323). This had enormous implications for the notion of territoriality within the Soviet Union, leading to sustained attempts to dissolve 'town-country' and 'region–region' differences within the country. In this 'grand plan' for the development of the Soviet state, little attention was paid to spatial variations in society and culture. The key factor that underpinned this drive was the vision of a homogeneous, and therefore 'legible', Soviet communist territory.

Second, the creation of a homogeneous state territory could occur through negotiation between the centre and the periphery. Ogborn (1992), for instance, has argued that many of the acts adopted in Britain during the nineteenth century as a means of creating a more homogeneous British territory were developed through a process of negotiation between Parliament in London and the various localities. Acts such as the Police Bill of 1856 and the development of poor relief were fine tuned as a result of political bargaining between centre and periphery. In no sense here was a homogeneous British territory created through an uncontested imposition of the norms of the core of the British state on its periphery. Building on this, we can think of a third mechanism for a state, eager to govern its territory in a more effective manner, namely that of federalism. In a federal system, some powers are delegated to various regions within the state whereas others are maintained at a federal scale. As such, it can be said that federal states represent somewhat of an emasculated form of territoriality: though some political and economic rights and responsibilities exist at a territorialised federal scale – thus forming a relatively homogeneous state space – others vary from province to province creating a mosaic form of territoriality (see Box 2.4).

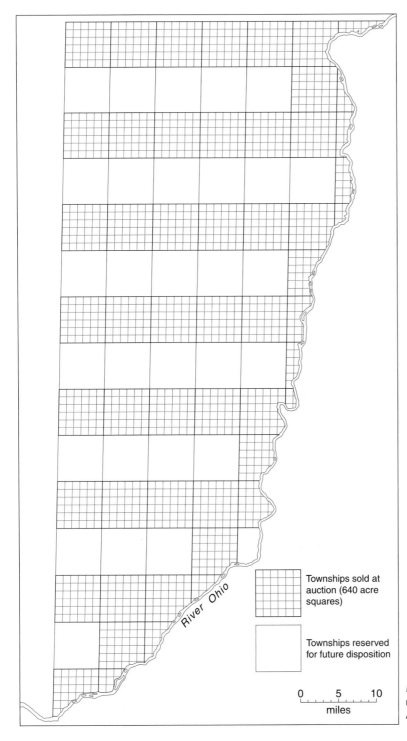

Townships sold at auction (640 acre squares)

Townships reserved for future disposition

River Ohio

0   5   10
miles

Figure 2.3 The imposition of a rational territoriality in North America (after Sack 1986)

BOX 2.4   TERRITORIALITY AND FEDERALISM IN THE US

One of the best examples of a federal state, incorporating a limited territorial homogeneity at a national scale, is the US. European expansion into North America between the sixteenth and nineteenth centuries progressively signalled a radical re-evaluation of the meaning of space compared with indigenous peoples' understandings of the land in which they lived. Mirroring Scott's ideas regarding the state's need to make society legible, many charters were granted to Europeans that described their newly acquired lands in a precise territorial manner. As Sack (1986: 136) notes, the charter granting land to William Penn in 1681 helped to define his land in a delineated manner. He was granted the land 'bounded on the East by the Delaware River, from twelve miles distance Northwards of New Castle Town unto the three and fortieth degree of Northern Latitude . . .'. At one level, therefore, there were considerable efforts to fashion a territorial basis for the new state that was forming in the US. At a grander political scale, however, there were many debates regarding the exact way in which this new state should be governed. The main sticking point was the potential size of the new state, and the belief in the difficulty of governing a large state in a democratic manner. After independence in 1776, Federalists, such as Madison and Jefferson, saw a federal state structure as one that offered the most democratic form of governance. According to these influential individuals, the smaller (provincial) states were favoured as the primary sites of governance, with the federal state scale limited to sectors that could not be governed at the scale of the individual state. In practice, this meant that the federal state's responsibilities were limited to defence and foreign affairs. As Jefferson argued (in Sack 1986: 148), 'it is not by consolidation, or concentration, of powers, but by their distribution, that good government is effected . . . were we directed from Washington when to sow, and when to reap, we should soon want bread'. The federal structure adopted in the US from this period onwards has, of course, profound implications for the nature of territoriality within the federal state. Federalism in the US, due to the devolution of considerable powers to various (provincial) states, has led to a limited territorialisation of power at the federal state scale, and an uneven pattern of territorial governance throughout the federal state.

Key reading: Sack (1986)

One final way in which territory helps to give shape to the power of the state is through the promotion of a state nationalism, which engenders a sense of loyalty amongst citizens towards the state and its institutions. This is a theme we will discuss at much greater length in Chapter 5. It is important to note at this juncture, however, that nationalism, as an ideology, emphasises the link between a group of people and a certain territory. As such, territory can be viewed as the concept that unites the bureaucratic and impersonal organisation of the state and the political and cultural community of the nation. It is another important reason for exploring the state – and the nation – from a geographical perspective.

Challenging state territoriality

In the above discussion, we have attempted to show how the state has sought to consolidate its power over the long term. It has done so by promoting: territorialised forms of power; the systematic collection of information; the development of new ways of 'knowing' its population. And yet, such arguments can give the impression that the state is all-encompassing and omnipotent. We can be led to believe that the state is all-seeing and all-knowing and that its power is absolute. This is patently not the case. As Alison Mountz (2003: 625, original emphases) has argued, 'the state *is* powerful, but *not* all-powerful and knowing'. Our aim in this section is to illustrate various ways in which the state – and more specifically its territoriality – is, and has always been, challenged by a variety of different processes.

One key way in which we can think about the challenge to the state is with regard to structural factors or, in other words, deep-rooted institutions and processes that constrain the state's ability to reproduce itself. A significant example of such a challenge lies in the context of the states of the global south. As we discussed in the introduction to this chapter, a number of authors have shown that there are fundamental structural inequalities between the states of the south and their counterparts in the north, largely as a result of the legacy of the empires that historically structured world politics (e.g. Flint and Taylor 2011). Primarily as a result of the imperial legacy, the leaders of southern states have sought to promote a political formation for their state that mirrors that of the various European metropolitan states. So, for instance, these new states have, on the whole, been organised territorially, and have incorporated a number of functions that are predominant in European and North American states (see Chapter 3). Vandergeest and Peluso (1995), for instance, have examined how state leaders in Thailand have sought to promote territoriality as a more effective way of controlling the country's people and resources. This project has unfolded in three main contexts: first, through the extension of a territorial form of civil administration into the rural areas of Thailand; second, through the promotion of survey-based land titles as the legitimate way to acquire and own land and, third, through the demarcation of certain lands as 'forests', which can then be acquired and exploited directly by the state.

Crucially, many southern states have faced considerable problems in achieving these goals. We do not argue, here, that northern states do not face significance challenges. We do argue, however, that the problems facing southern states are of a different magnitude to those experienced in their northern counterparts. Significantly, a number of the major issues facing southern states revolve around notions of territoriality. Vandergeest and Peluso (1995) have shown the difficulties experience by the state in Thailand, both in

promoting a uniform system of survey-based land titles and in defining large tracts of land as 'forest' to be exploited by the state. Major problems also arise with regard to southern states' attempts to control their more peripheral lands. Frontier regions within these states can often act as regional power bases for armed groups. We can think, for instance, of the *mujaheddin* rebels in Afghanistan, whose power base lay in the west of the country, the area farthest removed from the capital, Kabul. Other examples include the Tamil guerrillas located in the northern Jaffna peninsula of Sri Lanka and the southern regions of India.

The extent and depth of these problems has led some political commentators, along with influential organisations, to argue that 'southern states . . . have no choice but to reduce their expectations of what, as modern states, they can do' (Hawthorn 1995: 141). At one level, such an argument is morally distasteful since it opens the door to the potential neglect of the predicament facing southern states by the states and organisations of the north. We disagree with this position on ethical grounds. At a more theoretical level, however, the quotation raises interesting issues regarding the spatiality and temporality of state forms, and this is a theme we return to in the conclusion to this chapter.

Another important example of the structural challenges facing the state is the set of neoliberal processes associated with globalisation. Much is being made of the fundamental organisational changes that are affecting the state's functions. These changes are largely attributed to the forces of globalisation. As political, economic, cultural and social processes gravitate towards a broader, global scale, then the independence and autonomy of individual states throughout the world are allegedly being undermined. Importantly, these processes possess a territorial dimension. What we mean by this is that globalisation is often portrayed as something that is undermining the territorial integrity of states. If the states of the modern period can be described as 'power containers' (Giddens 1985), then there is some scope to argue that these containers are pitted with holes and are leaking badly (see Taylor 1995). We discuss the impact of globalisation on the state at greater length in Chapter 9.

A second set of reasons, which explain how and why the territoriality of the state has been challenged, revolve around the ability of individuals and groups in civil society – or, in other words, individuals and groups that lie outside state organisations – to question and sometimes undermine state authority. Even in states that seek to promote a wholly centralised notion of political power, there are numerous instances of the ability of citizens to challenge their power. France offers one example of this process. In his now-famous book, *Peasants into Frenchmen*, Eugene Weber (1977) discussed the efforts of the French state to mould its citizens into one community of people. Though this is linked to the creation of a coherent French nation, we argue that it is also symbolic of an effort to forge a common and uniform territoriality for the French state. Importantly, however, this project was not altogether successful. Even during the nineteenth and twentieth centuries, people living in the various localities in France were able to challenge the centralising tendencies of the French state, for instance, by preserving their own languages, dialects and customs. Another common way of challenging the territoriality of the state is through the act of smuggling and this is a theme we discuss in Box 2.5.

BOX 2.5 SMUGGLING AND STATE TERRITORIALITY: THE BORDER BETWEEN GHANA AND TOGO

The practice of smuggling is intimately bound up with state territoriality. By definition, smuggling depends on the organisation of the state: without state borders, smuggling would not have a reason to exist since it is the act of illicitly transporting or trading goods, services or information across borders that constitutes the criminal offence of smuggling. In one way, therefore, smuggling helps to reinforce the importance of state boundaries. At another level, of course, smuggling demonstrates how porous state boundaries have always been. It is a practice that challenges the notion of a clearly defined and homogeneous state territoriality. We can see this process at work at the border between Ghana and Togo. During the colonial period, when Ghana and Togo formed part of the British and French empires, respectively, attempts were made by the British to control the export of contraband from Ghana to its eastern neighbour. These included the construction of customs posts along key routes between the two countries and the implementation of monetary fines for those caught with contraband. These efforts were at best half-hearted and there was a general acceptance that some illicit trade between the two countries was inevitable. Indeed, it was welcomed, since it brought some degree of financial and social stability to the border region. Under British control, therefore, it was accepted that their control of the state of Ghana was only partially constituted along territorial lines. With independence, considerable efforts were directed towards 'hardening' the border between the two countries. Far harsher measures were adopted in order to discourage smuggling, including introducing the death penalty for those caught smuggling gold, diamonds or timber. This situation did not last long, however, mainly due to the ever-increasing levels of smuggling between the two states, and the evidence for the complicity of state officials in its practice. After a revolution in Ghana in 1982, the state realised that more lenient measures were needed to deal with smuggling, and it once again became a practice that was unofficially 'condoned' by the state. In this example, we see the uneasy relationship that exists between smuggling and the state: at one and the same time, they help to reinforce and challenge each other. As Donnan and Wilson (1999: 105) so aptly put it, 'smuggling . . . both recognises and marks the legal and territorial limits of the state and, at the same time, undermines its power'.

Key reading: Donnan and Wilson (1999).

Such attempts to contest the state – and its territoriality – can also occur in more subtle and mundane ways. Work on the state's interaction with the rural poor in India has shown how the state, and its impact on society, is conditioned by a range of state and non-state actors, each of which help to define the (territorial) form that state government takes (Corbridge *et al.* 2005).

The above research, as well as showing how people in civil society regularly contest the Indian state, also clearly demonstrates the way in which state employees themselves alter or reshape the state's organisations. The inherent danger with such work on India, of course, is that it can reinforce prejudices concerning the necessary corruptive qualities of states in the south. We argue, rather, that the practices and identities of state personnel reaffirm, contest and challenge state forms in *all* states. This is a third context, therefore, in which the state may be challenged: through the practices and identities of state personnel (Jones 2007). A number of authors have maintained that the state is a 'peopled organisation', rather than being 'an insulated domain of anonymous policy-makers' and 'authorless policy conventions' (Peck 2001: 451). In other words, we need to appreciate that state organisations are not staffed by faceless automatons; nor are state policies written by faceless bureaucrats. We need to consider, therefore, the way in which state personnel may facilitate and challenge the (territoriality of the) state.

A concrete example of the way in which state personnel can reproduce the states can be found in the context of some themes discussed earlier in this chapter; namely the way in which the growing consolidation of the early modern state was predicated on its war-making needs and capabilities (Giddens 1985: 112). Rather than thinking about this process in an abstract sense – in terms of the waging of wars between states or in terms of the development of new kinds of military technology – we can, alternatively, consider the role played by people in a number of different contexts. At one level, certain key individuals – Maurice of Nassau in particular – were influential innovators in the whole process of military modernisation and, in a related context, state consolidation that characterised the early modern period. Significantly, Maurice of Nassau's innovations led to the creation of a two-tier military machine that was based on different configurations of military agents and structures. While the upper echelons of the army, populated by members of the gentry, were based on a high level of knowledge of various administrative tasks – ones that were supported through the creation of military academies within the various European states – the lower reaches were characterised by a group of relatively deskilled ordinary soldiers. What is key, in this respect, is the effort made to shape the actions of individuals – whether through attendance at specialist military academies or through a prolonged process of training on the parade ground – and the subsequent impact of these new identities and skills on the conduct of an increasingly 'modern' form of warfare. Agents and structure were, thus, tied into closely articulated ties of interdependence and reproduction. The broader significance of this growing coordination of the armed forces within the early modern period, we would argue, was that it acted as a template for the later consolidation of the state. Notions of hierarchy, orderliness, rules and bureaucracy were easily translated from the battlefield into the emerging organisations of the state (see Gerth and Mills 1970).

At the same time, by conceding that the state is fundamentally peopled, then we also need to think about the ways in which the state may be challenged from within. The sociologist Bob Jessop (1990a: 269–70) has argued that the state

> has no power – it is merely an institutional ensemble; it has only a set of institutional capacities and liabilities which mediate that power; the power of the state is the power of the forces acting in and through the state. These forces include state managers as well as class forces, gender groups as well as regional interests, and so forth.
>
> (Jessop 1990a: 269–70)

If we take this claim seriously, then we need to think about the way in which the identities of state personnel – whether defined in relation to 'class forces', 'gender groups', 'regional interests' or any other kind of badge of identity – complicate the ways in which the state unfolds over time and space. An example of the impact of state personnel on the emergence of state policies is provided in Box 2.6.

Timing and spacing the state

In this chapter we have stressed the changing nature of the state, over time and over space. States in Europe back at the beginning of the modern period were relatively puny organisations, with little ability to affect the lives of their citizens. With the process of state consolidation, this situation gradually changed as states became more sophisticated and powerful in their degree of infrastructural control. Importantly for us as geographers, we can link this shift to a growing territoriality of the state: governing a territory came to be viewed as the most efficient way of governing a given population. Contemporary globalisation – amongst other processes – is allegedly challenging the functions and the territorial integrity of the state. As far as we are concerned, the jury is still out regarding the effect of globalisation on the territoriality of the state. These debates, however, help to reinforce the notion that the state is not a static organisation, and is always undergoing changes with regard to its territoriality and, as we shall see in Chapter 3, with regard to its functions.

Similar arguments can be made concerning the spatial variation of state forms. This was true for much of the modern period and is especially true of the contemporary state. The main difference, especially since the end of the Cold War, has existed between northern and southern states. Partly as a result of their recently created status as independent states, and

---

BOX 2.6   STATE PERSONNEL AND THE CONSTRUCTION OF HUMAN SMUGGLING IN CANADA

During 1999, the Canadian Federal Government intercepted four boats carrying immigrants being smuggled from Fujian province in China to Vancouver Island. 599 migrants were present on these four boats. 549 claimed to be refugees from China and, of these, 20 received refugee status. The majority were eventually deported back to China in 2000. The story became a key source of public debate in Canada. Mountz (2003: 623) notes, for instance, that the 'boat arrivals provoked a shrill response . . . among the Canadian public'. What is most interesting in the context of this chapter, however, was the response of Canadian state personnel to the episode. Following in-depth ethnographic research in the federal department of Citizenship and Immigration Canada (CIC), Mountz was able to chart the differing interpretations of the episode made by different state employees. Bureaucrats working in the federal capital of Ottawa, for instance, were able to construct the illegal immigrants in particularly negative terms. These state personnel characterised the immigrants as a 'criminal group', which should be ejected without delay (ibid.: 634). Other state personnel, who had come into more regular contact with the migrants – as prison workers in British Columbia, for instance – were far more ambivalent in terms of their discursive constructions of the group. As Mountz (ibid.: 635) notes, one state employee spent many hours on a aeroplane with some of the Chinese migrants as they were repatriated in 2000. By talking at length with the Chinese migrants, this employee re-evaluated his attitude towards the migrants: 'such employees experienced points of identification [with the migrants], rather than distance and abstraction'. Here, different attitudes arose amongst different workers concerning the issue of illegal immigration and, more importantly, concerning the sanctity of the state's borders. Such work illustrates the way in which differing state personnel interpret the state – and its territoriality – in contrasting ways.

Further reading: Mountz (2003).

partly as a result of their lack of economic viability, many southern states are struggling to maintain their territorial integrity. This again draws our attention to the changing nature of state form from one part of the world to another. A number of thorny questions arise with regard to this geographical patterning of state forms. Is the 'northern' form of the state necessarily the model for 'southern' societies? Is there a moral obligation on 'northern' states and societies to interfere in repressive state forms that exist in 'southern states'? Has the unstable state form of 'southern' societies the potential to undermine the stability of states elsewhere in the world? These are not easy questions to answer, but they deserve our sustained consideration and action, not only as political geographers, but also as ethical and moral citizens of the world.

Further reading

The best starting point for an explanation of the significance of territoriality to the state is Mann (1984), 'The autonomous power of the state: its origins, mechanisms and results', *European Journal of Sociology*, volume 25, pages 185–213. A good introduction to the changing nature of the state, especially in territorial terms, can found in Anderson (1996) 'The shifting stage of politics: new medieval and postmodern territorialities', *Environment and Planning D: Society and Space*, volume 14, pages 133–53. This paper discusses the changing character and importance of territoriality for the state over the long term and provides an outline of the increasingly tangled territorialities of the state under globalisation.

An examination of the contemporary significance of territoriality to the state, and the associated impacts of globalisation on state territoriality, can be found in two interrelated papers by Peter Taylor; see P. J. Taylor (1994) 'The state as container: territoriality in the modern world-system', *Progress in Human Geography*, volume 18, pages 151–62; P. J. Taylor (1995) Beyond containers: internationality, interstateness, interritoriality, *Progress in Human Geography*, volume 19, pages 1–15.

Fewer academics have attempted to chart the difficulties faced by southern states in promoting a territorial form of state power. An interesting study, however, can be found in P. Vandergeest and N. L. Peluso (1995) 'Territorialization and state power in Thailand', *Theory and Society*, volume 24, pages 385–426.

The state's changing forms and functions

The previous chapter discussed various aspects of the state's form. We focused on the historical development of the state and emphasised the significance of the state's efforts to control space as territory. As such, Chapter 2 concerned the internal political geographies of the state. The aim of this chapter is to complement this discussion by pressing the 'timing and spacing' button to focus on the changing forms and functions of the state. We specifically examine the changing institutional forms and functions of the capitalist state by drawing on a *régulation* approach to political economy and the state. The term 'political economy' is frequently used to discuss the interrelationships that exist between economic, social and political processes, which are forged through power relations as 'moving parts' (see Peet and Thrift 1989). The *régulation* approach has a neo-Marxist take on political economy that stresses the ways in which capitalism is managed through state, economy and society 'interactions' (Florida and Jonas 1991). Box 3.1 introduces capitalism and summarises the differences between Marxism, structural Marxism and neo-Marxism approaches to political economy. This chapter, therefore, suggests that state institutional forms and functions can be explored in relation to the ways in which states are embedded or 'integrated' into different economic, social and political processes; and vice versa.

The political geographer David Reynolds, writing during the early 1990s, remarked upon the increasing number of geographers drawing on approaches such as *régulation* theory to understand 'the behaviour of states as economic and geopolitical actors at a variety of territorial scales' (Reynolds 1993: 389). This research paradigm was considered important, because it was

taking political geography into new territory and giving progress to its intellectual development. Indeed, it took the British sociologist Bob Jessop five volumes to catalogue this work (Jessop 2001a) and later summarise its overall impact in no fewer than 479 pages (Jessop and Sum 2006).

We discuss the state's changing forms and functions from this perspective by first focusing on the origins of *régulation* theory and analysing its main arguments. Because the *régulation* approach is a challenging set of literatures, we then use case studies of how regulationist authors have applied this approach in their work and in doing so we draw out the changing institutional forms and functions of the state. The case studies analyse the *régulation* approach in relation to: economic and industrial geography; the geographies of scale within the context of neoliberalism; state intervention through modes of governance; the dynamics of urban politics and citizenship; the purported shift from the welfare state to the workfare state; and last, the cultural political economy of crisis management.

A rough guide to the *régulation* approach

The *régulation* approach emerged from a particular strand of French thinking during the early 1970s. Researchers at the Centre for Mathematical Economic Forecasting Studies Applied to Planning (CEPREMAP) in Paris were faced with an interesting set of problems that could neither be resolved through conventional economic planning nor be explained using existing theories of political economy (Jessop 1990b). Between

BOX 3.1   MODELS IN POLITICAL ECONOMY

What is capitalism?

Societies have moved through four different organising systems: primitive accumulation (bartering economies); antiquity (based on slavery); feudalism (supported by serfdom); and capitalism. Capitalism refers to a social and economic system that is divided into two classes: those owning the means of production (land, machinery, and factories, etc.) and those selling labour power. Under the capitalist mode of production, labour power is exploited to provide surplus value (or profit) and capitalists compete for this profit through a system that necessitates the 'accumulation of capital'.

Marxism: a critique of political economy

Karl Marx advocated an approach to political economy that he called 'historical materialism'. This materialist concept of history captured the shifting relationships between state, economy and state through struggles between opposites. This position was initially a critique of the model of political economy used by classical economists, such as Adam Smith and David Ricardo, whose work focused on production and exchange as somewhat isolated relationships. For Marx, a critique of political economy starts with property relations within different modes of production and then explores the relations between individuals in this context. Volume 1 of *Capital* took forward these concerns through an analysis of the commodity form, the nature of labour-power as a commodity, the labour process, the working day and alienation under capitalism. Volume 2 of *Capital* discusses the role of finance and money under capitalism. Volume 3 of *Capital*, which was completed after Marx's death by colleague Frederick Engels, focuses more on economic reproduction. Further volumes were planned to examine the state and other aspects of the capitalist mode of production.

Structural Marxism: a critique of classical Marxism

This was dominant in the 1960s and 1970s and had close relationships to political practice, especially in France. The leading thinker in structural Marxism, Louis Althusser, challenged what he saw as the technical and economic determinism within Karl Marx's thinking. Althusser introduced non-economic levels of analysis into a Marxist framework – such as consciousness and politics – and these were critically seen as 'relatively autonomous' because they formed an 'over-determined social structure'. This approach has also been termed 'base-superstructure' analysis, where historical materialism becomes a way of tracing out the connections between the main social elements. More recently, social scientists, drawing on the work of US economists Stephen Resnick and David Wolff, have revisited some of this thinking. Geographers such as J.K. Gibson-Graham use the term 'anti-essentialism' to reject economic determinisms of all kinds: they escape capitalism through developing anti-capitalist spatial analysis and anti-capitalist political strategies.

Neo-Marxism: rebel sons of Althusser

This mode of thinking originated, first, as a return to some of the principles of Marx's *Capital*. Authors such as Ernest Mandel and Paul Baran insisted on the necessity of rates of profit and labour theories of value as keys to studying the depressions of the 1970s. Another brand of neo-Marxist has been associated with development theory, and found in the work of André Gunder Frank. A third neo-Marxist strand can be located in critical theory associated with the Frankfurt School and more closely associated with the systems theory analyses of Jürgen Habermas and Claus Offe. A fourth strand challenges base-superstructure analysis and what is seen as the automatic reproduction of capitalism. Two distinct neo-Marxists groups argue that social action is situated within, but not reduced to, structural contexts. The social structures of accumulation school (SSA), seeks to explain the role of political and economic institutions in the making of capitalism. This is a North American approach found in the work of David Kotz, Michael Rich and colleagues. Another group of scholars answers this question by developing a regulation approach to growth and crisis, which uncovers 'mediating mechanisms' that help to bring about conflict resolutions under capitalism. Key regulationist authors are Michel Aglietta, Robert Boyer, Bob Jessop and Alain Lipietz.

Key readings: Jessop (1997a); Lipietz (1988); Peet and Thrift (1989).

the late 1960s and early 1970s, the 'Fordist' consensus (after the car manufacturer, Henry Ford) – based on economic planning, mass production, structured international financial systems and full employment – began to disintegrate. An international division of labour was emerging with newly industrialising countries, at the same time as widespread industrial unrest and declining productivity in developed capitalist economies, and France experienced stagflation (the coexistence of unemployment and inflation). Despite attempts to resolve this, state intervention exacerbated national economic instability.

Set within this context and also reacting against the structural Marxism of the 1970s (see Box 4.1), regulationists offered an analysis of socio-economic change that tried to understand the importance of rules, norms and conventions at a number of spatial scales (local, regional, national and supra-national) in the mediation of capitalism. Regulationists explore the regulation of economic life in its broadest sense, acknowledging that capitalist development does not possess its own 'self-limiting mechanisms' or follow an 'exclusive economic logic' (Aglietta 2000). Regulationists argue that socially embedded institutions and their networks, expressed as a series of 'structural forms', are critical to the endurance of capitalism, despite contradictions and crisis tendencies. The initial work of Michel Aglietta captured a concern with the roles played by trade unions, financial institutions and, perhaps most importantly, the state and its

changing institutional forms and functions under capitalism (Aglietta 1978).

This thinking has been extended by others through research on 'modes of regulation' (Lipietz 1988), 'modes of social regulation' (Peck and Tickell 1992) and 'social modes of economic regulation' (Jessop 1994) – terms that capture, amongst other things, the different institutional forms and functions of the state. For Robert Boyer (1990), the mode of regulation denotes five levels of analysis under capitalism: the wage relation, or wage-labour nexus; forms of competition and the enterprise form; the nature of money and its regulation; the state and its forms and

Regime of accumulation

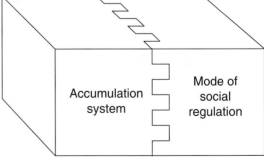

Figure 3.1 Regime of accumulation

Source: Redrawn from Peck and Tickell (1992: fig. 1), copyright 1992, with the permission of Elsevier

functions; and the international regime. When these act in tandem, a period of stable growth known as a 'regime of accumulation' is said to exist. Figure 3.1 depicts this relationship and Box 3.2 summarises some key terms in *régulation* theory.

Aglietta's research on the US national economy between 1840 and 1970 identifies five regimes of accumulation, each associated with a particular mode of state intervention, complementary economic system and form of state territoriality. Discussing the US after the Civil War, for instance, Aglietta talks about the importance of a territorial ideology called the 'frontier principal', which secured economic growth based primarily on agricultural production and the creation of urban commercial centres. This involved the 'domestication of geographical space' by those charged

---

BOX 3.2 THE *RÉGULATIONIST* VOCABULARY

Regime of accumulation (RoA)

This is used to denote a macro-economically coherent phase of capitalist development. There are connections here with 'long-waves' of economic growth, which emphasise technological phases of development (Marshall 1987). The regime of accumulation, however, is not reduced to purely techno-economic concerns: the RoA is forged through the 'structural coupling' of accumulation and regulation and this develops through 'chance discovery', involving trial-and-error experimentation.

Accumulation system

This explores the production–consumption relationship, whereby the individual decisions of capitalists to invest are met by demand for their 'products' through the market place. Convergence between production and its ongoing transformations and the conditions of final consumption can provide the basis for an RoA.

Mode of regulation

At one level this captures the integration of political and social relations, such as state action and legislature, social institutions, behavioural norms and habits, and political practices. For the purposes of undertaking research, modes of regulation can be unpacked as: the wage relation; forms of competition and the enterprise system; money and its regulation; the state and its forms and functions; and the international regime. The effectiveness of these institutional forms and their interrelations varies over time and across space.

Collectively these three terms allow regulationists to analyse the economy in its 'integral' sense, i.e. they are concerned with the social, cultural, and political context in which economic reproduction occurs. The spatial aspects of this are often expressed through:

Mode of societalisation

A term used to discuss the pattern of institutions and social cohesion, or the spatial patterning of regimes of accumulation.

Key readings: Jessop (1992); Tickell and Peck (1992).

with conquering territory and building railroads in line with a model of capitalism fostered on mobility and mutual competition (Aglietta 1978).

The *régulation* approach, then, is not restricted to Fordist analysis – a common mistake made by critics of this approach (see Brenner and Glick 1991). This said, it is more common for authors to use the example of Fordism – whose regime of accumulation can be analysed as a system supporting a virtuous model of production and consumption, and a mode of regulation consisting of: labour relations fostered on collective bargaining; the nationalisation of monopolistic enterprises; the creation and maintenance of national money; and the domination of the Keynesian national welfare state (Jessop 1992). Fordism also had a particular spatial pattern, or 'mode of societalisation', and links are frequently made between mass production and large-scale urbanisation in North America and Western Europe (Esser and Hirsch 1989; Brenner

1998) (see also Box 3.3). Last, given that the *regulation* approach is concerned with analysing the 'institutional infrastructure around and through which capitalism proceeds' (Tickell and Peck 1995: 363), this infrastructure varies within and between nation-states, according to the different sets of economic, social and political circumstances. Geography matters and Table 3.1 highlights the key national variants of Fordism. In each case, the state has a different institutional form and performs different functions to underpin models of economic development and also instigate social and cultural change.

What comes after Fordism?

Debates within the *régulation* approach have been preoccupied with what comes after the Fordist regime of accumulation. At the other end of the spectrum, a post-Fordist camp claims that flexible specialisation

Table 3.1 Variants of Fordism

Type of Fordist regime	Characteristics of coupling	Examples
Classic Fordism	Mass production and consumption underwritten by social democratic welfare state	United States
Flexi-Fordism	Decentralised, federalised state. Close co-operation between financial and industrial capital, including facilitation of interfirm co-operation	West Germany
Blocked Fordism	Inadequate integration of financial and productive capital at the level of the nation-state. Archaic and obstructive character of working-class politics	United Kingdom
State Fordism	State plays a leading role in creation of conditions of mass production, including state control of industry. *L'état entrepreneur*	France
Permeable Fordism	Relatively unprocessed raw materials as real leaders of economy. Private collective bargaining but similar macro-economic policy and labour management relations to classic Fordism. 'Bastard Keynesianism'	Canada, Australia
Delayed Fordism	Cheap labour immediately adjacent to Fordist core. State intervention played key role in rapid industrialisation in the 1960s	Spain, Italy
Peripheral Fordism	Local assembly followed by export of Fordist goods. Heavy indebtedness. Authoritarian structures coupled with movement for democracy, attempts to emulate Fordist accumulation system in absence of corresponding MSR	Mexico, South Korea, Brazil
Primitive Taylorism	Taylorist labour processes with almost endless supply of labour. Bloody exploitation, huge extraction of surplus value. Dictatorial states and high social tension	Malaysia, Bangladesh, Philippines

Source: adapted from Tickell and Peck (1995: table 1)

BOX 3.3 FORDISM AND POST-FORDISM

Fordism and post-Fordism can be analysed under three main headings: the relations of production, the socio-institutional structure and geographical form.

Fordism

Relations of production

Mass production, economies of scale, large firms and monopolistic competition, product and job standardisation

The socio-institutional structure

Collective bargaining through trade unions, demand management by the state and mass consumption through the welfare state

Geographical form

The manufacturing belt of the United States and the zone of industrial development in Europe stretching from the Midlands of England through North-West France, Belgium and Holland, to the Ruhr of Germany, with many outlying districts

Post-Fordism

Relations of production

Niche small batch production, economies of scope, small and high-tech firms, specialised products and jobs ('knowledge-based' workers)

The socio-institutional structure

Individualised bargaining and decline of trade union activity, supply-side state intervention and selective consumption through welfare privatisation

Geographical form

'New industrial spaces', such as Route 128, Silicon Valley and Orange County (North America), Baden-Württemberg (Germany), Emilia-Romagna (Italy) and Cambridge (Britain).

Key reading: Allen Scott (1988a, 1988b); Jessop (1992); Brenner (1998).

or flexible accumulation is emerging. This draws on developments taking place in industrial districts across North America and Western Europe, which have a model of economic growth built on flexible small firms and specialised high-technology production (Scott 1988a, 1988b). This is often supported by observations on localised modes of regulation. Sebastiano Brusco and Enzio Righi, for instance, draw attention to locally based institutions in Modena (North Italy), which have helped to forge a consensus around flexible economic growth (Brusco and Righi 1989; Dunford and Greco 2006). In extreme instances, authors such as Michael Piore and Charles Sabel selectively deploy regulationist language to push these localised observations further as one-region-tells-all scenarios. Flexible specialisation is presented as an economically and socially sustainable regime of accumulation (see Piore and Sabel 1984). Box 3.3 summarises some of the key characteristics of post-Fordism and compares these with Fordism. Debates on the 'new regionalism' in economic and political geography have been continuing these debates and exchanges (see Lovering 1999; MacLeod 2001).

At the other end of the spectrum, there are those sitting in the *after*-Fordist camp, which sees the contemporary stage of capitalism as still-in-crisis and not representing a new regime of accumulation. No prediction is made concerning the successor model to Fordism because regulation theory does not make any claims about the future (Peck and Tickell 1995). Their research focuses, amongst other things, on how the state's forms and functions have been changing in the *after*-Fordist era (see Brenner *et al*. 2003; Jessop 1994; Jones 1999; Moulaert 1996). These authors criticise post-Fordist accounts for generalising from a limited number of local case studies and over-emphasising the successes of this model to create sustainable and equitable growth (Amin and Robins 1990; Lovering 1990). By using *régulation* theory and *régulation* approach case studies, the remainder of the chapter discusses the changing institutional forms and functions of the capitalist state to get behind some of these ongoing debates and show their value for doing political geography.

Régulation approaches to the state: six examples

Economic and industrial development

The relationship between the state's institutional forms and functions and the economy is discussed in research by Sean DiGiovanna (1996). This analyses the roles played by 'institutionalised compromises' in the development of regions, using the *régulation* approach to compare the institutional basis and development of three economies (Emilia-Romagna in Italy, Baden-Württemberg in German and Silicon Valley in the US). These regions are selected because they broadly correspond to the model of industrial districts suggested by 'flexible specialisation theory' – where industrial clustering occurs within sectors associated with electronics, aerospace and high technology in general (see Krätke 1999). DiGiovanna's research reveals differences in both the institutional foundation and economic trajectory of each region, which results from characteristics within the mode of social regulation.

Box 3.4 summarises DiGiovanna's (1996) argument by analysing the three regions as different 'systems of regulation'. The first institutional form covers regional industrial relations and the structure of the labour market. DiGiovanna details how employers and employees relate to each other within the three regions, especially through skills, development and training policies. The second institutional form captures market-based relations and forms of competition. Attention is drawn to the size of firms and sub-contracting networks. Again, relationships are different in the three regions. Baden-Württemberg is characterised by large firms, whereas Emilia-Romagna and Silicon Valley rely more heavily on small firm alliances. The third institutional form deals with consumption regimes – such as market relations, inter-firm transactions, the different spatial structures of the firm and product flows within the regional economy. Striking differences exists between the three regions.

Last, and perhaps most importantly, DiGiovanna (1996) discusses the state as an institutional form capable of intervening in the economy to provide the necessary atmosphere for economic development. The state's forms and functions are very different across the three regions – an important argument that has been explored further in the work of Storper and Salais (1997) and Dunford and Greco (2006). In Baden-Württemberg the state, for instance, plays an important role in managing the production system and its industrial relations. In Emilia-Romagna, the

BOX 3.4 INSTITUTIONAL FORMS AND THE THREE INDUSTRIAL DISTRICTS

Emilia-Romagna

Wage labour nexus

Labour supply is segmented by employer and there is a clear divide between large firms (the primary sector) and small firms (the second sector). The primary sector is unionised and offers security, whereas the second sector is more precarious.

Forms of competition

Competitive advantage is secured from large numbers of specialist small firms, often working together on joint marketing and technology acquisition.

Consumption regime

Many firms are dependent on decisions made outside the region and few products are consumed in the region.

Role of the state

Local government is socialist or communist and heavily involved in social reproduction. There is a reluctance to be involved in industrial relations and the exploitation of labour often goes unchecked.

Baden-Württemberg

Wage labour nexus

A corporatist model with high rates of unionisation in large and small firms and the determination of wages by federal and regional patterns of negotiation. Security occurs by a commitment to training and skills development.

Forms of competition

Large firms dominate subcontracting relationships and lead developments in training and technology acquisition. Small firms are highly competitive and are hesitant to collaborate.

Consumption regime

Components are produced by smaller firms and consumed by larger firms within the region. Suppliers often follow firms in Baden-Württemberg to foreign locations.

Role of the state

The *Land* regional government is proactive in education, training and networking developments. The federal government is also supportive of research and development and technology diffusion. Struggles often occur between these two scales of regulation.

Silicon Valley

Wage labour nexus

Based around a bifurcated labour market model, where highly educated scientific workers often operate alongside low-skilled production workers (often women and immigrants). Unionisation rates are low and violations of health and safety standards are not uncommon.

Forms of competition

Dynamic strategic alliances exist between relatively small designers and customised equipment producers. This is augmented by extensive subcontracting between firms and collaboration on research, development and technical innovations.

Consumption regime

Products are created largely for large and small manufacturers in world-wide markets. Many products are aimed at niche markets, which are not accessed by large manufacturers.

Role of the state

Local governments are not primary players in Silicon Valley outside dealing with housing and environment concerns. Instead, economic development is fostered by venture capital institutions, founding firms, universities and the more informal networks of social capital.

Key reading: DiGiovanna (1996).

state's role is focused more on social reproduction and is more 'paternalistic'. Silicon Valley is an interesting example of the 'knowledge-based economy', where highly skilled workers are the source of innovation and economic success. National government expenditure on research and development and defence supports many of the organisational structures within this 'modern quicksilver economy' (Leadbeater 2000), whereas local-level government deals more with housing and environment concerns. Because California is a leading contributor to global warming, through high levels of car ownership and traffic congestion, environmental regulation is becoming increasingly important in this region. Moreover, private and increasingly authoritarian forms of non-state governance have emerged to manage an increasingly disparate suburban economic and political landscape (see Ekers *et al.* 2012).

Geographies of scale: local modes of social regulation

An interesting application of *régulation* theory can be found in the work of Jamie Peck and Adam Tickell. This examines the state's changing forms and functions in relation to economic, political and social processes, by incorporating the geography of scale into *régulation* theory (Peck and Tickell 1992, 1995). For these geographers, spatial scales – such as regions and localities— are fluid and actively produced, as opposed to being fixed and static. Scale is also relational; scales such as 'the local' should be analysed in relation to 'the region' and 'the nation', as well as in relation to each other (the process of combined and uneven development). Moreover, in the event of being produced, spatial scales can constrain some forms of activity and enable others to exist (see MacKinnon 2011; Smith 2003). By using the example of England's South-East region, a key space within Britain's response to globalisation, Peck and Tickell argue that political projects (in this case Margaret Thatcher's Conservative Party brand of neoliberalism) can mobilise geographical difference. They offer a regulationist reading of scale and uneven development by suggesting that modes of social regulation are mixtures of different 'regulatory systems', 'regulatory forms' and 'regulatory mechanisms' and these all operate at different spatial scales. Table 3.2 details these concerns, some of which point to the differently scaled forms and functions of the state.

Table 3.2 Regulatory forms and mechanisms at different spatial scales

Regulatory form/ mechanism	Spatial scale		
	Regional/local	Nation-state	Supra-national
Business relations (including forms of competition)	Local growth coalitions	State policies on competition and monopoly	Trade frameworks
	Localised inter-firm networks	Business representative bodies and lobbying groups	Transnational joint venturing and strategic alliances
Labour relations (including wage forms)	Local labour market structures and institutions	Collective bargaining institutions	International labour and social conventions
	Institutionalisation of labour process	State labour market and training policy	Regulation of migrant labour flows
Money and finance	Regional housing markets	Fiscal structure	Supra-national financial systems
	Venture capital and credit institutions	Management of money supply	Structure of global money markets
State forms	Form and structure of local state	Macro-economic policy orientation	Supra-national state institutions
	Local economic policies	Degree of centralisation/ decentralisation in state structures	International trading blocs
Civil society (including politics and culture)	Local trade union/ production politics	Consumption norms	Globalisation of cultural forms
	Gendered household structures	Party politics	Global political forms

Form of sub-national uneven development

Form of international uneven development

Source: reprinted from Peck and Tickell (1992), copyright 1992, with the permission of Elsevier

Peck and Tickell suggest that 'regional couplings' occur between accumulation and regulation and this gives rise to regional or 'local modes of social regulation'. This allows them to undertake a political geography of the South-East during the late 1980s, with this region representing a particular social structure within accumulation (Peck and Tickell 1992). Figure 3.2 details the South-East standard region, which covers the 'Home Counties' and the City of London. Peck and Tickell argue that in order to sustain a regime of accumulation uneven development needs to be contained and their research highlights the inability of Thatcherism to control growth, such that the South-East 'bubble' burst during the early 1990s. This region's model of economic growth was fuelled by a neoliberal ideology of 'individualism' and 'ownership' (see Box 3.5), which represented a challenge to the Fordist consensus of mass production, mass consumption and one-nation social democracy. This manifested itself as the consumer credit and mortgage boom of the late 1980s that, when combined with wage inflation resulting from skills shortages and recruitment difficulties, produced an overheated economy, and rapid and uncontrollable increases in

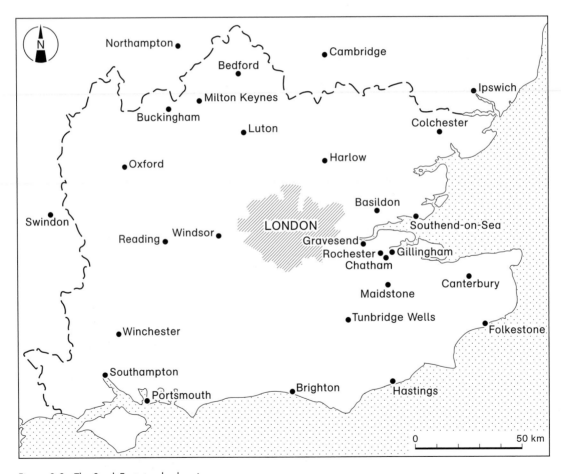

Figure 3.2 The South-East standard region

Source: adapted from Allen *et al.* (1998: map 2.1)

house prices. Peck and Tickell point out that these problems occurred partly because 'appropriate mechanisms for the regulation and reproduction of the economy had not been set in place' (1995: 35). This region suffered a regulatory deficit and this raises questions about the sustainability of post-Fordism.

Political geographies of the local state

The work of Mark Goodwin and Joe Painter has been important for developing links between the *régulation* approach, the local state and local politics. Goodwin's research has focused on the changing institutional forms and functions of the local state and how these can act as both an 'agent and obstacle' to regulation (Goodwin *et al.* 1993). Goodwin has argued that local states are products of uneven development: they have historically attempted to ameliorate the worst effects of socio-spatial polarisation by providing – through housing, education, and transport, etc. – the local means for securing collective consumption (Duncan and Goodwin 1988). Building on this, Goodwin discusses the ways in which *regulation*, its codes, and decision-making procedures, occur not in a national territorial vacuum, but through sub-national state agencies, which 'are often the very medium through

---

BOX 3.5   CHARACTERISTICS OF NEOLIBERALISM

Neoliberalism is a political philosophy stressing six central concerns

1   Liberalisation – promoting the free market.
2   Deregulation – reducing central state intervention and direct control.
3   Privatisation – selling-off nationalised and state controlled parts of the public sector.
4   Re-commodification – packaging remaining parts of the public sector to behave on a commercial basis.
5   Internationalisation – stimulating globalising market forces.
6   Individualisation – creating the opportunities for entrepreneurial activity within high-income earners.

Elements of this were initially realised in the 'new right' political strategy of *Thatcherism*, taken as the period of British political economy from 1979 to 1997 and covering the British Conservative Party under the leadership of Margaret Thatcher and John Major. According to Ray Hudson and Allan Williams:

> The Thatcherite project was above all else, an attempt radically and irrecoverable to redefine the relationships between the state, economy and society, and to break out of the old social democratic consensus of One Nation politics.
>
> (Hudson and Williams 1995: 39)

These characteristics can also be applied closely to US under especially the Reagan administration and public sector restructuring in New Zealand throughout the 1990s.

Neoliberalism has become increasingly dominant across the globe in the twenty-first century and for authors such as Harvey (2005) and Peck (2010), writing on the 'new political governance', neoliberalism restores class power goes hand in hand with authoritarianism, and dramatically realigns democracy.

Key readings: Allen *et al.* (1998); Hudson and Williams (1995); Brenner *et al.* (2010); England and Ward (2007); Harvey (2005); Peck (2010).

which regulatory practices are interpreted and delivered' (Goodwin *et al.* 1995: 250). Painter and Goodwin explore the notion of 'regulation as process' – where the institutional forms and functions of the state are not only seen as being associated with trying to secure stability; they are also concerned with managing fluidity, flux and change, which is 'constituted geographically' (Painter and Goodwin 1995). This emphasises the 'ebb and flow' of regulatory processes across time and space by using a 'modified version' of *régulation* theory that can explore the plethora of new institutions emerging in the local state. New institutions often incorporate business sector elites to undertake the delivery of economic development and have a specific, rather than a multi-functional policy remit, operating through territories smaller than those of local government (Peck 1995). Set within the context of a neoliberalist shift from 'managerialism' to 'entrepreneurialism' (Harvey 1989b), local authorities – key players under the Fordist mode of regulation and underwriters of many of its consumption norms – now operate within a system called 'local governance'. As we have detailed in Figure 3.3, governance captures the broader concern with how the local state is managed not only through elected local government but also through

central government, a range of non-elected organizations of the state (at both central and local levels) as well as institutional and individual actors from outside the formal political arena, such as voluntary organizations, private businesses and corporations, the mass media and, increasingly, supra-national institutions, such as the European Union (EU). *The concept of governance focuses attention on the relations between these various actors.*

(Goodwin and Painter 1996: 636, emphasis added)

Table 3.3 summarises this work on new developments taking place in local governance, which should *not* be read as applying only to Britain. New local 'sites of regulation' are common across many developed economies (see Brenner 2004; Brenner and Theodore 2002). Local governance, then, allows political geographers to present the local state as a system of

regulation that involves different actors and regulatory practices: sometimes this is based on government and at other times governance is the norm. It is, therefore, not accurate to talk about a binary shift from local government to local governance. Many have made this mistake (see debates in Valler *et al.* 2000) and have fallen into the same trap as those offering post-Fordist forms of analysis (see above). Governance present does not presuppose government past. Furthermore, developments need to be related to processes occurring *outside* the local state to assess the effectiveness of the new institutions. Painter and Goodwin use the term 'local regulatory capacity' to probe such issues, discuss the impact of these shifts within Sunderland (in the North East of England), and claim that there is little evidence of new institutional state forms and their functions providing the necessary mechanisms for stabilising a new mode of regulation (see Box 3.6). Sunderland has a 'deficit in local regulatory capacity' and some state forms and functions are 'clearly counter-regulatory' (Painter and Goodwin 2000). This last point has been explored further by others, who suggest that the state forms and functions become modified to deal with policy problems created by previous rounds of state intervention (Jones 2010). The German state theorist Claus Offe called this situation the 'crisis of crisis-management' (Offe 1984) and in some circumstances this raises issues of political legitimacy that can ultimately threaten the state's operation. This is discussed further in our next case study.

Urban politics, citizenship and legitimacy

This work on the local state's changing functions raises important issues of urban politics, especially in relation to citizenship. Ade Kearns, for instance, talks about the consequences of the shift towards multi-agency modes of delivery in the local state for 'senses of belonging to a community that lies at the heart of existential citizenship' (Kearns 1995: 169). Under Fordism, local government was important in creating 'certainty' and 'clarity': it was the main regulatory mechanism operating within the local state. With the arrival of non-elected local agencies, often drawing their personnel

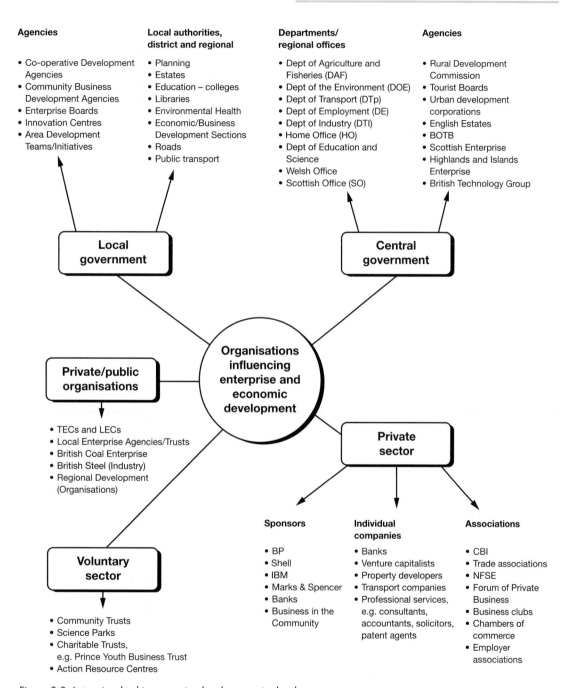

Figure 3.3 Actors involved in promoting local economic development

Source: redrawn from Richardson and Turok (1992: fig. 2.1)

Table 3.3 New developments in local governance

Sites of regulation	Local governance in Fordism	New developments
Financial regime	Keynesian	Monetarist
Organisational structure of local governance	Centralised service delivery authorities Pre-eminence of formal, elected local government	Wide variety of service providers Multiplicity of agencies of local governance
Management	Hierarchical Centralised Bureaucratic	Devolved 'Flat' hierarchies Performance-driven
Local labour markets	Regulated Segmented by skill	Deregulated Dual labour market
Labour process	Technologically undeveloped Labour-intensive Productivity increases difficult	Technologically dynamic (information based) Capital-intensive Productivity increases possible
Labour relations	Collectivised National bargaining Regulated	Individualised Local and individual bargaining 'Flexible'
Forms of consumption	Universal Collective rights	Targeted Individualised contracts
Nature of services provided	To meet local needs Expandable	To meet statutory obligations Constrained
Ideology	Social democratic	Neoliberal
Key discourse	Technocratic/managerialist	Entrepreneurial/enabling
Political form	Corporatist	Neocorporatist (labour excluded)
Economic goals	Promotion of full employment Economic modernisation based on technical advance and public investment	Promotion of private profit Economic modernisation based on low-wage, low-skill, 'flexible' economy
Social goals	Progressive redistribution/social justice	Privatised consumption/active citizenry

Source: adapted from Goodwin and Painter (1996: table 2)

from outside the locality and driven by service agreements and market ethics, Kearns highlights the emergence of 'confused citizenry' and somewhat diluted senses of place (ibid.: 169). For Kearns, fundamental tensions exist in neoliberal local governance between service-based and citizen-orientated strategies (Kearns 1992, 1995). We discuss this further in Chapter 4.

Using the example of Los Angeles, Purcell's work also considers the complex links between the changing forms and functions of the state, citizenship and political legitimacy (see Purcell 2001, 2002a). Working within the *régulation* approach, Purcell presents a 'consciously political conception' of the state to draw attention to bottom-up state-citizen relations that can challenge the state's functions. Although the state ultimately has political authority within its given territory (see Chapter 2), this is derived from the collective willingness of its citizens to be ruled. The state, then, is involved in a careful balancing act, set around what Purcell calls 'mutual expectations': citizens expect the state to meet, or at least perceive that it can meet, their obligations in return for territorial allegiance.

BOX 3.6 'LOCAL REGULATORY CAPACITY' IN SUNDERLAND

Fordism in Sunderland

Sunderland represents the model of 'blocked Fordism' as outlined by Tickell and Peck in Table 3.1. This locality in England's North-East – one of the five districts that constituted the former Metropolitan County of Tyne and Wear – was dominated by heavy industry (mainly shipbuilding and coal), had a dominance of male full-time employment, was regulated by high-levels of unionised workers and was unwritten by a welfare state system that provided a social wage. Fordist local governance also provided large-scale public housing, further supporting this model of economic growth. During the era of state modernisation in the 1960s, the Washington New Town introduced interventionist planning and considerable central government resources, which temporarily absorbed the crisis tendencies of Fordism.

Sunderland and the crisis of Fordism

As with most resource-based regional economies, Sunderland went through a period of intense economic restructuring during the 1970s. The heavy industry of the past gradually disappeared and was replaced by an expanding service sector and Washington New Town attracted small-scale manufacturing that offered low-skilled work. By the early 1990s, two-fifths of the male population of working age had no direct income from a job, whereas half of women worked in low-skilled and often part-time jobs. Sunderland was very much becoming an industrial wasteland.

Is Sunderland post-Fordist?

During the 1990s, Sunderland experienced an entrepreneurial city council through a shift from single agency approaches to partnership-based organisations. This had the potential to provide the institutional basis for a sustained period of post-Fordist growth. The Tyne and Wear Development Corporation, Sunderland City Challenge, Sunderland City Training and Enterprise Council, The City of Sunderland Partnership and Sunderland Business Link were the key players in economic development and were largely involved in supply-side interventions to promote the locality to inward investors and increase the skills of the unemployed in a shrinking labour market. These institutions promoted Sunderland as 'the advanced manufacturing centre of the North', but without co-ordinated demand-side intervention policies, these institutions did not possess 'regulatory capacity' to intervene in the locality and regulate the contradictions of *after*-Fordism. The local state is driven by agendas and funding regimes determined *outside* the locality. The City of Sunderland Council and the various partnership organisations also have little impact on wage relations and the norms of collective consumption: they dealt with fire-fighting the consequences of *after*-Fordist economic decline, rather paving the way for post-Fordist high-tech prosperity.

Key reading: Painter and Goodwin (2000).

Purcell demonstrates this through research in Los Angeles and Box 3.7 summarises how the state changes its interventions in response to state–citizen tensions. LA contains 3.8 million people and is an incubator for 'secession movements' – groups that have turned their back on mainstream political parties and prefer to pursue more unconventional ways of making themselves heard (Purcell 2001, 2002b). Purcell gives examples of such movements in the San Fernando Valley, a 'microcosm of the twentieth-century suburban America' (Purcell 2001: 617), focusing on an organisation called Valley VOTE (Voters Organized Together for Empowerment). Also by drawing on the Staples Center Project – a previously rundown area of LA that is now the 'entertainment centre of the world' and home to basketball, ice-hockey and football teams – Purcell explores the tensions between the 'competition state' (Cerny 1997; Fougner 2006), focused on economic competitiveness strategies such as inward investment, and the maintaining of political legitimacy. The politics of economic development thus relates to defending a form of growth *and* preserving state-citizen relations (Purcell 2002a and 2002b).

Towards workfare states

Since the middle of the twentieth century, the welfare state has dominated the political landscape of North America and Western Europe. During the last twenty years and set firmly within the context of globalisation, advanced nations have been addressing the problems associated with economic decline and spiralling

---

BOX 3.7   CRISES OF LEGITIMACY IN THE LOS ANGELES LOCAL STATE

VOTE is a coalition of Valley business interests and Valley homeowners' groups. Both are influential forces: in some cases homeowners' groups have 2,000 members. Although there are conflicting agendas within the coalition – with homeowners wanting controlled growth and development and the business community advocating *laissez-faire* land use policy and low tax – they agree on key reasons for secession. Both sides feel that the City of Los Angeles (local government) is too large to be responsive to local needs and has been 'short changing' the Valley in terms of providing the necessary level of services for collective consumption. Interesting examples here are struggles over the ownership of water, with the urban population being privileged over suburban interests in the Valley region. After many struggles, the City of Los Angeles was forced to launch an independent charter reform commission, which recommended a rewriting of the charter for public services to defuse organisations such as VOTE.

In the example of the Staples Center Project, a local commercial real estate agent, who at the same time was an advisor to the Mayor, was offering tax breaks and nominal rents deals (worth $70 million and 25 per cent of the arena's estimated cost) to attract developers to this area. The rather clandestine processes at work here were uncovered by a populist councillor, Joel Wachs, who started a group, Citizens Against Secret Handouts (CASH). This argued that the Mayor's office should offer less costly incentives to develop the site. Because Wachs's concerns were not taken seriously within City government, Wachs balloted citizens in a move that would require a voter referendum, enforced by city law, for any new sports facility. An anti-arena movement quickly developed and it appeared that Wachs' efforts would derail the project. The developers avoided this by renegotiating a deal whereby they would absorb the costs of the project, with minimal costs being picked up by the City government. The 'city chose to assuage the discontent among its citizens rather than meet the imperatives of economic development' (Purcell 2001: 308).

Key readings: Purcell (2001, 2002a, 2008, 2013); and http://www.secession.net/

public expenditure by restructuring the institutional forms and policy functions of the welfare state. For *regulation* theorists, welfare state restructuring entails the displacement of 'passive' with 'active' forms of labour market regulation (see Jessop 2002; Peck 2001). Within active labour market regulation, the 'work ethic' is being used to reconfigure the universal rights and needs-based entitlements to welfare that characterised the state's historical commitment to full employment and social rights for all citizens. The term 'workfare' – literally meaning welfare + work – is becoming increasingly dominant in the political vocabulary as a means of securing a 'new paternalist' relationship between the state and its subjects (Mead 1997). Workfare introduces strict behavioural requirements and new social responsibilities to encourage the unemployed to become more employable and job-ready through compulsory participation in training and education programmes. The workplace is also presented as the best means of avoiding poverty through slogans such as, 'I fight poverty, I work'. Workfare is frequently legitimised as a reaction to economic globalisation (see Chapter 9) through the need to secure labour market flexibility as the basis for competitiveness and based on these changes to the state's functions Jessop suggests that we are moving from a Keynesian Welfare State to a Schumpeterian Workfare State. This also has implications for the state's institutional form (see Box 3.8).

BOX 3.8 FROM KEYNESIAN WELFARE STATES TO SCHUMPETERIAN WORKFARE STATES

Jessop highlights a new era for the institutional forms and functions of the state that is associated with the shift from Keynesian welfare to Schumpeterian workfare. In more recent work, this is expressed as a movement from Keynesian welfare national states (KWNS) to Schumpeterian workfare postnational regimes (SWPR).

Keynesian welfare states

After the British political economist John Maynard Keynes . . .

Function of the state

The KWNS supported full employment through demand management, provided public infrastructure to support mass production and consumption, and ensured mass consumption through collective bargaining and the expansion of welfare rights.

Form of the state

The national scale was used for state intervention in economic and social policy-making, with local as well as central modes of delivery.

Schumpeterian workfare regimes

After the Austrian political economist Joseph Schumpeter . . .

Function of the state

The SWPR supports supply-side innovation and competitiveness through promoting open economies and subordinates social policy to the needs of competitiveness by pushing down wages and promoting low skilled employment.

Form of the state

The national scale is no longer the dominant scale for state intervention, with the emergence of devolved local and regional 'partnerships' and networks.

Key reading: Jessop (1994, 2002, 2008, 2014).

Workfare began in the buoyant labour markets of North America (such as California), where since the 1960s state governments have been experimenting with mandatory work and training programmes to reduce the welfare caseload. This increased throughout the 1970s and the 1988 Federal Family Support Act required state governments to provide mandatory work or training activities for welfare recipients as the condition of receiving benefits. Based on local success stories – such as Riverside, California (see Box 3.9) – workfare was claimed to be a national success and was transferred across North America through think-tanks and political advisors (Peck 2001). Welfare state restructuring is, therefore, an excellent example of what McCann and Ward (2011) have termed 'mobile policies' – the facilitated movement of policies across geographical space. Workfare became intensified under the Personal Responsibility and Work Opportunity Reconciliation Act, which replaced the sixty years old Aid to Families with Dependent Children programme (AFDC) with block-granted welfare payments to the state level and also introduced a time-limited unemployment benefit system. Signing this Act in August 1996, President Clinton famously argued this was about 'ending welfare as we know it'. Critics highlight the impacts that this strategy is having on the plight of welfare children and the deepening of America's economic problems (Baratz and White 1996).

BOX 3.9   BORN IN THE USA: WORKFARE IN RIVERSIDE, CALIFORNIA

The Riverside Greater Avenues for Independence (GAIN) model concentrates on moving welfare recipients into work as rapidly as possible and with minimum costs. Evaluations have shown it to be successful in driving down costs and accelerating the process of labour market re-entry, although there is no evidence that it can lift participants out of poverty. The Riverside model has excited much interest across North America and Western Europe, because of its 'pure' workfare appeals. It is a no-frills, high-volume, low-cost way of enforcing work participation and work disciplines. Peck's research uncovers a 'new mode of labour discipline' that seeks to conscript the poor into low-wage, or contingent, work. Peck discusses the various strategies used by officials in Riverside and draws attention to the consequences that these have on local labour markets.

Key reading: Peck (1998).

BOX 3.10   RESISTING WORKFARE

Workfairness

Workfairness is a New York-based organisation of workfare workers and their supporters that emerged as a response to New York's Work Experience Program (WEP). Workfairness has been campaigning for a better deal for workfare workers and has challenged the 'new paternalist' ideology of blaming the unemployed for their position in society.

> People on welfare have been stereotyped, maligned in the media, and made into scapegoats for the politicians, the rich and powerful to target. The truth is that Workfare mothers get up in the morning just like any other worker, they see their children are cared for, and they go to work. Workfare workers work very hard, and they are proud of the work they do. They don't want to be cheap replacements for their friends and neighbors fortunate enough to have union wage jobs. Workfare workers want permanent jobs at union wages. They want to join unions. They want respect, dignity and equality. These are the things that WORKFAIRNESS, and others are trying to fight and win.

Key reading: Holmes and Ettinger (1997); http://www.iacenter.org/workfare.htm

Community Voices Heard

Community Voices Heard (CVH) is an organisation of low-income people, mostly women on welfare, working together to improve the lives of the poor in New York City. It is run by low-income people on welfare. CVH uses public education, public-policy research, community organising, leadership development, political education and direct action issue organising, to campaign around issues such as 'welfare activism'. In the late 1990s, CVH lobbied New York City politicians to ensure that welfare reform moved people out of poverty by creating jobs, job training, education and childcare. CVH has also been developing grass roots leadership among women on welfare to recognise their power and potential to impact public policies that impact on their daily lives.

Key reading: http://www.cvhaction.org/

Welfare state restructuring in North America has not gone unchallenged at the local level and new spaces have been opened up for contesting the state through political activism within civil society. Organisations such as Workfairness and Community Voices Heard in New York have been lobbying since the mid-1990s for regulatory standards to minimise the exploitation of labour and to get workfare workers unionised. Box 3.10 describes these organisations, Figure 3.4 details a Workfairness anti-workfare leaflet, and Plate 3.1 is a protest against workfare in New York.

These developments are not isolated to North America. Ivar Lødemel and Heather Trickey (2000) and the OECD (1999) highlight similar trends occurring across Western Europe, but in doing so reveal subtle differences in the changing institutional form and function of the welfare state. Authors such as Gøsta Esping-Andersen (1990) and Evelyne Huber and John Stephens (2001), attribute geographical

DON'T CUT
OUR
FOOD STAMPS

RESTORE EMERGENCY ASSISTANCE TO PEOPLE ON PUBLIC ASSISTANCE
REAL JOBS NOT WORKFARE !

COME TO THE **PROTEST**

ON THE STEPS OF **CITY HALL**

FRIDAY APRIL 16th **at 12:00 PM**

(Take the east-side trains to the City Hall stop, and west side trains to Chambers St.)
Join our supporters, speak up for yourself, *IT'S YOUR RIGHT*

Mayor Giuliani is cutting all food stamps to people on public assistance between the ages of 18 and 50 after three months beginning April 1, 1999. This cruel act will deprive more than 25,000 people of food. To make matters worse, Giuliani is also ending emergency assistance to poor people who are evicted. City Hall refuses to support a jobs bill for WEP workers and people on public assistance.

The mayor can reverse the food stamp cut off, the elimination of emergency assistance to people in need, and sign legislation creating jobs instead of slave labor workfare. Come to the City Hall Protest to demand that Giuliani **STOP TREATING PEOPLE LIKE DIRT**!

For more information, or if you need assistance to get to City Hall on April 16, call:
WORKFAIRNESS *an organization of WEP workers, people on public assistance, and there*
supporters - 39 West 14th St. #206, NY, NY 10011 **Phone (212) 633-6646** Fax (212) 633-2889

Union Labor Donated

Figure 3.4 Workfairness leaflet

Source: reprinted by courtesy of Workfairness, New York

differences to the role of different interest groups that can influence those holding power within the institutions of the welfare state.

The arrival of the 'new paternalism' in the UK, for instance, has been more recent. Due to trade union pressures, workfare was resisted during the 1980s and the post-war labour market settlement remained more or less intact until the 1996 Jobseeker's Allowance required a strict 'agreement' between the 'jobseeker' and the state, as a condition for the receipt of benefits

Plate 3.1   Anti-workfare protest, New York City, April 1999

Courtesy of Martin Jones

Jones H 1996). With the arrival of the New Labour administration in 1997, elements of workfare – presented as 'welfare-to-work' – have been evident in the various New Deal programmes. These require participation in a series of 'options' in return for welfare benefits. Priority is given to immediate placement in the labour market to embed the work ethic at the earliest opportunity (Peck 2001). Gordon Brown, at the time Britain's Chancellor of the Exchequer, famously argued that: 'Rights go hand in hand with responsibilities, and for young people offered new responsibilities . . . there will be no third option of simply staying at home on full benefit doing nothing' (Brown 1998: 1). A fascinating study of Britain's changing work–welfare regime at this time has been undertaken by Peter Sunley, et al. (2006), which points to how policy makers need to be aware

of how institutional spaces and labour market conditions interact to produce local knowledges.

The New Deal was renamed the 'Flexible New Deal' in October 2009 and replaced in the summer of 2011 by the UK Coalition Government's 'Single Work Programme'. The effect has been to align the UK more closely to the US model noted above, with similar resistance politics being displayed (see Box 3.11). Authors such as Linda McDowell (2005) have considered the complex gender implications of this shifts in state form and function and MacLeavy (2011: 355) has argued that

the reorientation of state assistance towards work, coupled with the proposed simplification of working-age benefits and tax credits, is argued to present a particular challenge to the financial

BOX 3.11 BOYCOTT WORKFARE

Boycott Workfare is a UK-wide grassroots campaign, formed in 2010 by people with experience of workfare and those concerned about its impact, to end forced unpaid work for people who receive welfare. According to their website, 'workfare profits the rich by providing free labour, whilst threatening the poor by taking away welfare rights if people refuse to work without a living wage'. Boycott Workfare aims to 'expose and take action against companies and organisations profiting from workfare; encourage organisations to pledge to boycott it; and actively inform people of their rights'. Claimed not to be a 'front for any political party, or affiliated with any political party', those who share the views of Boycott Workfare are encouraged to find a group near to their location. This ranges geographically from Birmingham (Birmingham Against Cuts) to Edinburgh (Edinburgh Coalition Against Poverty) and Leeds (Leeds Unemployment Action Group).

Key reading: http://www.boycottworkfare.org

security and autonomy of women, signaling the end of the process of modernizing the welfare system that was forged around the single earner family model in the period of post-war austerity.

In contrast with the neoliberal approaches of the US and UK, research undertaken in Denmark has suggested that a welfare-*through*-work strategy is being deployed and this retains many of the state's welfarist labour market functions (see Box 3.12).

The cultural political economy of crisis management

Bob Jessop has suggested that the *régulation* approach needs to be combined with other approaches, such as state theory and discourse analysis, to get a better handle on political economy (Jessop 1995). Working

BOX 3.12 WELFARE STATE RESTRUCTURING IN DENMARK

Denmark has adopted a 'welfare-*through*-work' model. Due to the power of the labour movement and the pressures exerted against the state by gender movements, this retains some key social policy functions. First, social partnerships have been strengthened in respect to the delivery and implementation of welfare. Second, financial planning and decision-making has been decentralised to regional-based institutions. Third, the unemployed have been given rights to counselling, an individual action plan and, more importantly, access to a comprehensive package of leave schemes including job training, education and childcare. A key aspect of this model is a work-sharing scheme called 'job-rotation', whereby the unemployed are recruited and given direct job training experience in posts vacated by (predominantly) unskilled workers, who in turn are given the opportunity to update their training and education. The unemployed receive both work experience at trade union negotiated rates as well as additional vocational training. This benefits the firm by providing the sustainable basis for an up-skilled workforce, without loss in employment. This initiative was promoted in 14 different countries, through a European Union funded transnational programme, but is under threat by work-first neoliberal labour market strategies.

Key reading: Etherington and Jones (2004a, 2004b); Torfing (1999).

for two decades with Ngai-Ling Sum, Jessop and Sum have advocated a post-disciplinary approach to political economy that takes a 'cultural turn' seriously (see Jessop and Sum 2001, 2006; Sum and Jessop 2013). Calling this Cultural Political Economy (CPE), they look at 'semiotic aspects' of political economy, i.e. not just the linguistic and narrative turn (culture, language, visual images, media, etc.), but the construction of inter-subjective meanings (researched using techniques such as discourse analysis, content analysis and conversational analysis). CPE demands attention to 'sense and meaning-making', as this does not reduce culture to language or discourse. Culture is redefined to encompass the ensemble of social processes by which meanings are produced, circulated and exchanged. This allows political geographers to look at the inter-connected and co-evolution nature of the economic, the political and the social-economic relations more broadly – through the creation of 'economic imaginaries'. Imaginaries shape the interpretation of crises and the responses thereto. Four 'regulation as process' (see above) research agendas follow from this (Jessop and Sum 2001: 98–99):

- How are objects of economic regulation and governance created and how do they survive?
- How are the actors/institutions and their modes of calculation constituted and how do they interact to produce these objects in both discursive and extra-discursive fields of action?
- What are the specific practices for creating narratives and stories about knowledge in specific institutional contexts in order to reinforce and reproduce these objects over time and space?
- What are the struggles around these objects and are they successful in challenging projects and strategies?

Sum and Jessop (2013) put this to work on the post-2007 'free fall' economy (Stiglitz 2010) – that moment that started with the sub-prime mortgage crisis, tipped into the bailout of US banks, over time triggering the economic fiscal crisis of the US state (evidenced by the October 2013 US budget crisis and government shutdown) and has begun to spread across the globe in what Harvey calls 'episodes of meltdown' (Harvey 2010), or what we might want to call the 'new political geography of austerity' (see Box 3.13).

BOX 3.13 CRISIS RECOVERY IN THE NORTH ATLANTIC FINANCIAL CRISIS

The so-called global financial crisis offers a good opportunity to test the CPE approach. It is far more complex, multi-dimensional and multi-scalar than its simple label implies, and has unfolded very unevenly around the globe. A CPE approach explores which interpretations get selected as the basis for private and public attempts to resolve the crisis. This is not reducible to narrative resonance, argument force or scientific merit, but also depends on important structural, agency-based and technological selectivities of political geographical knowledge.

The dominant crisis interpretation in liberal market economies after the initial emergency measures is that this is a crisis in finance-led accumulation or, at most, in neoliberalism. As such it can be resolved through a massive, but strictly temporary, financial stimulus, recapitalisation of the biggest (but not all) vulnerable banks, (promises of) tighter regulation and a reformed (but still neoliberal) inter-national economic regime. This will allegedly permit a return to neoliberal 'business as usual' at some unfortunate, but necessary, cost to the public purse, some rebalancing of the financial and 'real' economies and, in the medium term, cuts in public spending to compensate for the costs of short-term crisis management.

One reason for the lack of popular mainstream mobilisation against the crisis and these measures in the heartlands of neoliberalism may be the widespread belief that 'everyone' is to blame because of generalised 'greed' based on the financialisation of everyday life in the neoliberal economies. This implies that the house bubble and financial meltdown were due to excessive consumption rather than unregulated, profit-oriented supply of loans, and also distracts attention from the explosive growth of global finance through derivatives. A more significant account, especially in the USA, 'blames' China for its exchange rate policy, sweated labour, excess savings and so forth, and, accordingly, demands that it bears a significant part of the burden of economic restructuring in the immediate post-crisis period. The crucial sites of this crisis interpretation and crisis management following the outbreak of crisis in 2006–08 have been the USA and the international financial institutions (IFIs) that it dominates and influences. The IMF, World Bank, G-77 have become 'discursive sites' in which sense and meaning are created in the formulation of knowledge on the cause of crisis, its form and crisis resolutions. In the process, and under US influence, institutions and 'experts' can be proposed as the appropriate practitioners of this knowledge, which is then transferred from one scale to another and presented as universally applicable.

Building on the notion of 'discursive sites', CPE encourages an explanation that sees this as a North Atlantic financial crisis (NAFC) because it originated in the USA and UK, spread through contagion effects elsewhere in North America and Europe and, in part, China and to other export-orientated emerging economies, and, more recently, acquired a new dynamic through the eurozone crisis. The NAFC has a complex aetiology and, just as labelling it as global distracts attention from its origins in a particular accumulation regime in the world market, labelling it as financial distracts attention from other mechanisms that led to its complex emergence. The NAFC began to develop well before it attracted general attention in 2007–08 and is the product of the interaction of at least five processes: 1) the global environmental, fuel, food and water crisis; 2) the decline of US hegemony, dominance and credibility in the post-Cold-war geopolitical order; 3) the crisis of a global economy organised in the shadow of ongoing neoliberalisation; 4) a range of structural or branch crises in important sectors (such as automobiles and agriculture); and 5) the crisis of finance-dominated accumulation regimes that emerged in a few but important contingent economic spaces. Each process has its own political geographical logic, each interacts with the others and, collectively, they are influenced by specific local, regional, national and macro-regional factors that ensure that crisis tendencies are always geographically specific rather than instances of global crisis tendencies. Lastly, there are unevenly distributed capacities for crisis management and crisis mismanagement and it is CPE's research future to document these.

Key reading: Sum and Jessop (2013); Jessop (2013).

Summary

This chapter has provided an overview of the uses made by the *régulation* approach to capture the changing institutional forms and functions of the capitalist state. We have only scratched the surface of political geographical enquiry here. This approach to political economy is an ongoing research paradigm method and should not be read as fully finished or complete. As Aglietta puts it, 'We must speak of an approach rather than a theory. What has gained acceptance is not a body of fully refined concepts but a research programme' (Aglietta 2000: 388). The *régulation* approach doesn't have all the answers but it asks some interesting political geography questions. Tickell and Peck have highlighted five important 'missing-links' – more work on modes of social regulation; more research on leading-edge motors of growth; consideration of how and why economies change; more attention to spatial scales of analysis; and heightened consideration for consumption issues (Tickell and Peck 1992) – all of which remain important concerns for political geographers.

Political geography students might wish to consider a recurring criticism levelled against this approach to political economy, the regulationist defence and the ongoing extensions to *régulation* theory through approaches such as Cultural Political Economy. It is often stated that *régulation* theory tends to insert a divide between the economy, which is bracketed as a black box and simultaneously cast as a key protagonist, and the cultural and political realms (see Graham 1992; Jessop and Sum 2006; Jones 2008; Lee 1995). In reply, regulationists claim that the economy is constructed, reconstructed and institutionalised *through* social, cultural and political relations (Bakshi *et al.* 1995; Sum and Jessop 2013). For uncovering further changing institutional forms and functions of the state, it is also suggested that mileage can still be gained from developing a 'regulationist state theory', which draws on Bob Jessop's work on states as mediums and outcomes of territorially-distinct political strategies and policy projects (Jessop 1997b, 2014). We discuss this further in Chapter 12 on public policy and political geography.

Further reading

The subjects covered in this chapter are wide-ranging and there are several avenues for further reading.

The political economy approach is discussed further from a wide range of perspective in: Noel Castree (1999) 'Envisioning capitalism: Geography and the renewal of Marxian political economy', *Transactions of the Institute of British Geographers*, volume 24, pages 137–158; J.K. Gibson-Graham (1996) *The End of Capitalism (as we knew it): A Feminist Critique of Political Economy* (Blackwell); Alain Lipietz (1988) 'Reflections on a tale: The Marxist foundations of the concepts of accumulation and regulation', *Studies in Political Economy*, volume 26, pages 7–36; Richard Peet and Nigel Thrift (1989) 'Political

economy and human geography' in R. Peet and N. Thrift (eds) *New Models in Geography: The political economy perspective* (Unwin Hyman); Martin Jones (2008) 'Recovering a sense of political economy' *Political Geography*, volume 27, pages 377–399.

The papers by – Adam Tickell and Jamie Peck (1992) 'Accumulation, regulation and the geographies of post-Fordism: Missing links in regulationist research', *Progress in Human Geography*, volume 16, pages 190–218; Gordon MacLeod (1997) 'Globalising Parisian thought-waves: Recent advances in the study of social regulation, politics, discourse and space', *Progress in Human Geography*, volume 21, pages 530–553; Joe Painter and Mark Goodwin (1995) 'Local governance after Fordism: Investigating the uneven development of regulation' *Economy and Society*, volume 24, pages 334–356; and David Valler, Andrew Wood, and Peter North (2000) 'Local governance and local business interests: A critical review, *Progress in Human Geography* volume 24, pages 409–428 – all provide very good overviews on the development of regulation theory and its applications in human geography.

Those wanting to trace the intellectual origins of regulation theory should consult books by Boyer (1990) *The Regulation School: A Critical Introduction* (Harvard University Press); Aglietta (2000) *A Theory of Capitalist Regulation: The US Experience*; Robert Boyer and Yves Saillard (2002) *Régulation Theory: The State of the Art* (Routledge); two key papers by Bob Jessop (1997a) 'Survey article: The regulation approach' *The Journal of Political Philosophy*, volume 3, pages 287–326, and (1997b) 'Twenty years of the (Parisian) regulation approach: The paradox of success and failure at home and abroad' *New Political Economy*, volume 2, pages 503–526; and Bob Jessop's (2001a) *Regulation Theory and the Crisis of Capitalism: Five Volumes* (Elgar).

The debates on Fordism and Post-Fordism are covered in the excellent edited book by Ash Amin (1994) *Post-Fordism: A Reader* (Blackwell); the collection of essays brought together in Michael Storper and Allen Scott's (1992) *Pathways to Industrialization and Regional Development* (Routledge); David Harvey's (1989a) *The*

Condition of Postmodernity (Blackwell); and papers by, Frank Moulaert and Erik Swyngedouw (1989) 'Survey 15: A regulationist approach to the geography of flexible production systems', *Environment and Planning D: Society and Space* volume 7, pages 327–345; and Michael Webber (1991) 'The contemporary transition', *Environment and Planning D: Society and Space*, volume 9, pages 165–182. Some of these debates have been revisited within the context of neoliberalism in a collection of essays edited by Neil Brenner and Nik Theodore (2002) *Spaces of Neoliberalism: Urban Restructuring in North America and Western Europe* (Blackwell) and Neil Brenner, Bob Jessop, Martin Jones, and Gordon MacLeod (eds) (2003) *State/Space: A Reader* (Blackwell).

The majority of our case studies cover the broad area of local and regional economic development and those wishing to further explore the comparative nature of this work should consult a number of excellent sources: Neil Brenner (2004) *New State Spaces: Urban Governance and the Rescaling of Statehood* (Oxford); David Harvey (1989b) 'From managerialism to entrepreneurialism: The transformation of urban governance in late capitalism', *Geografiska Annaler*, volume 71B, pages 3–17; Susan Clarke and Gary Gaile (1998) *The Work of Cities* (University of Minnesota Press); Micky Lauria (ed.) (1997) *Reconstructing Urban Regime Theory: Regulating Urban Politics in a Global Economy* (Sage); Norman Walzer and Brian Jacobs (eds) (1998) *Public-Private Partnerships for Local Economic Development* (Praeger); Helga Leitner (1990) 'Cities in pursuit of economic growth', *Political Geography Quarterly* volume 9, pages 146–170; Andrew Wood (1998) 'Making sense of entrepreneurialism', *Scottish Geographical Magazine* volume 114, pages 120–123; and Peter John (2001) *Local Governance in Western Europe* (Sage); and Gordon MacLeod and Martin Jones (2011) 'Renewing urban politics', *Urban Studies*, volume 12, pages 2,443–2,472.

Place, participation and citizenship

As the previous chapters have discussed, for the state to effectively govern society it needs to break down its territory into smaller geographical units and to operate through a hierarchy of local and regional governments and agencies. In a democracy, the offices and institutions of the 'local state' are not appointed agents of the central state installed to impose the state's diktat on the local population, but rather they are democratic institutions in their own right, accountable to local citizens and offering opportunities for participation, either through elections or through forms of direct engagement and active citizenship. Indeed, there has historically been a close association between 'the city', citizenship and democracy. The word 'citizenship' is etymologically connected to the Latin word for city, *civitas*, and the concept of citizenship implying not only a sense of belonging to a city, but also a set of rights and responsibilities exercised at the scale of the city evolved in the ancient city-states of Greece and Rome and the self-governing cities of mediaeval Europe (Isin 2002). In modern times, citizenship came to be more commonly associated with the nation-state and with the legal codification of rights and responsibilities in relation to the state (see Box 4.1); however, it can be argued that the contemporary restructuring and rescaling of the state discussed in Chapter 3 is once again shifting more emphasis on to how citizenship is constructed and practised at a local level. Accordingly, geographers such as Lynn Staeheli (2011) have proposed that research should focus less on categories and definitions of citizenship and more on questions about how citizenship is constructed and contested, and the sites through which citizenship

formation takes place (see also Desforges *et al*. 2005; Kurtz and Hankins 2005).

Considering the formation and practice of citizenship at the local scale intrinsically leads us into issues around the exercise of power within local communities and the local state. Citizenship can empower individuals to act politically in their cities and communities, but it can also be used to disempower and exclude marginal groups. Questions about who has power in localities and how power is exercised have long been a topic of interest for political scientists, sociologists and geographers. In 1950s and 1960s America, the 'community power debate' pitched social scientists who argued that power was concentrated in the hands of exclusionary local elites (notably Hunter 1953), against a more pluralist model in which the power of local leaders was checked and influenced by the interests and voting behaviour of the wider population (Dahl 1961). Both approaches were critiqued by neo-Marxist scholars who emphasised the position of local government as part of the state, and contended that the 'local state' served primarily as an instrument of class politics, reinforcing the interests of capital by providing services that supported the reproduction of labour and distracting political opposition into inconsequential local disputes (Castells 1983; Cockburn 1977).

The apparent dismissal of local politics in the Marxist concept of the local state was in turn critiqued by a series of studies that argued that local political cultures and local government institutions could make a difference in shaping conditions for capital accumulation and thus the competitiveness of a city-region in the national and global economy (Cooke 1989;

BOX 4.1 CITIZENSHIP

Citizenship codifies the relationship between the individual and the state. At one level, citizenship is a mark of belonging – our national citizenship is a sign of the nation-state to which we 'belong'. This is a legal notion of citizenship, which we acquire either through birth or through application, and which then defines certain legal rights that we enjoy as citizens and certain legal responsibilities that we must perform as citizens. The right to vote and the responsibility to pay taxes are examples of this. At a second level, however, citizenship exists through its *practice* in ways that may extend responsibilities and restrict rights beyond the legal framework. For example, the practice of citizenship within a particular local community may be about helping out that community through, for instance, various types of voluntary work. Equally, members of some minority groups may find that their *de jure* citizenship rights are in practice compromised by, for example, racist or homophobic attitudes.

Concepts of citizenship are not static but have changed over time. The idea of citizenship first developed in the city-states of ancient Greece and Rome and related to rights and responsibilities concerned with city government. With the rise of the nation-state from the seventeenth century (see Chapter 5) citizenship was re-imagined at the national scale and in emerging liberal democracies such as Britain and the United States the emphasis was placed on the *political rights* of citizens, such as the right to vote, right to free speech and the right to assembly, and *economic rights* such as the right to own property. During the twentieth century, the development of the 'welfare state' introduced new *social rights* of citizens, such as the right to education and the right to healthcare, especially in Europe. More recently, it might be argued that the social rights of citizens have been diluted by state restructuring, whilst more emphasis has been placed on the responsibilities of citizens, particularly as *active citizens* involved in volunteering and community self-help.

Some observers have also argued that the primacy of the nation-state as the scale at which citizenship is defined and performed is becoming weakened. As discussed in this chapter, state restructuring and the concept of active citizenship have placed renewed emphasis on the community as the place where citizenship is practised; whilst ideas of a national collective interest have been challenged by the rise of more individualistic, atomistic notions of citizenship that emphasise individual rights and the freedom to make lifestyle choices within self-defined communities of interest. At the same time, rights and responsibilities have been up-scaled in the concept of *global citizenship* that separates citizenship from the state (see Chapter 9); whilst intensified global mobility means that an increased number of people hold dual national citizenship, occupy shadowy positions as non-citizens in their country of residence, or conceive of their citizenship in multi-layered ways evoking not only national identity, but also race, gender and religion.

Key readings: Desforges *et al.* (2005), Kurtz and Hankins (2005), Staeheli (2011).

Cox and Mair 1988; see also Jones and Woods 2013). This led to the emergence of 'urban regime theory', which proposed that in order to maintain stable conditions for capital accumulation, local regimes are formed that draw together coalitions of institutions, interest groups and political leaders around the pursuit of particular goals (Lauria 1997). Such regimes are understood as contingent in that they must respond and adapt to changing social, economic and political circumstances (both local and external) and therefore can evolve in their membership and strategy. As there are a number of different strategies that local regimes can adopt (for example, they can be entrepreneurial or anti-development), local factors can shape the form of regimes and subsequent policy outcomes. The urban regime theory approach was exemplified by Stone's (1989) study of Atlanta, Georgia, which documented the shifting influence of different regimes in the post-war era and the consolidation of power by a coalition of white business leaders and black middle class community leaders that found consensus on a broadly pro-business agenda whilst marginalising radical black community groups, job advocates and white neighbourhood and preservationist groups.

Urban regime theory gained popularity during the 1990s as it was perceived to resonate with the early stages of neoliberal state restructuring in the 1980s and 1990s that had sought to strengthen private sector involvement in local government by encouraging partnerships between municipal authorities and business groups (Cochrane *et al.* 1996; Tickell and Peck 1996). Yet, as Macleod (2011) observes with reference to the UK, 'the extent to which the bewildering array of partnerships and quasi-autonomous agencies . . . assembled across Britain throughout the past two decades can plausibly be analysed as proper growth coalitions or urban regimes remains a moot point' (p. 2,637), and the ongoing mix of reforms that have combined privatisation, voluntary and non-profit sector engagement and initiatives to invigorate neighbourhood governance have arguably splintered the urban political sphere in a way that some commentators have described as 'post-political' or 'post-democratic' (Swyngedouw 2011a, 2011b; see also Macleod 2011).

This chapter explores the new landscape of governance and citizenship at the local scale and addresses questions of power, participation and exclusion. The next section charts the promotion of a 'new localism' as part of a changed strategy of governmentality and examines how the new framework of local governance has created opportunities for both 'defensive localism' and 'progressive localism'. The final section then concentrates on citizenship. It discusses how ties between community and citizenship are being reshaped by wider social and economic trends, but also how communities continue to serve as key sites in which citizenship is constructed, negotiated and contested.

Localism and local governance

All countries, except the very smallest, have structures of local government that sit beneath the tier of the nation-state. However, the average size, responsibilities, autonomy and constitutional position of local governments all vary significantly between countries. In some countries, such as France and Italy, local government is closely aligned with local identity and municipalities or communes have both high levels of autonomy and a long history of stable existence. In other countries, including much of northern Europe as well as Brazil and South Africa, local governments play a key role in welfare provision but are subject to central state direction and periodic reorganisation; whilst in the United States, the function of local government is largely focused on economic development (Stoker 2011). In some countries, including the United States, the division of rights and responsibilities between local and central government is constitutionally defined, but in others, notably Britain, central government has complete freedom to restructure, reorganise and even abolish local government. This was demonstrated in 1984 when the Conservative national government in Britain abolished the metropolitan governments of several large city-regions, including London, which were controlled by the opposition Labour party.

Accordingly, any discussion of shifting relations between central and local governments must be careful to avoid generalisations and acknowledge that national political contexts are important in shaping developments. Nevertheless, a number of broad global trends can be observed, reflecting the place of local governance in changing modes of governmentality. As discussed in Chapter 2, *governmentality* refers to the means by which society is made governable. Whereas in Chapter 2 we were primarily concerned with the technologies and techniques of governmentality that enable the state to exercise power across a dispersed territory (such as the collection of statistics), here we are more interested in governmentality as a rationality for imagining how the state should operate in relation to society, and particularly how it organises 'apparatuses of security' such as health, education, social welfare and economic management systems (Dean 1999).

The precise form of organisation of these systems, and the extent to which access to the services they provide is positioned as a social right of the citizen, depends upon the *political rationality* adopted by a state at a particular time. Advanced industrial democracies (including Western Europe, North America, Australia and New Zealand) have experienced a transition in the dominant political rationality from 'managed liberalism' in the mid-twentieth century to a new 'advanced liberal' or 'neoliberal' rationality at the start of the twenty-first century (Rose 1993). Managed liberalism imagined society as a coherent entity at the national scale, with national economic planning and welfare defined as a universal social right to be uniformly delivered across the nation through state intervention. This suited the needs of a Fordist mode of economic production, which required a healthy and educated work force, economic stability and a mass consumer society, but it also substantially expanded the role of the state and its expenditure. As capitalist economies moved towards post-Fordist modes of production in the 1980s, which emphasised flexible accumulation and the exploitation of competitive differences between regions in a free market, so managed liberalism was supplanted by advanced liberalism or neoliberalism. In addition to cutting the functions and cost of the state, the neoliberal mode of governmentality has also re-imagined society, replacing national planning with a strategy of 'governing through communities' that 'does not seek to govern through "society"; but through the regulated choices of individual citizens, now construed as subjects of choices and aspirations to self-actualization and self-fulfilment' (Rose 1996: 41).

Although the 'communities' in the 'governing through communities' approach need not be geographical in nature – they could, for example, be communities of interest, or ethnic or religious communities – the rationality has become identified with a shift in government thinking that has involved, for example, 'a redefinition of the problem of social exclusion *as a problem of local origin* and of the challenge of local regeneration as a challenge for local actors'

(Amin 2005: 615). In other words, economic development is now regarded as best planned at a local or regional scale and driven by local business interests; whilst social problems such as poverty, homelessness and anti-social behaviour are represented as being embedded in local communities and best tackled by local volunteers and non-profit organisations. It is in this sense that the approach has also been labelled the 'new localism' (Amin 2005; Elwood 2004; Imrie and Raco 2003).

The new localism has been operationalised through policies and programmes such as City Challenge, Local Strategic Partnerships and the New Deal for Communities in Britain (see also Box 4.2), Empowerment Zones, Enterprise Communities and Community Development Block Grants in the United States (Elwood 2004), and LEADER and URBAN in the European Union. These policies and programmes adopted and promoted governmental techniques, including partnership working, citizen participation initiatives, neighbourhood governance and active citizenship that have expanded the scope of local involvement beyond the institutions of the conventional local state. The practice of governing has hence been localised but also dispersed and now rests with a complex array of groups, associations, partnerships and initiatives in networks of community or local *governance* (Stoker 2011; Swyngedouw 2005).

Furthermore, the new localism has frequently reached beneath the level of traditional local authorities such as cities and counties to focus on new spatial scales of organisation such as the neighbourhood. In Britain, the National Strategy for Neighbourhood Renewal, launched in 1998, promoted the formation of neighbourhood forums and adoption of neighbourhood plans as a mechanism for regenerating deprived urban areas, and similar neighbourhood governance initiatives have been employed as part of urban policy in Canada, Denmark, France, the United States and elsewhere (Bailey and Pill 2011; Davies and Pill 2012). As Bailey and Pill (2011: 930) note, 'the neighbourhood has particular attractions to policy makers because it is manageable in size and has many "taken-for-granted" attributes of sociability, familiarity, and convenience in providing services and

BOX 4.2   THE NEW LOCALISM, THE THIRD WAY AND THE BIG SOCIETY IN ENGLAND

The introduction of neoliberal rationalities of governmentality and the development of new networks of local governance have been promoted in England by successive governments from both the centre-right and centre-left. The Conservative government of Margaret Thatcher (1979–1990) was generally regarded as a centralising government for its restructuring of local government and its removal of powers and autonomy from elected local authorities, however it also sowed the seeds for new localism in its transfer of responsibilities from the state to the private and voluntary sectors and its championing of active citizenship. For example, responsibility for social housing was shifted from local councils to private, non-profit or tenant-owned housing associations, whilst a 'neighbourhood watch' scheme was introduced to involve local residents in crime prevention. Moreover, business was given a stronger role in economic development and local governance, especially on Urban Development Corporations and Training and Enterprise Councils.

The approach was continued by Thatcher's Conservative successor as Prime Minister, John Major (1990–1997), but it was the New Labour government led by Tony Blair (1997–2008) and Gordon Brown (2008–2010) that most clearly articulated the new localism as part of its 'Third Way' philosophy of combining free market economics with social responsibility. In this context, localism was promoted as a means of delivering social benefits from business-led regeneration and a mechanism for targeting efforts to address spatial and social inequalities. Flagship programmes included the New Deal for Communities (launched in 1998), the Neighbourhood Renewal Fund (launched in 2000) and the Neighbourhood Management Pathfinder Programme (launched in 2001). Collectively, these and other programmes invested over £2bn in urban regeneration targeted at deprived communities generally identified from statistical measures in the Index of Multiple Deprivation. These schemes involved local residents in drawing up community or neighbourhood plans, coordinated by Local Strategic Partnerships (LSPs) that included representatives from the public, private and voluntary sectors. LSPs were established in all local authority areas in England and charged with preparing Community Strategies that would set out a vision for the area and outline an action plan for coordinating the work of various stakeholders towards achieving outcomes (the Local Area Agreement).

Although these flagship programmes focused on urban regeneration, the principles and techniques of new localism were applied by New Labour across the full scope of government, from education to the environment. As Amin (2005: 616–7) observed, 'the targeted, neighbourhood-level approach is common in efforts to tackle crime and youth alienation, disability, child poverty, poor housing and wasted neighbourhoods, health and care, and a host of employment and opportunity initiatives'.

The Conservative–Liberal Democrat Coalition government, formed in 2010 by David Cameron, has continued the promotion of localism under the banner of the 'Big Society', which has placed more emphasis on entrepreneurship and volunteering. Whilst presented as a transfer of power from government to local communities, charities and social enterprises, the Big Society has been criticised by political opponents as providing cover for cuts in public services. A 'right' for charitable trusts and voluntary bodies to take over services provided by local councils was introduced by the Localism Act 2010, which also enabled registered community organisations to draw up and implement neighbourhood plans.

Key readings: Amin (2005); Imrie and Raco (2003); Raco *et al.* (2006).

generating data'. Yet, this attraction is based on questionable assumptions about the ease of defining neighbourhoods, the degree of interaction of local residents and the identification of shared neighbourhood 'interests'. Bailey and Pill's analysis of neighbourhood forums in west London, for example, observed that neighbourhoods contain a variety of groups and that the multifaceted characteristics of neighbourhoods were often overlooked in the interest of achieving defined policy goals. As such, they concluded that although community engagement in the forums had produced some positive outcomes, they had 'limited impact in challenging deep-seated inequalities' (Bailey and Pill 2011: 940).

The effectiveness of the new localism has also been critiqued in the context of Australian rural development policy, which in the 1990s began to emphasise the importance of building self-reliant communities that had the capacity, vision and motivation to drive their own regeneration (Cheshire 2006; Herbert-Cheshire 2000). State subsidies for agriculture and direct state intervention in major infrastructure developments were replaced by initiatives such as 'community builder programmes', community leadership retreats and conferences with titles like *'Positive Rural Futures'*. Supporters of this strategy celebrated it as an *empowerment* of local communities, because it removes decision-making from the arbitrary processes of a distant state and places it in the hands of local citizens. However, critics have argued that it is more accurately a privatisation of responsibility:

> For those who advocate the self-help approach to rural development, its empowering potential for rural people is a fundamental strength. In contrast, for those authors who are more critical of the underlying intentions in governmental discourses of self-help, empowerment represents little more than a rhetoric to obscure the true extent to which power remains (increasingly) in the hands of political authorities. Whichever side of the debate individual authors might take, the main issue for local people, perhaps, is not so much the intent behind discourses of self-help – that is, whether government policies are actually constructed around

genuine notions of shared ownership and control or not – but rather, how those forms of empowerment are actually played out at the local level; whether individuals themselves feel empowered by the process or whether, as is suspected, it is not so much control as the added burden of responsibility that is being devolved.

> (Herbert-Cheshire 2000: 211–212)

The new localism might therefore be more accurately defined as 'the decentralization of responsibility, but not power, from the national to local level' (Coaffee and Johnston 2005: 165). Individual communities may achieve positive outcomes from engaging with the opportunities presented by specific schemes and programmes – as illustrated below – but as a strategy of governmentality the purpose of the new localism is not to empower local communities, but to enable the state to retain power whilst reducing its costs and liabilities. Swyngedouw (2005) calls this the 'Janus face' of the new local governance: on the one side facilitating new forms of participation and citizen engagement, but on the other side introducing new technologies of regulation and discipline and eroding the democratic character of the local political sphere.

Geographies of localism and the local trap

Localism, by definition, allows geographical variations in the practice of governance, the delivery of public services and the outcomes of regeneration initiatives. As localism has devolved responsibility for tackling social and economic problems to communities themselves, so outcomes of initiatives will depend on the capacity of individual communities to act effectively, and this capacity in turn is influenced by a number of factors.

First, as the programmes and projects undertaken by local governance are defined and regulated by the state, capacity to act can be constrained or enabled by the degree of autonomy afforded and the resources made available. Davies and Pill (2012) note this in a comparison of neighbourhood renewal schemes in Bristol, UK, and Baltimore, USA. In Bristol, funding had been secured from a number of government programmes, including New Deal for Communities, and managed according to programme requirements. These held that projects were led by local people, though Davies and Pill noted that in practice the autonomy of community members to shape projects had been eroded by changes in funding programmes. Furthermore, whilst the programmes provided resources for regeneration work, the continuity of funding was vulnerable to changes in government policy, limiting opportunities for long-term planning. In Baltimore, neighbourhood renewal schemes were run by the City Council, but with a strong pro-business ethos, with an emphasis on encouraging investment in neighbourhoods by private sector partners. This model permitted little room for community involvement in governance, but encouraged local volunteers to undertake activities such as clearing litter and mowing grassed areas. The emphasis on investment opportunities also meant that the most deprived neighbourhoods tended to be ignored, such that the approach arguably contributed to a widening of spatial inequalities in the city.

Second, as the last observation from Baltimore implies, capacity to act is also constrained by the strength of social capital in the community, by which we mean the ties between local residents, the existing structure of community organisations and the resources that might be mobilised from within the local social economy (see Box 4.3). Amin (2005), for example, contrasts the social economies of four British cities. In Bristol and Tower Hamlets, London, he observes a vibrant range of social enterprises and projects, with significant community involvement, facilitated by

the presence of a class of experienced and peripatetic social entrepreneurs, proximity to work opportunities in the formal labour market, commitment among providers, clients and intermediaries such as the local authorities and funding bodies to the sector as a non-mainstream economic sphere, and a heterogeneous social structure containing the disadvantaged as well as community activists, a proactive middle class, and other agents of civic engagement.

(Amin 2005: 621)

In Glasgow, however, top-down regeneration projects had produced a narrower range of activities, with limited community involvement; whilst in Middlesborough, historic dependence on large industrial employers meant that there was no base of grassroots

---

BOX 4.3 SOCIAL CAPITAL

Social capital refers to the worth and potential that is invested in social networks and contacts between people. The term is intended to be analogous to 'economic capital' – or financial resources – and 'human capital' – or the skills and attributes of individuals. It has had a number of different precise usages, with some theorists, like Pierre Bourdieu, employing the level of the individual to describe the resources contained in an individual's social network, and others speaking of social capital in a collective sense to describe the sum value of networks and interactions in a society. Robert Putnam, the American political scientist who has popularised the term in recent years, tends towards the latter position. He further makes a distinction between two main types of social capital. The first, *bonding capital*, refers to social networks that bring closer together people who already know each other. The second, *bridging capital*, refers to contacts that bring together peoples or groups who did not previously know each other. Bonding social capital helps to promote community solidarity, whilst bridging social capital assist in enabling communities to access external resources.

Key readings: Putnam (2000); Mohan and Mohan (2002).

activities and social entrepreneurs. Similarly, Woods *et al.* (2006) noted in a study of small town regeneration projects that the most successful tended to be in communities with professional middle-class residents who could contribute skills in project management, accountancy, fundraising, proposal writing and so on.

Third, effective community mobilisation requires strong and coherent leadership, as well as inclusive engagement. Woods *et al.* (2006) also highlight the example of an English town in which the impressive list of local projects on paper disguised disagreements and splits among key community activists, and a second town where a lack of clear leadership and debates over the need for regeneration had limited participation in rural development schemes. Some government programmes have expressed aims to build local social capital and capacity to act, especially within more deprived communities, yet promoting effective community participation requires cultivation of the confidence and energy of local residents, as demonstrated by Jupp (2008, 2012) in research on neighbourhood activism in Stoke on Trent, England. Here community activists were motivated by concerns about the deprivation of their neighbourhoods, but faced obstacles in knowing how to engage with officials, balancing activism and family life, and from emotional exhaustion. Achieving successful, inclusive community involvement, Jupp argues, requires not formal meetings, but support for everyday spaces and seemingly banal activities, such as coffee mornings, in which residents can feel comfortable and through which connections and relationships can be developed over time.

Collectively these factors mean that the greatest beneficiaries of the new localism might not be the most deprived communities, but rather communities that already possess relatively high social and economic capital. As such, the new localism has been criticised for potentially widening social and spatial inequalities and for weakening the social rights of citizens. In a similar vein, critics have argued that the new localism fails minorities by enabling local elite groups to further exclusionary initiatives and policies (Parvin 2009). This *defensive localism* combines the capacities afforded by local autonomy with a commitment to preserving local identity, which may be defined in terms that discriminate between social groups in terms of race, class or religion.

The small town of Hazleton, Pennsylvania, for example, became the first municipality in the United States to use its constitutionally defined local autonomy to introduce local anti-immigration legislation in 2006, the Illegal Immigration Relief Act (IIRA). This introduced a fine of $1,000 per day for landlords who rented accommodation to undocumented immigrants, and established powers to revoke the licences of businesses employing undocumented immigrants. As Steil and Ridgley (2012) describe, the legislation followed a period of deindustrialisation and economic decline, with falling average incomes and business closures, that had coincided with increased Latino immigration and visibility, such that 'long-term native-born residents began to feel not only that their city was economically marginalized in relation to the region and the nation, but also that they themselves were being marginalized in the city of their birth as their control over local public spaces and retail establishments declined' (Steil and Ridgley 2012: 1032). Thus, although the legislation targeted illegal immigrants it was perceived as an assault on the minority Latino population as a whole (the first version of the ordinance had also declared English to be the official language of the town). This perception was reinforced by the rhetoric surrounding the introduction of the legislation, which presented Hazleton as an 'all-American small-town' and as 'one community of legal, hardworking, law-abiding, taxpaying citizens' (ibid.: 1037). The effect of the act, however, was to fracture the community by polarising its residents:

> By dividing Hazleton's population into 'citizens' and 'illegals' and enlisting the native born in its enforcement, the IIRA created new identities and transformed previously existing personal relationships . . . [I]nterviewees reported that the ordinance did not reflect tensions in Hazleton as much as it created them.
>
> (Steil and Ridgley 2012: 1040)

The promotion of neighbourhood governance within large cities can similarly reinforce social divisions, as England (2008) records in a study of Seattle. Initiatives in the city such as Good Neighbor Agreements and an Environmental Design programme were intended to involve local residents in tackling social problems such as alcohol abuse and crime, but in being framed through neighbourhood territories they intrinsically differentiated between 'neighbours' and 'outsiders' through the policing of public space. Thus, England describes the use of Good Neighbor Agreements by neighbourhood associations to address perceived problem drinking by making agreements with bars and alcohol vendors that restricted the type of alcohol sold, placed obligations on owners to maintain the appearance of the building and introduced other standards of 'good neighborliness', in return for a promise of local custom. However, England also observes that the agreements have often been used to target businesses that serve African-Americans in predominantly white neighbourhoods, and in effect displace problems of chronic public inebriation into less affluent and organised communities.

It is evident, therefore, that the reality of the new localism does not always live up to the political discourse of community empowerment. In this way, Purcell (2006) warns against the 'local trap' – the equation of local with 'good' and preferred presumption of the local over non-local scales. Although initially coined with respect to development studies (Purcell and Brown 2005), the concept of the 'local trap' captures problematic assumptions inherent in both policy and academic writing on urban and rural governance:

> The local trap in the urban democracy literature is founded on the assumption that devolution of authority will produce greater democracy. It is assumed that the more localized governing institutions are, the more democratic they will be. More specifically, the assumption is that the more autonomy local people have over their own local area, the more democratic and just decisions about that space will be.
>
> (Purcell 2006: 1925)

As Purcell argues, and as the examples discussed above have illustrated, these assumptions can be contested on a number of grounds: localisation is not the same as democratisation, and can lead to tyranny and oppression; 'community-based development' is not the same as 'participatory development' as local-scale control does not necessarily result in greater public participation; and as communities can be found at all scales, there is nothing intrinsically superior about the geographically defined 'local community'.

Yet, it would also be erroneous to dismiss the new localism as inevitably ineffective or reactionary. Brownlow (2011) contends that local communities are not 'easily duped', but can find opportunities for insurgent political performances within neighbourhood renewal programmes; whilst Featherstone *et al.* (2012) call for a 'progressive localism' to counter what they identify as the 'austerity localism' of the current UK government. The possibility for subversive progressive localism is demonstrated by Brownlow with the case of Cobbs Creek park in Philadelphia, which was restored during the early 2000s by volunteers from the local African-American community. Brownlow notes that the local community had largely abandoned the park and showed little previous interest in organised activity in the park, were critical of the park's managing body and had little involvement in the design or governance of the restoration project – and consequently, might be expected to have little motivation for providing unpaid labour for restoration work. Yet, many hundreds of residents did volunteer, using the scheme as an implicitly political activity to, on the one hand, critique the previous management of the park and the resource-distribution decisions of the city authorities, and, on other hand, to articulate a sense of community responsibility and identity. As such, 'the Cobbs Creek volunteers effectively capitalized upon a devolved opportunity to reclaim long-absent rights of *state citizenship*, all the while demonstrating the responsibilities of *community citizenship*' (Brownlow 2011: 1269).

Citizenship, community and place

The new localism has reasserted the community as the site in which citizenship is practised. In this it follows a long historical association of community and citizenship that rests on the claim that 'democratic community (the community to which citizenship would normally be thought to be most relevant) is rooted in some form of commonality' (Staeheli 2008: 8). However, as demonstrated by several of the cases discussed in the previous section, this association is problematic on several counts.

First, the places in which we live are rarely socially homogeneous, and attempts to define community by characteristics such as ethnicity, language, religion, culture and class, as well as to police 'appropriate' forms of activity and behaviour within a place, invariably creates insiders and outsiders, such that individuals with equivalent legal citizenship may experience differentiated abilities to practise citizenship in relation to a community.

Second, even in relatively homogeneous neighbourhoods there may be a lack of commonality between residents. Surveys have shown that everyday interactions between neighbours have decreased over time, and public spaces such as churches or sports clubs no longer function as uniting focal points for neighbourhood life (Putnam 2000). Neither does living next door to someone indicate a relationship of care, or shared set of values or interests. As Painter (2012: 524) notes, 'neighbours, after all, can be hostile as well as friendly, indifferent as well as interested, passive as well as active'. The conflation of 'community' and 'neighbourhood' in the new localism misunderstands the relation between these two concepts.

Third, communities are not necessarily geographical, and communities based on ethnicity, religion, gender, sexuality, work, lifestyle or interests are arguably becoming more significant both as sites of social identity and interaction, and as vehicles for citizenship formation and practice. Staeheli (2008: 16) gives the example of the Arab-American community in the United States, which by acting as a focal point for cultural and political activity has created a new kind of political subject 'located in a complicated geography of home and citizenship that is simultaneously grounded in more than one place, as well as in a metaphorical, diasporic home'. Online social media have helped to facilitate translocal communities, as well as creating spaces for new social and political mobilisations that have no geographical root, but have also enabled the stretching of traditional geographical communities, with individuals participating in community life from afar (Bernal 2006; Cammaerts and Van Audenhove 2005).

Fourth, as both 'community' and 'citizenship' are socially constructed concepts their meaning varies between individuals. Thus, attempts to mobilise active citizenship in community-based initiatives can expose differences in understanding and act as a catalyst for debates over community values and the expectations of citizenship. Staeheli (2008), again, provides an example of a crime prevention initiative in Spokane, Washington State, which organised citizen volunteers in neighbourhood patrols through neighbourhood offices known as COPShops. For the programme organiser, the COPShops were 'sites of community building and education; they helped residents become active and responsible citizens' (Staeheli 2008: 6), yet for some residents they represented a top-down model of citizenship and helped to impose an exclusionary idea of the community.

Accordingly, communities cannot be regarded as unproblematically fostering citizenship, nor can citizenship be unproblematcally considered to build communities. Rather, communities – and especially geographical communities – are contexts in which citizenship is negotiated, contested and performed, with multiple possible outcomes. As Staeheli observes,

> Citizenship is constructed in and through the contradictions of community. This is the terrain that is negotiated as particular constructions of citizenship are put forward, contested and changed through time. Citizenship and community, therefore, are always unsettled. In that sense, they are always a problem. They are problems to be

engaged and worked with, rather than to be dismissed, ignored or condemned. They are part of the agonism of democracy and citizenship, as riven with problems as these may be.

(Staeheli 2008: 18)

The performance of citizenship through communities is not only manifest in political acts or participation in governance initiatives, but is more routinely enacted in the interactions of everyday life – taking children to school, shopping at a corner store, drinking in a local bar – through which the entitlements and responsibilities of citizenship are continuously and implicitly played out (Staeheli *et al.* 2012). It is through these performances that 'outsiders' gain acceptance in local communities and consolidate their citizenship status. Pine (2010), for example, discusses the position of immigrant Dominican grocery-store owners (*bodegueros*) in Philadelphia. Whilst the *bodegueros* claimed that they were considered as 'invaders' by the local community – including by some community leaders – Pine demonstrates that by anticipating and adapting to the service needs of local residents, the *bodegueros* had come to be not only accepted as part of the community, but also to be providers of public spaces 'where the *bodegueros* chatted with the customers and discussed local political issues with whoever was around' (Pine 2010: 1118). As such, 'they "performed" the actions of communitarian citizenship, even while expressing a sense of not belonging' (ibid.).

In some cases, however, more overt political organisation is required by marginalised citizens to assert their rights and challenge conventions of community. For example, Bosco *et al.* (2011), describe the organisation of Latina immigrant women in a city in southern California in a Neigborhood Action Group (NAG) that operated on the one hand as a self-help group providing support for members, but on the other hand as an advocacy group representing community needs to city leaders and involved in direct action projects. Significantly, by organising as a geographical community, a *Neighborhood* Action Group, rather than as an ethnic or women's organisation, the women signalled their rights as residents, even though many had not

obtained US citizenship or faced obstacles of language in engaging with the state.

Both these cases – the Dominican shopkeepers in Philadelphia and the Latina women in California – can be seen as examples of what Henri Lefebvre calls the 'right to the city' (Lefebvre 1968; see also Harvey 2008; Purcell 2002b). This is not a legal right ascribed by citizenship of a state, but rather evokes a more innate practice of urban citizenship – the right of residents to occupy urban space to live, work and play, to represent the city and to participate in its production and governance, including in decisions over land use and community resources. It involves the collective power of urban citizens to make and remake cities, performing 'a right to change ourselves by changing the city' (Harvey 2008: 23). The 'right to the city' is hence central to struggles over the performance of urban citizenship, as the next section discusses with a case study of Istanbul.

The 'right to the city' in Istanbul

Istanbul is one of Europe's fastest growing and most multicultural cities, where European secularism and consumerism mixes with conservative Islamic values. Tensions between these two tendencies are part of the politics of everyday life in the city, and especially issues of urban development and the use of public space. They gained international attention in early summer 2013, when demonstrators against the redevelopment of a public park occupied the central public space of Taksim Square for several days before being violently dispersed by police, but have a much longer history, shaping the claims made to the city by different social groups. Anna Secor (2004) and Dikmen Bezmez (2013) narrate stories that explore this struggle from the perspectives of two groups.

Secor (2004) follows the experience of Kurdish women in the city. The women are Turkish citizens, internal migrants from south-eastern Turkey displaced by economic hardship and political insurgency. Yet, their Kurdish ethnicity, language and culture marked them out as different. Whilst some celebrated the multicultural nature of the city and a feeling of freedom in public spaces such as Taksim Square, others found that their Kurdish identity compromised their de facto citizenship, especially in education and finding employment. As one woman told Secor, 'For us, this place counts as a foreign place. This place is also our homeland, but we feel that we are in a foreign place' (ibid.: 359). Some of the women responded by trying to become anonymous, not revealing their Kurdish identity, but others had become more assertive of their identity as Kurds *and* Turkish citizens, for example, by wearing Kurdish clothes: 'Last year, I was wearing traditional Kurdish clothes, and one week I walked around like that . . . . They stopped me in the street and asked. "What country are you from?" I said, "I am from this country, this land, but I am a Kurd and because of this I wear the clothes of a Kurdish woman"' (Kurdish woman quoted by Secor, 2004: 363). Through such small, everyday acts, the Kurdish women articulated their citizenship and claimed their right to the city.

Bezmez (2013) in turn focuses on the politics of disability in Istanbul. As in most modern cities, people with disabilities in Istanbul have had their de facto citizenship rights compromised by discrimination and inaccessibility, with a disabled activist movement asserting their right to the city. However, Bezmez notes that since the 1990s a number of initiatives had brought support for people with disabilities into the mainstream of local government in the city, with groups such as the Accessible Istanbul for Everybody Platform expressly employing the language of urban citizenship to talk about the rights of disabled people. Yet, Bezmez argues that this development is not the outcome of a claim for people with disabilities' right to the city, but rather was an expression of the religious charity-based approach to disability of the ruling moderate Islamist AKP party. Morever, Bezmez suggests, appropriating the relatively uncontroversial issue of disabled rights allowed the city government to demonstrate its liberal values, whilst avoiding the claims associated with politically more difficult categories of ethnicity, gender and sexuality. As such, Bezmez echoes Purcell's (2006) warnings about the 'local trap' in revealing the complex politics that can underlie struggles for urban citizenship.

Summary

This chapter has examined the association between citizenship, community and place. Historically, the notion of citizenship was closely tied to the city as a place, bestowing civic rights and responsibilities. From the eighteenth century onwards, however, citizenship became understood in terms of membership of a nation-state, and the rights of citizens extended over time from political rights – such as the right to vote – to social rights – such as the right to healthcare – especially as states adopted 'managed liberalism' as a mode of governmentality to underpin Fordist capitalism. More recently, the 'advanced liberal' or neoliberal mode of governmentality has promoted a rationality of 'governing through communities' in which the identification of citizenship with local communities has been re-asserted. Sometimes referred to as 'the new localism', this approach has been adopted in different variations in different countries, but has commonly involved the engagement of private and third-sector partners in local governance and the formation of new initiatives for neighbourhood governance. The model has also depended on the participation of local residents as 'active citizens' in local governance, taking on responsibility for certain activities in their communities that were previously undertaken by the state.

However, critics have argued that the 'new localism' has been a devolution of responsibility more than a devolution of power, with control still resting with the central state. Mark Purcell has similarly warned against falling into 'the local trap' by presuming that governance structures are more democratic or accountable if they are organised at the local scale. Rather, critics point out the limited autonomy afforded by the new localism, the uneven geography of its outcomes, and its proneness to be captured by reactionary interests and used to discriminate against minority groups. For these and other reasons, communities might be more accurately regarded as contexts in which citizenship is negotiated, contested and performed, with multiple possible outcomes. Citizenship is hence articulated and tested through everyday interactions as different groups seek to assert their 'right to the city' – that is their freedom to live and work in a geographical place and to be involved in decisions about its future.

Further reading

The relationship between citizenship and community, and the study of this relationship in geography, has been discussed in a number of papers by Lynn Staeheli. These include Staeheli (2008) 'Citizenship and the problem of community', in *Political Geography*, 27, 5–21; Staeheli (2011) 'Political geography: where's citizenship?', *Progress in Human Geography*, 35, 393–400; and a co-authored paper, Staeheli *et al.* (2012) 'Dreaming the ordinary: daily life and the complex geographies of citizenship', *Progress in Human Geography*, 36, 628–644.

For more on the history of citizenship, see Engin Isin (2002) *Being Political: Genealogies of Citizenship*, whilst the transition to an 'advanced liberal' mode of governmentality is discussed by Nikolas Rose (1993) 'Government, authority and expertise in advanced liberalism', in *Economy and Society*, 22, 283–299. The 'new localism' is critically analysed, especially in a UK context, in Ash Amin (2005) 'Local community on trial', *Economy and Society*, 34, 614–633, and Mark Purcell warns against the 'local trap' in Purcell (2006) 'Urban democracy and the local trap', *Urban Studies*, 43, 1921–1941.

The case study of anti-immigrant localism in Hazleton, Pennsylvania, is described in Steil and Ridgley (2012) 'Small town defenders: the production of citizenship and belonging in Hazleton, Pennsylvania', *Environment and Planning D: Society and Space*, 30, 1028–1045, and the case study of Kurdish women in Istanbul is discussed by Secor (2004) 'There is an Istanbul that belongs to me: citizenship, space, and identity in the city', *Annals of the Association of American Geographers*, 94, 352–368.

The political geographies of nationalism

I was walking down this street in Mold, north Wales [in the UK] – years ago now – with some friends from my home town in Llanelli in south Wales. We were laughing, playing about and talking to each other in Welsh. A group of locals came down the street, and after hearing our accents, came up to us quite aggressively . . . obviously looking for a fight. They tried to taunt us by calling us 'Cymry plastic' or 'plastic Welsh'. Obviously, we didn't fit in with their ideal type of Welsh person. We weren't local and we spoke south-Walian Welsh. That was enough for them. Anyway, we ran in to the nearest pub and managed to get out of any trouble.

This tale, narrated by Rhys Jones, one of the authors of this book, helps to illustrate a number of crucial themes relating to the geographies of nationalism. First, it demonstrates the close relationship that exists between place and nationalism. In this story, certain places can be seen to represent the Welsh nation more effectively than others: north rather than south Wales, Mold more so than Llanelli. It shows the way in which places help symbolise and anchor nationalist discourse. Second, the story shows us the subdivisions that exist within the allegedly homogeneous entities of nations. We usually think of nations as coherent 'imagined communities' (Anderson 1983) of people who follow the same customs and speak the same language. Indeed, this is one of the main ideological foundations of nationalism; to encourage us to believe that it is possible to draw boundaries around homogeneous groupings of people. Rhys Jones's experiences in north Wales would seem to undermine this notion. Even though he and his friends we born in the same country,

and spoke the same language, as the other group of men, they were still thought of as being somehow different in nature. In this example, these imagined boundaries between the two groups were constructed along place-based and linguistic lines. To be born in a different region or a different town, or to speak a different dialect, led to the construction of imaginary boundaries between the two groups of people. Finally, the tale illustrates the close link that often exists between language and nationalism. The ability or inability to speak a language can often be used as a means of defining who actually belongs to, and who is excluded from, a nation. In this case, the ability to speak a language was not deemed to be enough of a badge of national identity, since the Welsh language had to be spoken in a particular way in order to gain membership of the Welsh nation.

These are all key themes that need to be explored when discussing the geographies of nationalism, and we shall focus on each of these during the course of the chapter. The first section will introduce the different kinds of theories that have tried to explain the formation of nations. Following on from this, we then proceed to explore the significance of the key geographical concepts of place, landscape and territory for nationalism. Finally, we focus on the way in which nationalism is contested and complicated in a variety of different ways.

Reproducing nationalism

So what do we mean by a nation? Similar to comments made regarding the state in Chapter 2, it is often the case that our membership of nations – as individuals and as communities of people – makes it difficult to think about them in an objective way. This is even more of an issue since it is nigh-on impossible for any individual to escape from the grasp of nationalist discourse. Gellner (1983: 6) notes the prevalence of such a discourse when he argues that it is as if 'a man (sic) must have a nationality as he must have a nose and two ears'. When nations are commonly viewed as 'natural' phenomena like this, it becomes difficult critically to analyse their form and function. A further problem arises when attempting to distinguish between states and nations. In the first part of this section, our aim is to define what we mean by nations and emphasise the difference between states and nations.

According to Anthony Smith (1991: 14), a prominent writer on themes of nationalism, a *nation* should be viewed as a 'named human population sharing an historic territory, common myths and historical memories, a mass, public culture, a common economy and common legal rights and duties for all members'. Although definitions of nations display slight variations in the themes that they emphasise (see below), on the whole they follow the general principles laid down by Smith. Smith's definition draws our attention to a number of important themes. First, and crucially, all nations should possess a geographical referent in their claims to a particular territory. It is difficult to imagine a nation that does not claim access or control over a certain territory and it is this feature of nations that acts as the main justification for studying nations from a geographical perspective. We will argue in subsequent sections that territory is only one geographical theme that is of importance to nations. Others, such as place and landscape, are also highly significant. Second, Smith's definition stresses some of the key cultural aspects of nations. Nations are said to possess a common 'public culture'. They also emphasise common historical memories that help to engender a sense of loyalty towards the nation. It is these common cultural characteristics that enable the members of a nation to imagine the existence of this large-scale, yet close-knit, community of people, even though they will never meet all other members of the nation

(Anderson 1983). It is these cultural themes that also help us to distinguish between states and nations. Whereas states are organisations that seek to control a particular territory – similar to nations – they do so in a relatively impersonal manner. States, *per se*, do not attempt to stress the cultural commonalities that exist within and between the state's citizens. Nations, on the other hand, are conceived of as communities that are characterised by common ties of custom, history and culture and a connection to a certain territory. It is when both of these institutions combine, of course, that the powerful organisation of the nation-state is formed. Third, Smith emphasises the role that certain legal and economic processes play in forging a nation. Once again, this draws us near to definitions of the state (see Chapter 2) and helps to illustrate once again some of the common elements that are said to characterise states and nations.

If nations are common communities of people that share certain cultural attributes and a particular territory, then we need to think of *nationalism* as an ideology that seeks to promote the existence of nations within the world. Furthermore, an important element within nationalism is the belief that every nation should possess its own sovereign territory or state. The ideology and political practice of nationalism therefore seeks the ideal political and territorial scenario of the *nation-state*, in which every citizen of the state is a member of the same nation. Thinking of this in geographical terms, nation-states represent political geographies in which the boundary of the nation coincides with the boundary of the state (Gellner 1983: 1). Obviously, in a world characterised by continuous flows of people, the ideal of the nation-state is precisely that – an ideal that can never be achieved. Indeed, it has been famously argued by Mikesell (1983) that the only example of a nation-state in the contemporary world is Iceland. Unfortunately, the difficulties in achieving the goal of the nation-state do not stop states, nations and minority groups from trying – sometimes peaceably and sometimes violently – to reach the ideal. Examples such as the attempts to create an independent Quebec through referenda and the terrorism of ETA, the Basque separatist movement in Spain, demonstrate the salience of such processes within contemporary political geography.

We now possess an understanding of what nations are, along with some of the ideologies that are linked to them. The key question that has exercised the minds of social scientists and historians in the field of nationalism is 'How are nations formed and continue to exist?' To put it another way, 'How are nations reproduced?' We want to distinguish here between two major categories of theories of nationalism. A set of classical theories of nationalism has sought to examine the longer-term processes, which have contributed to the emergence both of nationalism as an ideology and of specific nations. These classical theories have been challenged in recent years by the emergence of various academics that have sought to understand the social construction of nationalism.

Classical theories of nationalism

Classical theories of nationalism are numerous, and draw on a broad range of political, economic and cultural viewpoints. Following the work of Smith (1998), we can conceive of them as existing in two broad groups (see Table 5.1 below).

In the first category are those theories that see nations as communities of people that have some discernible roots in the pre-modern period (in other words before approximately 1500). These theories

often posit a link between nations and ethnic communities of people. Most problematic in this category are those so-called primordialist theories of the nation that argue that nations have always existed, from time immemorial as it were. Nations are therefore not produced, as such, since they are seen to represent essential qualities – linked to race, blood, language, religion or ethnicity – that are inherent in communities of people. These can appear as quite naïve interpretations of nations since they posit that certain definable and homogeneous communities of people have always existed in the world. Halvdan Koht (1947: 278), for instance, perceived a late-twelfth-century uprising in Italy as an early display of Italian nationalism against the Germanic peoples who were occupying their country. Koht argued that the Italians succeeded in rallying their people by proclaiming that the German language was similar to the 'barking of dogs and the croaking of frogs'. Clearly, here, we see an attempt to view a medieval conflict as an early instance of nationalism. It is doubtful whether this is the case since it is unlikely that the communities of people that lived in Italy at this time displayed the characteristics common in modern nations (though see Reynolds 1984). More seriously, these primordialist theories are also potentially dangerous ideas since they can be used to legitimise reactionary and racist attitudes towards 'outside' people and influences that allegedly undermine the 'purity' and 'integrity' of the nation.

Table 5.1 Theorising the reproduction of the nation

Theorising the roots of nations in pre-modernity
• Primordialist theories: the nation is an essential element of the human make-up and has always existed
• Perennialist theories: the nation is a product of modernity but has its roots in earlier pre-modern ethnic communities

Theorising the nation as a product of modernity
• Nationalism as 'high culture': industrial development requires a literate work force and state education systems 'educate' citizens regarding nationalist ideals
• The socio-economic development of nationalism: 'peripheral' countries mobilise their populations, using nationalism as a means of competing against the 'core'
• Nationalism and the state: nations are formed as a result of the creation of state bureaucracies
• Nationalism as ideology: nationalism is created as a means of escaping the isolation and social disruption caused by processes within modernity

Source: after Smith (1998)

Many instances of historical and contemporary nationalism – especially during times of conflict – have witnessed exclusionary and essentialist attitudes towards people who do not 'fit in' with the norm of the nation. In recent times, we can think of ethnic cleansing in the former Yugoslavia as a particularly repugnant example of attempts to rid the nation of 'impure' or 'undesirable' elements (e.g. Mirkovic 1996).

Less problematic are the so-called perennialist theories promoted by academics such as Anthony Smith (see especially 1986). In his attempt to chart the ethnic origins of nations, Smith has argued that 'nothing comes from the nothing' (1996: 386): in other words, it is extremely difficult to create a nation out of a group of people who do not feel any sort of communal feeling towards each other. Rather than seeing nations as totally modern fabrications (see below), Smith argues that the more successful nations have been based on early ethnic communities of people, or *ethnies*. In doing so, Smith tries to emphasise the long history of nations. The processes and institutions of modernity – industrial development, capitalism and the state – impact on ethnic communities in various ways. Some ethnic communities are dissolved as a result of these processes as the languages, cultures and customs of a particular *ethnie* are adopted as the norm for a number of adjacent ethnic communities. This process possessed distinctive geographical undertones since it often involved an imposition of the language spoken in the core of the nation on its peripheral regions. We can think, for instance, of the adoption of the language of the Île de France – the area around Paris in France – as the official language of the French nation, and the subsequent attempts to dilute and extinguish the other so-called vernacular languages – Breton, German and Basque – spoken in other parts of the country. Smith's ideas certainly make us think of the long-term development of nations out of earlier ethnic communities and they can act as a powerful antidote to the more explicitly modernist theories of the formation of nations.

Modernist theories view nations as communities of people that have come into being as a result of various processes that happened in the modern period. To a greater or lesser extent, they view the nation as a new creation, actively produced by means of the processes and institutions of modernity. As such, these theories tend to focus on the civic quality of nations. Anthony Smith (1998) has classified them into four main categories of theories. We briefly discuss each category in turn. First, some theories emphasise the role played by industrial development and state education systems in the formation of national 'high cultures'. Gellner (1983) argued in his later writings concerning the important role of state educational systems in the creation of a 'high' culture necessary for successful industrial development. In this period there is a need for individuals who understand the same language. Communication must be made possible throughout the whole territory of the state. It is this need to create an impersonal society in which all individuals speak the same language, are educated by the same state education system and are, to a large extent, interchangeable one with another, that explains the formation of nations. State education systems replace 'low' cultures – meaning localised and traditional customs, languages and cultural norms – with the 'high' culture, meaning a 'standardised, education-based, literate culture' (Smith 1998: 32). These shared, literate cultures are the bases for national sentiment. A famous example of this process relates to the different histories portrayed in Estonian school textbooks before and after its independence. While still part of the Soviet Union, the Soviet Union was portrayed favourably since it helped to liberate Estonia from German rule towards the end of the Second World War. After independence, however, school textbooks in Estonia portrayed the Soviet Union in a negative light, seeing it as a repressive political and cultural force within Estonian history (Smith 1996). This example clearly shows the key role played by state educational systems in moulding nations (see Box 5.1). Rather than representing age-old cultures and customs, as primordialists argue, Gellner argues that nations are in fact created out of new, state-based cultures, languages and customs. As such, 'nations are functional for modern society' (Smith 1998: 35) meaning that they are indispensible to it.

BOX 5.1   SCHOOLS AND NATIONALISM IN SINGAPORE

Kong (2005) argues that schools seek to promote two types of discourse as part of a nation-building project within Singapore. The first is based on a series of ideologies and practices that further a multiracial and multicultural vision of Singapore, which are said to lead to a process of national integration. The aim of this discourse is to promote a common experience for all schoolchildren in Singapore, in which 'multicultures occupy and interact in common space on terms specified by the state' (ibid.: 621). The second discourse is based on the construction of state schools as 'sites of modernity, in which students are provided with education aimed at enabling them to participate effectively in the modern economic life of the country' (ibid.). The economic relevance of state education is of especial significance in this respect. A fascinating aspect of Kong's work is the way in which she is able to show how madrasahs – schools embedded in the Islamic religion – offer an alternative understanding of the role of education within Singapore. The primary consideration of madrasah schools is to produce the Muslim religious leaders of the future but they are also viewed as sites of a more 'holistic' type of education, in which moral and religious issues are deemed to be of equal importance to purely academic training (ibid.: 622). Such work shows how different groups of people – based on religious beliefs in this case – can alter and refract the nation-building project associated with state education systems.

Key reading: Kong (2005).

Second, and to some extent linked to the first set of theories, are socio-economic models of the formation of nations. Characteristic of these types of theories is the work of Tom Nairn (1977), who views nations as political and cultural entities that are formed as a result of socio-economic processes of the world economy. In a nutshell, Nairn argues that nationalism derives from the uneven development of capitalism. By this, he means that the success of the capitalist process in certain Western states after 1800 depended on their exploitation of cheap labour and resources in the periphery. Nairn views nationalism as a means by which political and cultural leaders in the periphery seek to mobilise their populations against the imperialism of the core, much in same way as a sense of nationalism can be said to help the players in a sports team to compete against an opposing team. Efforts were made, therefore, to encourage popular forms of nationalism in the periphery. Nairn uses the spread of nationalism in Germany and Italy during the nineteenth century, as they sought to oppose the capitalist might of the UK and France, as an illustration of this process. There are weaknesses with Nairn's model, for

instance his inability to present an explanation of the way in which political and cultural leaders in the periphery succeeded in promoting nationalisms within the population that they ruled. Similarly, there is firm empirical evidence that the first nationalisms appeared in the core, rather than the periphery. According to Smith (1998: 53) it appeared in England, Britain, France and America before Germany, the alleged first state to promote a sense of nationalism. We can, therefore, take only so much from Nairn's model. It offers some insight into the relationship between capitalism and nationalism but it displays too many weaknesses in order to be considered as a truly useful explanation of the formation of nations and nationalism.

Third, we need to consider political explanations of the formation of nations and the development of nationalist ideologies. Here, the development of the modern, bureaucratic state is used as an explanation for the formation of nations. Anthony Giddens (1985), for instance, has argued that nations can only exist with regard to states. What is key here, therefore, is the consolidation of the state as a bureaucratic organisation that extends its administrative control outwards from a core to defined boundaries (see Chapter 2). The significance of this is two-fold: it brought a singular and uniform administrative rationality to all citizens of the state, and it also imposed a fixed boundary on the national community of the state. In a few rare instances, this could lead to the creation of a singular nation within the boundaries of the state, but it more often led to cultural and national reactions against bureaucratic control within particular states. Some support for Giddens's ideas appears in the context of African states. Here, we see clearly the influence of states on national communities during the post-colonial era. Leaders of these states have tried to forge a national community to match the boundaries of their administrative control, but these boundaries have also acted as the frame of reference for secessionist and independence movements within these countries. A painful example of this process is the Democratic Republic of Congo where a number of armed revolutionaries, especially in the eastern half of the country are challenging the territorial integrity of the state.

Other authors have emphasised the way in which the state's need to wage wars effectively – and specifically the process of military conscription upon which this effort was based – has affected the kinds of identity felt by the state's citizens. Eugene Weber (1977; see also Mann 1986; Tilly 1975) described a process whereby 'peasants' in various parts of France were gradually turned into 'Frenchmen' (sic). An important contributor to this process was the conscription of young men into the French armed forces. Conscription in France began in 1798 and young men were forced during the nineteenth and much of the twentieth century to commit between one and eight years of their lives to the armed forces. Alan Baker has described the broader significance of this process:

> Conscripts were, in effect , attending a school for the patrie [the country], which distanced them from their pays [region]. Military orders were, of course, given in French; regimental allegiances were actively cultivated, erasing local attachments and replacing them with a sense of national identity. Regimental schools taught conscripts not only how

to read, write and count, but also what it meant to be a French citizen.

> (Baker 1998: 195–196)

The main problem with political accounts of nationalism, which focus on themes such as a state's bureaucracy and military conscription, is their inability to deal with more cultural forms of nationalism. Not all nationalist movements seek to create their own independent state, and may rather seek to propagate forms of nationalism that seek to improve the moral and cultural well-being of the nation. In many ways, the nationalism of the Maori people of Aotearoa/New Zealand fits into this model. Maori are more concerned with ideas of cultural well-being, socio-economic development and the protection of the natural environment than they are with the need to create and independent Maori state (see Pawson 1992; Bery and Kearns 1996). As such, political explanations of the nation fail to account for the more diffuse form of nationalism experienced in this context.

Fourth and finally, are the ideological interpretations of Elie Kedourie (1960; 1971) who views nationalism as a system of beliefs similar to religion. Here, nationalism is viewed as a system of beliefs that is promoted by groups of intellectuals as a means of making sense of the fundamental changes that affect societies as a result of the processes of modernity. Indeed, the nation's immemorial history plays an important role in anchoring the life-worlds of individuals who are lost within the modern maelstrom. Importantly, Kedourie grounds the growth of nationalism in two specific periods and places: an early nineteenth-century Central Europe that was experiencing large-scale political and social change and Africa and Asia during the late-nineteenth and twentieth centuries. The key impetus for the development of nationalism, in this regard, was the colonial process that operated in particular countries. Traditional forms of living were destroyed, and local intellectuals were marginalised from colonial bureaucracies and reacted to this process of marginalisation by engaging with ideals of nationalism that diffused from Western states. For instance, Kedourie (1971: 42–3; see also B. Anderson 1983: 195) describes the role of the Greek intellectual, Adamantios Korais, in helping to frame Greek history within Western conceptions of nationalism. Having experienced life in Revolutionary France, Korais began to comprehend Greek history as a nationalist history of the Greek people. This intellectual enabled Greeks to begin to view their place in the world as one that was structured by the new secular religion of nationalism: instead of experiencing the *anomie* and isolation of modernity, they began to view themselves as members of a durable Greek nation. According to Kedourie, therefore, nationalism should be viewed as a political and cultural opiate (Smith 1998: 103) that enables individuals to make sense of the world in which they live.

Here, then, are the classical theories that have sought to explain nationalism as a political ideology. The categories Smith (1998) has constructed on this broad-ranging literature should be viewed as definitive in nature. It is clear, for instance, that there are many overlaps between the various categories. There is much of value in these theories. They provide a sense of the large-scale socio-spatial changes that have characterised the past 200 or so years – whether in the form of capitalist processes, state formation, changing forms of rationality – and the way in which they have affected group identities. These theories also begin to indicate important geographies that are associated with the reproduction of nationalism, most notably the way in which these processes attempt to reify the significance of a national territory as the spatial basis of nations, along with their emphasis on particular kinds of social and spatial mobility as an important contributor to the emergence of nationalism. And yet, these theories have been criticised in recent years on a number of different fronts by a loose band of social constructivists and we turn to examine their arguments in the following sub-section.

The social construction of nationalism

Classical theories of nationalism have been criticised for their tendency to view nations as ontological social categories. Nations, within classical theories, have been viewed almost as agents in their own right, possessing the ability to affect social and spatial change

(Brubaker 2004: 2–3, 8). Authors such as Craig Calhoun (1997: 3) have sought to counter such a viewpoint by arguing that nationalism should be viewed as 'a "discursive formation", a way of speaking that shapes our consciousness'. Rogers Brubaker (2004: 9, original emphasis), too, has maintained that a nation 'is a key part of what we want to explain, not what we want to explain things *with*'. He elaborates on this idea as follows:

> Ethnicity, race, and nation should be conceptualized not as substances or things or entities or organisms or collective individuals – as the imagery of discrete, concrete, tangible, bounded, and enduring 'groups' encourages us to do – but rather in relational, processual, dynamic, eventful, and disaggregated terms. This means thinking of ethnicity, race, and nation not in terms of substantial groups or entities but in terms of practical categories, situated actions, cultural idioms, cognitive schemas, discursive frames, organizational routines, institutional forms, political projects and contingent events.
>
> (Brubaker 2004: 11).

The 'groupness' of nations is something, therefore, that needs to be forged through continual and iterative practice and explained by academics rather than being taken as an ontological 'given' (see Box 5.2). Group-making is, therefore, a project that is based on a series of events – both extraordinary (Ignatieff 1993) and banal (Billig 1995). According to Calhoun (1997: 4–5), critical features of nationalist discourse or rhetoric, which help to constitute a nation as a 'group', include notions relating to boundaries, indivisibility, sovereignty, an 'ascending' notion of legitimacy, popular participation, direct membership, culture, temporal depth, common descent and territoriality. While the exact combination of these discursive underpinnings of nationalist discourses may vary from one nation to the next, they are all key features of a broader discourse of nationalism. Nations, in this respect, are communities of people that are imagined through a series of nationalist discourses (Anderson 1983), rather than being actual communities of people, which are purposive in character.

BOX 5.2 'MAKING GROUPS' IN TRANSYLVANIA

Brubaker's (2004) work on the Transylvanian town of Cluj has elaborated on the production, circulation, reworking and contestation of nationalist discourses and the related attempts to make distinct Romanian and Hungarian groups. By examining the particularities of place within Cluj, among other things, Brubaker is able to demonstrate the highly contingent consumption of nationalist discourses within the town. Despite the active production of nationalist discourse by a nationalist elite within the town – most notably, its Romanian mayor, Gheorghe Funar – the definition of Romanian and Hungarian groupness within Cluj has been low (ibid.: 23). The contrast between Cluj and the neighbouring town of Târgu Mureş has been marked, where a different kind of localised politics of place has led to an extreme hardening of nationalist identities. In Brubaker's work, therefore, we see how nationalism should be viewed as a contingent group-making project. Despite similar attempts to define Romanian and Hungarian groups in two seemingly similar towns, the results have been significantly different in the two places.

Key reading: Brubaker(2004: 20–27).

Classical theories of nationalism have also been criticised by social constructionists for their tendency to portray nations as unproblematic and homogeneous social categories. The focus within classical theories of the nation is to examine the production and circulation of dominant forms of nationalist discourse. Classical theories of the nation conveniently 'forget' the differing engagements of various marginalised and underrepresented groups with the discourse of nationalism. Özkirimli (2000: 192), for instance, makes clear that '[n]one of these [classical] theories took account of the experiences of the "subordinated", for example the former European colonies and their post-colonial successors, or women, ethnic minorities and the oppressed classes.'

Influenced by a post-structural turn in the social sciences, contributions by various authors have sought to rectify this deficiency by demonstrating the multiple ways in which a dominant nationalist discourse can be complicated and problematised: Enloe (1993) and Yuval-Davis (1997), for instance, with regard to issues of gender; Bhabha (1990) and Chatterjee (1986) with reference to the 'Westocentric' (Yuval-Davis 1997: 3) aspects of classical theories of nationalism; and Billig (1995), whose focus has been on demonstrating the everyday and banal contexts within which nationalism

is reproduced. We return to these themes in the final section of this chapter, where we show how dominant nationalist discourses can be complicated and contested by a variety of social and spatial factors.

In the following section, we turn to more geographical themes by discussing the way in which the key geographical concepts of place, landscape and territory help to inform the character of nationalism.

The geographies of nationalism

Since approximately the 1980s, geographers have begun to examine the significance of spatial concepts for nations and nationalism. The key contribution that acted as a clarion call for political geographers to examine nationalism from a spatial perspective was that published by Colin Williams and Anthony Smith in 1983 (Williams and Smith 1983). In their paper on the 'The national construction of social space', they argued that much research needed to be carried out on the geography of nations and nationalisms. As a starting point for this project, they outlined eight different contexts in which geographical themes could inform our understanding of nationalism. These are noted in Table 5.2 below.

Table 5.2 Space, territory and nation: eight dimensions

Dimension of national space and territory	Recent examples
Habitat The nature of the environment and the soil helps to explain the location and nature of human communities. Following on from this, the members of the nation come closer to the ideal of the nation if they live close to the soil, in other words, in rural rather than urban areas	Jewish fundamentalists try to recapture the essence of the Jewish nation by creating new rural settlements in disputed territories
Folk culture Soil and environment lead to particular customs and social norms, e.g. daily, monthly and annual rhythms of life in rural areas. These become prized as peasant virtues and folk cultures. Efforts are made to avoid the cosmopolitan culture of the cities and recapture the rural roots of the nation	Nazism and the emphasis placed within this extreme nationalist ideology on the virtue of a folk culture of the Volk
Scale The size or scale of nations helps to position the nation in an international league table of nations. We can link this most clearly with the attempts made by nations to expand the reach of their territorial control. By doing this, the nation increases its own prestige	This was implicit in much of geopolitics in the first half of the twentieth century, and helps to explain the 'scramble for Africa' by European nations
Location We can think here of the uneasy relationship between a nation and other neighbouring nations. This can lead to warfare, which may affect the character of the nation. Also the distance between peripheral areas of the nation and the core can lead to uneasy relationships	For instance, the uneasy relationship between India and Pakistan has led to more virulent nationalisms in the two countries. We can also think of the border community of Kashmir that lies between India and Pakistan and which has been problematic for both nations
Boundary Crucial here is the idea of finding the 'natural' frontiers of the nation and the need to get away from artificially imposed borders	This idea has been important in the development of the French nation, which has professed an explicit need to defend its natural borders
Autarchy Land is viewed as a resource deposit for the benefit of a particular nation. Any struggle for land and independence is linked with the struggle for the use of national resources, first in the context of agrarian resources but also in the context of minerals	A good example of this was the attempt by the Kikuyu to wrest land from white control in Kenya. This process is being repeated in Zimbabwe
Homeland Territory is not a neutral term for the nation. It is the national homeland, the historic root of the nation	We can think of the importance of the homeland for Jews, something that sustained them during thousands of years of exile
Nation building The process of forming and maintaining nations involves improvements carried out within the nation's territory. it is this 'infrastructure of the nation' that turns a territory into a national territory	For instance, cities, communications networks, power stations, law and educational systems are all crucial to the construction of a national territory. A fine example of this is the communication node of Grand Central Station in New York

Source: after Williams and Smith (1983)

Williams and Smith's (1983) ideas are an extremely valuable way of understanding the relationship between geography and geographical concepts, and nationalism. Indeed, their ideas have been taken up by a number of political geographers who have brought new insights into the study of nationalism through their focus on geographical concepts, such as place, landscape and territory. It is to their work that we now turn.

Placing the nation

In this section we focus on the importance of certain places for nationalist discourse. By referring to place here, we are specifically concerned with place as locality, in other words, places that occur at small scales. Even though nationalism refers to an ideology that exists at a national scale, and draws members of a nation together as one common community of people, nationalist discourses always draw on specific places as sources of ideological nourishment.

At one level, we can think of generic places that help to sustain nationalist discourses. One particularly powerful type of place is the memorial to the dead of the nation (see Plate 5.1). These are individuals who have paid the ultimate price for their loyalty to their nation. Michael Heffernan (1995), for instance, has described in detail the discussions and, indeed,

Plate 5.1 The Anzac Memorial in Sydney, Australia

Courtesy of Rhys Jones

conflict that revolved around the commemoration of those members of the former British Empire who died during the First World War (1914–1918). After much wrangling a system was adopted whereby those who had died in the war were buried in cemeteries along the Western Front, mainly in northern France and Belgium, while monuments were raised along the length and breadth of the UK in remembrance of each community's human losses. Importantly, both the cemeteries and the monuments represent key symbols of the loyalty of thousands of individuals towards the British nation. The Arlington Cemetery in the US and the Arc de Triomphe in Paris are other clear examples of this memorialisation of the dead of the nation.

Perhaps most powerful, in this respect, are the tombs of unknown soldiers that exist within most states. Benedict Anderson (1983: 9) has argued that 'no more arresting emblems of the modern culture of nationalism exist than cenotaphs and tombs of Unknown Soldiers'. Because of the anonymity of the soldier that lies under the tomb, these tombs can come to represent the more general sacrifices that all individuals (should) make for their nation.

Memorials to the dead of the nation are one powerful type of place that helps to focus the nationalist sentiment of members of the nation, but others also exist. Parliament buildings and monuments (see Plate 5.2), for instance, embody the citizenship of

Plate 5.2   A statue commemorating Garibaldi, 'father' of the Italian nation
Courtesy of Rhys Jones

all individuals within a state, along with their membership of a nation. Other examples include national museums, which are seen as illustrations of the nation's historical development and are therefore a key method by which the nation can demonstrate its achievements to visitors, whether they are members of the same nation or others. The role of museums as generic places that play a key role in symbolising nations is elaborated upon in Box 5.3.

As well as certain generic places that are important within all kinds of nationalist discourse, we also need to consider the specific places that possess a significant meaning for particular nations. What is important here are the particularities of the histories and geographies of a given nation, ones which give meaning and value to specific places. In the remainder of this section, we discuss two brief examples of places that play significant roles in symbolising or inspiring particular nations.

Our first example is the Cathedral of Christ the Saviour in Russia. Dmitri Sidorov (2000) has examined the significance of this particular place for the Russian nation. Originally designed as a memorial to the great Russian victory against Napoleon's French forces in 1812, it was viewed as a way of memorialising the 'unprecedented zeal, loyalty to and love of the faith and Fatherland' (quoted in Sidorov 2000: 557). Obviously, therefore, from the very beginning the Cathedral was perceived as a means of symbolising the commitment of the Russian people to their nation. What is equally crucial, according to Sidorov, is that the Cathedral, since its construction, has reflected and symbolised broader political and national changes that have occurred in Russian/Soviet society. So, for instance, the original European design was quashed in the late 1820s because it did not tally with the new Tsar's desire for the Cathedral to reflect national Russian architectural forms. Similarly, its demolition in 1931, as a result of the Bolshevik revolution, was carried out as preparation for the use of the location where it stood as a site for the construction of the secular Palace of the Soviets.

BOX 5.3 FOLK MUSEUMS AND THE MEMORIES OF THE NATION

National museums are key sites for any nation because they enable a nation to represent itself to its members and to the world. Particularly important are the folk museums that exist within many countries. Seen as the repositories of the folk culture of the nation, they are viewed as key means of representing the essential cultural truths about a particular nation (see Williams and Smith 1983). A good example of a folk museum is the Skansen Open Air Museum, situated in Stockholm, the capital of Sweden (M. Crang 1999). Skansen, the world's first Open Air Museum, was opened in 1891 as a way of preserving elements of the Swedish folk culture. Crang has noted how Swedish intellectuals at the time displayed a great deal of interest in the cultural pasts of the Swedish folk. These folk (and mostly rural) cultures were perceived as the only links between a Swedish nation and its past, especially since the cultural make-up of the vast majority of the Swedish population was being transformed as a result of the processes of modernity. Skansen today comprises a number of costumed workers who 'inhabit' the houses and farms drawn from all parts of the country. Importantly for Crang (1999: 451) the folk culture preserved in Skansen in set up as 'a timeless, interior Other within the modern nation'. Skansen, and other folk museums, therefore, illustrate the importance of preserving a folk past for the modern political and cultural communities of the nation. They help to emphasise how contemporary nations are allegedly connected, through ties of culture, to the earliest incarnations of the nation in the dim and distant past.

Key reading: Crang (1999).

In other words, by replacing the Cathedral with the Palace of the Soviets, this one place in Moscow could be seen to represent the far broader changes occurring in Russian/Soviet nationalism as it shifted from one that emphasised a strong religious form of nationalism to one that was wholly secular in nature. As it turned out, the Palace of the Soviets was never constructed. After the accession of Mikhail Gorbachev in 1985, discussions began concerning the construction of a new Cathedral. Sidorov argues that this process, once again, has reflected broader political and national currents in Russia. Crucially, therefore, this one place in the centre of Moscow can be viewed a key symbol of Russian nationalism. Throughout its period as a construction project, it has illustrated some of the significant changes to have affected Russian society. Moreover, it has been viewed as a place that should help to symbolise and inspire the Russian/Society nation.

Our second case study appears in Louise Appleton's (2002) description of the significance of the *Saturday Evening Post*, weekly magazine published up until 1969, for the reproduction of nationalist discourse in the United States. The weekly magazine was extremely popular in the United States for much of the twentieth century, with between 15.4 per cent and 22.9 per cent of its population reading the *Saturday Evening Post* during the 1950s (ibid.: 424). As Appleton (ibid.: 423) notes, the magazine had been viewed since the late nineteenth century as a means of creating and disseminating 'nationhood in the aftermath of America's civil war'. The most significant aspect of Appleton's work, in this respect, is her effort to demonstrate how the magazine used the 'local scale' as a way of symbolising the American nation. Drawing on ideas such as the local scale as a 'microcosm of an exceptional nation', of places as 'crucibles of national life', of the nation as a 'patchwork quilt' and of places that were beginning to challenge the national ideal, Appleton demonstrates the way in which places and the local scale were used by the *Post*, both as representations of the American nation and, through its large circulation, as a way of reproducing an American national consciousness throughout the United States.

Place – whether thought of in generic or specific terms – is, therefore, critical to any understanding of nations and nationalism. Places help to symbolise and sometimes inspire the nation. In many ways, certain places become important elements of the national imagination. Another geographical concept that plays a significant role in national imaginations is the landscape.

Landscapes of the nation

Along with place, the most potent way of imagining the nation is through reference to particular landscapes. By landscape we mean not just the physical environment but also the meaning and values that are ascribed to them by individuals or communities. Nations tend to view particular types of landscapes as ones that represent the values or the essence of the nation. In this section, we focus on these landscapes at both a conceptual and more empirical level.

Generally speaking, nations tend to portray rural landscapes as ones that symbolise the nature of the nation. We can relate this to some of the themes raised by Williams and Smith (1983) in their classification of the various dimensions of state territory. Two, in particular, emphasise the important of rural landscapes to the nation. By referring to the habitat of the nation, Williams and Smith draw our attention to the stress placed by nations on the belief that members come closer to the nationalist ideal if they live 'on the soil'. What this means is that individuals are more likely to adhere to nationalist principles if they live in rural, rather than urban, areas. Folk culture is also important to many nations: the rhythms of life in rural areas lead to the formation of peasant lifestyles and these are prized as manifestations of the true character of the nation. In both of these contexts, therefore, rural lifestyles and rural landscapes are to be cherished by the nation (see Box 5.4).

As well as helping symbolise the purity of the nation, certain landscapes can be used as a means of emphasising the differences that exist between one nation and another. Rob Shields (1991: 182–99), for instance, has focused on the importance of the

BOX 5.4 FOLK CULTURE, HABITAT AND RURAL LANDSCAPES IN WALES AND IRELAND

We witness an excellent example of the promotion of these two elements – folk culture and habitat – in the politics of the Welsh Nationalist Party, Plaid Cymru, in the interwar period. Pyrs Gruffudd (1994: 69–70) has demonstrated how the party argued that Welsh people would have to 'return to the land' if they were to gain their rightful place as a moral nation. Significantly, part of this political strategy was based on a belief that the Welsh nation needed to live in rural areas so that they could avoid 'Anglicized metropolitan values'. It was also based on the belief that it was only in rural areas that the Welsh *gwerin* (folk), or in other words, the upholders of true Welsh national and moral values, lived. In many ways, Gruffudd's research echoes the work carried out on the national imagination of the West of Ireland within Irish nationalism (see Johnson 1997). Here, once again, for much of the twentieth century the West of Ireland was perceived as the main bastion of Irish national identity. This national imagination drew on linguistic geographies. The west was, by far, the least Anglicised and most Gaelic-speaking region of the island. It was also based upon the religious differences that were thought to exist between a Catholic west and a Protestant north. Equally important was a romanticised understanding of the rural lifestyles that existed there among the Irish 'folk'. Though these ideas have been thoroughly deconstructed by Irish commentators, they represent a set of discourses that are still of great relevance to popular understandings of Irish nationalism.

Key readings: Gruffudd (1994); Johnson (1997).

Canadian North for the constitution of Canadian nationalism. In one context, the Canadian North can be viewed in much the same way as Gruffudd and Johnson respectively view rural Wales and Ireland. For Shields (ibid.: 198), 'the 'True North' is a common reference 'point' marking an invisible national community of the initiated'. For instance, Canadian historian W.H. Morton notes that an understanding of the Canadian North is imperative if people are to fully comprehend the nature of the Canadian identity (in ibid.: 182). The rugged northern landscape – divorced from a southern, urbanised Canada – is seen to represent the essence of the Canadian nation. Shields argues, however, that the Canadian north is more than merely a symbol of the purity of the Canadian nation, for it helps to distinguish the Canadian topography and nation from that of the US. For example, much of the nineteenth-century literature that described the Canadian North as a significant factor in the formation of the Canadian nation was consumed by audiences in the US. In effect, the landscape of the Canadian North came to be used as a symbol of the necessary national difference that existed between the Canadian and US nations. The existence of this massive arctic hinterland in the north of Canada in many ways enabled the Canadian nation to identify with other northern nations and states, such as Norway, rather than with the US (ibid.: 198). In this example, landscape, as well as being a symbol of the nation, was also a signifier of the differences between neighbouring nations.

And yet, the role played by notions of landscape within nationalist discourse is never simple and uncontested. Echoing the arguments made earlier about the social construction of nationalism, it is clear that different groups residing within a particular national territory can emphasise the significance of different kinds of landscape as well as assigning different meanings to the same landscapes. Such themes come to the fore in Nogué and Vicente's (2004) research on the connection between mountain landscapes and national identity in Catalonia. They maintain that while 'mountains have wide and varied material dimensions . . . they are also endowed with a spiritual and symbolic dimension' (ibid.: 118) and this is certainly the case in Catalonia. Catalonia is a stateless nation in north-east Spain, which is centred on the city of Barcelona. Its integration into a wider Spain began during the fifteenth century and it has suffered from a lack of political liberty since this period. Largely as a result of the industrial and demographic vitality being experienced in Catalonia during the late nineteenth century – especially when compared with the stagnation existing in the rest of Spain during the same period – led to the emergence of a nationalist movement, entitled the *Renaixença*, within the province (ibid.: 121). Certain mountain landscapes were used for their symbolic value by this emerging nationalist movement; the Pyrenees, located to the north of Catalonia, and the mountain of Montserrat, located to the south-west of Barcelona, in particular, were viewed as signs of the 'purity and virginity' (ibid.: 122) of the Catalan people. The most significant aspect of Nogué and Vicente's research for the present argument is that differing interpretations of the status to be ascribed to the mountainous landscapes of Catalonia emerged during the early twentieth century. While the *Modernisme* movement emphasised the fact that the 'natural' Catalan environment was to be found in the mountains and that this landscape represented, in essence, an 'irrational' and 'violent' landscape and an 'untamed nature' (ibid.: 124), the *Noucentisme* movement stressed far more a Mediterranean model and a far more civilised and humanised account of Catalonia's mountainous landscapes. For this latter movement, rural and mountainous Catalonia could only be understood in relation to its urban centres; *Noucentisme* stressed a 'city-Catalonia as a model for country and landscape' (ibid.: 125). In this example, we see how a mountainous landscape, while being important for these two different nationalist movements, were scripted into nationalist discourses in different and contradictory ways.

Nation and territory: the homeland of the people

Nations, as is shown in Anthony Smith's definition, must 'share an historic territory'. Nations are, in effect, rooted in particular territories. Indeed, James Anderson (1988: 24) has eloquently argued that

> the nation's unique history is embodied in the nation's unique piece of territory – its 'homeland', the primeval land of its ancestors, older than any state, the same land which saw its greatest moments, perhaps its mythical origins. The time has passed but the space is still there.

In this quote, we see part of the significance of territory for the nation. A nation's territory helps it to commune with its past and to emphasise the strong links that have always existed between it and the land in which it now resides. A particularly striking example of this process exists in the context of the Jewish nation. During its long time in exile, and since the formation of the Israeli state, the territory of Israel has furnished the Jewish nation with much ideological support (Azaryahu and Kellerman 1999; Hooson 1994). Struggles over land between Jews and Arabs in contemporary Israel, therefore, do not merely represent attempts to increase the amount of land under one's control in a physical sense. They also represent ideological struggles for the control of the symbolic body of the nation (see Box 5.5).

Territory plays other important roles for a nation. In the first place, it has been argued that territory is the conceptual link between the nation and the state in the form of the nation-state (Taylor and Flint 2000: 233). Of course, the notion of the nation-state is the ideal of nationalism, the perfect political scenario in which the boundary of a state matches in an exact manner the geography of the nation. It is in these political contexts that both the nation and the state may feed off each other. The nation helps to legitimise the whole existence of the state, binding its citizens into an unswerving loyalty towards it. At one and the same time, the state exists to protect the members of the nation, ensuring that their national rights are promoted at the expense of the rights of the members of all other nations. In this way, territory can be viewed as the basis for the ideological and organisational marriage between the nation and the state.

BOX 5.5   THE FIGHT FOR TERRITORY IN THE MIDDLE EAST

Since the formation of the modern state of Israel in 1948–49, the status of Israelis and Palestinians has been a source of considerable debate in the Middle East and, indeed, on a global stage. Matters were exacerbated as a result of the 1967 war, following which Israel occupied the West Bank and the Gaza Strip. Newman and Falah (1997) maintain that the discourses of Palestinians and neighbouring Arab countries in the period between 1967 and 1974 emphasised the need to eject Israelis from the Middle East. The state of Israel was not recognised; rather there was a demand to re-create a Palestinian state that would extend throughout the whole of the territory allocated to Israel in 1948–49. From 1974 to 1988, the territorial context for struggle changed. Arab states began to contemplate a compromise position, in which Israel should withdraw from the occupied territories of West Bank and the Gaza Strip, while still maintaining control of its own allocated territories. Palestinians, under the leadership of the Palestinian Liberation Organisation (PLO), also began to accede to the right of the state of Israel to exist, although in less explicit terms than their counterparts in other Arab countries. The post-1988 period, has witnessed a more explicit acceptance of a 'two-state solution within Palestine . . . amongst both the PLO leadership and the Palestinian masses' (ibid.: 119). In this way, we witness how understandings of territory – and its role as a spatial foundation of nationalism – have constantly been re-evaluated in the Middle East, whether across time and space, or across different organisations. This pattern is likely to continue in the face of changing global attitudes towards Israel/Palestine and the actions of, amongst others, suicide bombers and Israeli settlers.

Key reading: Newman and Falah (1997).

Of course, boundaries are crucial elements in the constitution of territories and this fact draws our attention to a second important link between nations and territories. Daniele Conversi (1995) has argued that the boundaries between nations are critical elements in the constitution of the nations that exist on either side of the boundary (also Paasi 1996). For Conversi, nations are defined, at least in part, through a process of othering, in which the faults of neighbouring nations – whether real or perceived – are emphasised as a means of promoting the strengths and qualities of the 'home' nation. All nations engage with this act of othering to a greater or lesser degree. The US, in recent years, has attempted to contrast itself with the Japanese nation, most clearly with regard to economic and cultural practices. Nations from the neighbouring countries of Greece and Turkey have also symbolised each other's nations in negative ways. This situation has been further exacerbated by the conflict between the two countries over the national status of the island of Cyprus (see Box 5.2).

Of course, in the contemporary world of cultural globalisation, nations need not define themselves against other nations, but rather against the perceived growing importance of supra-national sense of group identity. In many ways, this mirrors David Harvey's (1989a) arguments regarding the tendency for contemporary groupings to try to preserve their own individuality and distinctiveness as communities of people. Michael Billig (1995: 99) has noted, for instance, how much of contemporary British nationalism is couched in terms of a need to preserve a distinctive British identity in the face of an ever-increasing 'Europeanisation' of identity within Europe. For instance, the then UK Prime Minister John Major announced in 1992 that he would 'never let Britain's distinctive identity be lost'. His attempt here was to defend a proud and valuable British national identity against what he and many others saw as the detrimental cultural and political influences emanating from European politicians and bureaucrats.

BOX 5.6   GREEK AND TURKISH NATIONALISM IN CYPRUS

The history of the struggle between Greece and Turkey over the island of Cyprus has its roots two main processes. First, the domination of the eastern Mediterranean by the Ottoman Empire for much of the modern period and the related repression of the Greek population. Second, the efforts of the British Empire to control the eastern Mediterranean as a bulwark against the growing power of Germany and Russia. With the collapse of the Ottoman Empire after 1918, there has been much conflict over the status of the island. In the period between 1918 and 1939, many Greek nationalists on Cyprus and the Greek mainland sought to gain independence for the island from British rule. The Turkish minority on the island were naturally wary of this prospect. The relationship between Greece and Turkey deteriorated further in the period sub-sequent to 1945, with much of the nationalist angst deriving from conflict over the status of Cyprus. Atrocities were committed towards Greeks living in Turkey and numerous riots took place on the island itself. The creation of an independent Cyprus in 1960 did not solve the issue and, indeed, four years later an all-out war between Greece and Turkey was only narrowly avoided. Today, it is only a large UN presence that prevents the escalation of hostilities between Greece and Turkey. During the whole of this period, there has been a steady deterioration in the relationship between the Greek and Turkish governments and peoples. To a large extent, both nations exist in opposition to one other.

Key reading: Calvocoressi (1991).

Here is another example, then, of the key role played by the nation's boundaries and territory in helping to shape the nature of the nation. Admittedly, Benedict Anderson (1983) has criticised this inter-pretation of nations as being too exclusionary, negative and regressive in nature. For Anderson, the fact that certain individuals may become naturalised members of other nations demonstrates that the boundaries of nations are not always fixed. In other words, the boundaries of nations may be viewed as places where national cultures and languages mix, rather than being places of exclusion and mutual denigration. Of course, such arguments resonate with recent debates within human geography regarding the necessity to view place as one which is open to outside influences, rather than being one that is closed to them (see Massey 1994). Laudable though these sentiments are, plainly, nationalist discontent and war are still grounded, in many cases, in the defamation and hatred of neighbouring nations.

In the three preceding sections, we have discussed the key role played by the concepts of place, landscape and territory in ideologies of nationalism. As a result of this discussion we argue that geography and spatial themes are at the heart of any understanding of the nation. We further reinforce this claim in the final section of the chapter, where we focus on processes that contest the nation.

Contesting the nation

The impression given in definitions of nationalism, and also in various nationalist ideologies, is that the nation is a coherent and stable community of people, to which its members demonstrate an unswerving loyalty. Within the ideology of nationalism – and also, to a large extent, in classical theories of nationalism – the nation is conceived of as an homogeneous group of people indoctrinated with the nation's ideals. The physical boundaries of the nation demarcate the territorial extent of a group of people totally unified in their love of the nation. All members of the nation contribute to this notion of a unified 'imagined community' of people (Anderson 1983).

Of course, it is possible to challenge such ideas. In empirical terms, we know that claims such as these are unlikely to mirror the reality of social existence, and that other spatial, scalar and social processes help to contest the ideology of nationalism. As we noted in an earlier section, there is an additional conceptual challenge to such ideas. Social constructivists have long argued that if nations and nationalism are constituted on the basis of discourses and practices – or, in effort, that nations are processes in motion rather than being fixed entities – then we also need to appreciate the way in which they are crosscut by other kinds of individual and group identity, whether relating to gender (Enloe 1993; and Yuval-Davis 1997) or race (Bhabha 1990; and Chatterjee 1986). Some of these themes come to the fore in the work of Sarah Radcliffe (1999) on feelings of national identity within a lower-middle-class neighbourhood of Quito, the capital of Ecuador. In her research, she found that women and men experienced 'different trajectories of affiliation' to the Ecuadorian nation (ibid.: 217). This meant, for instance, that women were more likely to express their sense of national identity through reference to ideas of love of nation, while men spoke in more neutral terms of a sense of obligation to the nation. Furthermore, there were significant differences in the way in which male and female Ecuadorians engaged with the *mestizo* racial category. Originally, this was the name of a mixed racial group of European and indigenous peoples, but especially in the post-war era came to symbolise 'an engagement in the urban, market-led and modernizing national society and an avenue for social advancement' (ibid.: 215). Importantly for Radcliffe, men and women in Ecuador engaged with this *mestizo* category in a different way to each other. On the whole, a lower proportion of women identified with the *mestizo* category, and those who did, did so in an ambivalent manner. Indeed, many women viewed themselves as white rather than *mestizo*, and identified with 'white' forms of language-use, dress-code and literature. For Radcliffe, therefore, women and men in Ecuador became entangled in Ecuadorian nation identity in different ways, as a result of their different engagements with issues of gender and race. This is one clear instance of the way in which different kinds of identity can contest a homogeneous vision of nationalism.

In another context, we need to think about the way in which different people in different places interpret or consume nationalist discourses in different ways. If we think place to be important in the shaping of our identity, then surely it will mean something slightly different to be a member of the American nation in Birmingham, Alabama, as opposed to Seattle, Washington. Or to be a member of the Brazilian nation in Rio de Janeiro, as opposed to a village in Amazonia. If this is the case, then we argue that nations are contested from within, for the simple reason that they incorporate numerous different places, and therefore, numerous different types of people with numerous types of national identity within their boundaries.

One way in which we can further such ideas is through reference to Michael Billig's (1995) concept of banal nationalism (see Box 5.7). He has urged social scientists to consider the banality of nationalism, or in other words, the way in which it is subtly reinforced on a daily basis as a result of small-scale, mundane and banal processes. Numerous social scientists – geographers among them – have examined the various contexts within which nationalism is reproduced on a day-to-day basis. Research has focused, for instance, on the symbols and imagery used on national currencies (see Gilbert and Helleiner 1999; Unwin and Hewitt 2001). Plate 5.3 illustrates another instance of banal nationalism, namely the inclusion of the words '*Je me souviens*' or 'I remember' on all car registration plates in Quebec, a slogan first carved onto the Quebec parliament building in 1883 and intended to prompt Quebecois to think of their distinctive francophone heritage and struggle for autonomy.

At the same time, the banal qualities of nationalism enable individuals and groups to contest dominant nationalist discourses and practices. Raento and Brunn (2005), for instance, have examined the significance of postage stamps as banal reminders of state or official nationalism. There can be few more banal signifiers of the distinctiveness or characteristics of a

BOX 5.7 BANAL NATIONALISM

Michael Billig has attempted to show how prevalent nationalism is in the established nation-states of the 'West'. Rather being something that exists 'out there' or something that is the characteristic feature of 'extremists', Billig argues that nationalism is something that pertains to 'our' identities in the 'West'. He has stressed, moreover, the way in which all kinds of nationalism are reproduced on a day-to-day basis for, as Billig (ibid.: 6) states, 'the world of nations is the world of the everyday'. Billig (ibid.: 93–127) demonstrates various ways in which nationalism is reproduced within nation-states. His main example revolves around the use of language used by political leaders and newspaper columnists. The use of personal collective pronouns – words such as 'we', 'us', 'our' but also, by implication, words such as 'them', 'they', 'their' and so on – illustrates the widespread belief amongst groups of people in the existence of their nation, their membership of it, and the existence of other groups of people who are members of other equally valid nations. The notion of banal nationalism, therefore, helps to show how the 'groupness' of nations (Brubaker 2004) is something that is socially constructed rather than being an ontological 'given' and is something that needs to be forged daily.

Key reading: Billig (1995).

Plate 5.3 The banality of nationalism: remembering the sacrifices made for Quebec

Courtesy of Rhys Jones

particular nation. Their 'mundane omnipresence . . . gives them considerable nation-building power and makes them exemplary tools of . . . "banal nationalism"' (ibid.: 145). The day-to-day use of stamps, at the same time, provides an opportunity for those opposed to state or official nationalisms to voice their opinions, given that 'any icon is a potential tool for resistance' (ibid). Raento and Brunn note, for instance, that many Finns during the late nineteenth century, when the

Russian empire's control over Finland was being enhanced, subverted the official nationalism of the state by placing Russian stamps upside down on the envelope. In this way, the banal qualities of stamps also provided an opportunity for individuals and groups to contest the dominant nationalist discourses being promoted in Finland during the period of Russian dominance.

A further significance of Billig's ideas, therefore, is that they help to emphasise the significance of localities as places within which the mundane and everyday processes that help to mould individuals as part of the nation occur. Research by Fevre *et al.* (1999) in Wales, for instance, has demonstrated how such a process of reproduction could take place at small scales. For example, for many people in north Wales, it is the processes that operate within the local housing market, in which Welsh-speakers cannot afford to compete with English newcomers, that helps to engender within them a sense of Welsh nationalism. Similarly, it is individuals' experiences in pubs and bars – where people with varying linguistic abilities meet each other, and where arguments may take place – that enable them to shape their own interpretation of their national identity. As individual members of nations, we can all recall experiences from school, university, on the street or when socialising, that have made us think about our role and place in our nation in particular ways. For Rhys Jones, for instance, his experiences on the streets of Mold in north Wales, forced him to re-evaluate his place within the Welsh nation. We argue that the logical outcome of these local reproductions of nationalism is the creation of nationalisms that vary slightly from place to place (see Edensor 1997; Jones and Fowler 2008).

The role played by localities in reproducing nationalism, of necessity, draws attention to ideas of geographic scale. We began the chapter by noting how classical theories of nationalism have tended to emphasise the national scale as the only appropriate scale at which to study nations. Gellner's (1983) work, for instance, seeks to highlight the growing importance of state education systems in obliterating 'low' cultures and replacing them with national 'high' cultures. Of necessity, Gellner's work speaks of national institutions, leading to a national set of processes, which help to create a national high culture. Giddens (1985), too, in examining the extension of state bureaucracies and means of administrative control outwards from a core to defined boundaries, and the subsequent impact of this process on senses of identity, tends to reify the national scale as the appropriate scale of enquiry. At one level, such a preoccupation with processes operating at the national scale is to be expected, given that nations are intimately connected to a national territory. And yet, recent work has shown how nationalism is actually produced through processes operating across a number of scales. We have already seen how recent work in Wales has examined the role played by networks of people within certain localities in producing nationalist discourses. Work by Catherine Nash (2002), has similarly shown how recent concerns about genealogical identities have problematised the scales over which nationalism is reproduced. She has examined how genealogical research amongst Irish immigrants to the New World has raised significant questions about Irish national identity and has 'stretched' it to include debates about: a global Irish diaspora, which

is trying to re-connect to its roots in (different parts of) Ireland; the genetic make-up of the body, which can be used to designate (and sometimes complicate) national identities.

These recent debates about the way in which nations can be reproduced at scales other than the national raises also important issues regarding the impact of cultural globalisation on nationalism. As we discuss in Chapter 9, the growing production, circulation and consumption of cultural messages at a transnational or global scale does not necessarily mean that nationalisms are being undermined within specific countries. Potentially, nationalism can be reproduced as a result of processes operating at any spatial scale, and therefore it would be unwise to predict the end of nationalism as a structuring principle for cultural communities. Nationalism is here to stay, at least for the foreseeable future, and as long as it still exists it will draw on geographical concepts of place, landscape and territory for its sustenance.

Further reading

The key starting point for geographical studies of the nation and nationalism is Williams and Smith (1983) 'The national construction of social space', *Progress in Human Geography*, volume 7, pages 502–18. This paper, which in many ways represents the beginnings of geographical studies of the nation, elaborates on different geographical themes that are intimately related to nationalism.

Numerous studies have examined the significance of place, landscape and territory for the nation. For a good discussion and case study of the significance of particular places for the nation, see Sidorov (2000) 'National monumentalization and the politics of scale: the resurrections of the Cathedral of Christ the Savior in Moscow', *Annals of the Association of American Geographers*, volume 90, pages 548–72. Whelan's discussion of the politics of monuments in Dublin, although couched in terms of colonialism, also illustrates the importance of monuments for cultural identity; see Whelan (2002) 'The construction and destruction of a colonial landscape:

monuments to British monarchs in Dublin before and after independence', *Journal of Historical Geography*, volume 28, pages 508–33.

The significance of particular landscapes for the nation has also been the source of much debate within geography. A good discussion of these themes can be found in Daniels' (1993) *Fields of Vision: Landscape Imagery and National Identity in England and the United States* (Polity Press). This book gives an account of the use of rural images as a way of inspiring members of a nation, particular during times of tension and conflict, for instance wars.

More recently, a number of authors have started to demonstrate the way in which nations are contested from within. A good example can be found in Radcliffe (1999) 'Embodying national identities: mestizo men and white women in Ecuadorian racial-national imaginaries', *Transactions of the Institute of British Geographers*, volume 24, pages 213–26. See also Jones and Desforges (2003) 'Localities and the reproduction of Welsh nationalism', *Political Geography*, volume 22, pages 271–92, for a discussion of how places-based identities can undermine notions of a coherent and uniform community of people within the nation.

Contesting place

Introduction

In the previous chapter we discussed the nation as the pre-eminent scale of political power and organisation in the modern era, demonstrating that nations are not pre-ordained naturally occurring entities, but are socially constructed and frequently contested. The same applies to other 'places' in which we encounter politics at levels beneath the nation-state: regions, cities, towns, neighbourhoods, streets and so on. 'Place' is a central concept in human geography, yet it can be slippery to define and describe. A 'place' is not necessarily a bounded territory that ends where another place begins. As relational theorists such as Doreen Massey (1994, 2005) have shown, places are unique entanglements of wider social, economic and political relations – or what Massey calls 'power-geometries' (Massey 2005: 64). This means that we cannot examine the politics of a place in isolation. Places are intimately connected to other places, and thus influenced by political actions and decisions occurring elsewhere, and by inequalities of power within translocal networks. A 'sweatshop' clothing factory in Bangladesh, for example, is tied to corporate headquarters and commercial shopping malls in Europe and North America, but with little power over the nature of the relationship. Equally, the precise ways in which different relations are combined in a place are dynamic and open to contestation – for example, in local opposition to a coffee chain opening in a small town.

In referring to geographical locations as 'places' we are further imbuing them with associations and meanings that make them more than a set of coordinates on a map or an address on an envelope. A place has an identity, a character, but this is socially constructed – it is imagined and brought into being through the exchange of ideas and perceptions between people and through representations in the media and popular culture. Yet, because place is socially constructed in this way, the meanings of places are always contested. Different people will perceive the same 'place' differently, and this will influence their opinion on the appearance of the landscape (and the impact of new building), the types of activities that are undertaken in the place (and those which should be prohibited) and the people that fit there (and those who do not). Moreover, as places become associated with particular social groups, religions or political movements, so the contestation of place can become a proxy for wider social or political struggles – for example, protestors breaking the windows of a bank as a symbol of global capitalism.

This chapter explores this theme from two broad perspectives. First, we look at the role of landscape in the promotion of particular discourses of place, and discourses of power within place, and at how symbolic landscapes can become the focal point for conflict. Second, we examine the interconnection between place, community and identity and how conflict over the development or representation of particular places is contested because of meanings conveyed about community identity.

Landscape and power

Landscape is the physical manifestation of place. Places as social constructs may exist as abstract ideas, on maps, or in written documents, but when we actually go to a specific place, or we see a place represented in photographs, art or film, what we are experiencing is the *landscape* of that place. As such, landscapes are frequently seen as symbolic of the meanings that people attribute to particular places. By landscape we are here referring to all the various components that make the visual appearance of a place, including the natural geomorphology, elements of cultivation such as trees, flowers, crops, gardens and parks, and the built environment of buildings, roads, paths, monuments and so on. Thus a city centre, a factory, a theme park or a rubbish tip are all just as much a 'landscape' as a bucolic pastoral visage that we might associate with, for instance, 'landscape painting'.

Moreover, landscapes are not just assemblages of natural and manufactured objects. Cosgrove and Daniels (1988) describe a landscape as 'a cultural image, a pictorial way of representing, structuring or symbolising surroundings' (p. 1), and if we follow this definition we can see that landscapes are full of social, cultural and political meaning.

Landscapes are *powerful* because of the role that they play in structuring our everyday lives. Some of that power results from the permanence of certain landscapes, and their ability to transcend history; and some results from the fact that landscapes are shared points of experience for large numbers of people who live in, work in or visit the same place. As such, points in the landscape can symbolise particular memories and meanings of place, including messages about power and politics.

We refer to landscapes that work in this way as *landscapes of power*. A landscape of power operates as a political device because it reminds people of who is in charge, or of what the dominant ideology or philosophy is, or it helps to engender a sense of place identity that can reinforce the position of a political leader. Landscapes can express power by emphasising the gap between the 'haves' and the 'have nots' – for example, through the contrasting landscapes of rich and poor neighbourhoods – and they can also become sites through which such relations of power and oppression are resisted. As Sharon Zukin (1991) remarks in her book *Landscapes of Power*, 'themes of power, coercion,

and collective resistance shape landscape as a social microcosm' (p. 19).

What landscapes of power do and how they do it

Broadly speaking, we can identify four main functions of landscapes of power. First, they show who is in charge. Think, for example, of the castles of mediaeval Europe. As well as being important military installations, their size, construction and position also served as a reminder to local people of the power of a particular baron or king (Gould 2006). When King Edward I of England conquered Wales in the thirteenth century he ordered the construction of a series of castles, less to ensure military security – they could have little practical effect in controlling a dispersed upland population – than to symbolise the dominance of the English. More recently, 'company towns' such as Hershey in Pennsylvania, and Port Sunlight and Saltaire in England, were not just acts of social philanthropy, but also served as constant reminders to the workers they housed of their complete dependence on a single company and, usually, a single industrialist (Dinius and Vergara 2011).

Second, landscapes of power remind people of dominant ideologies or economic interests. An explicit example of this was the ubiquity of the red star on public buildings in communist states, but the physical layout of the landscape and the prominence of certain buildings can also covey this message. The dominance of Christian culture in Europe, for example, was historically symbolised by the centrality, size and extravagant design of cathedrals and churches; whilst the modern skyscrapers of the financial districts in London, New York, Chicago, Tokyo and other cities, symbolise the power of contemporary capitalism (Bradford *et al.* 1999; Willis 1995; Zukin 1991).

Third, landscapes of power broadcast a statement about the status of a place – and send a signal to rival cities or countries. In the late nineteenth century, for example, the new industrial cities of England and Wales engaged in highly competitive programmes of public building, erecting large and elaborate city halls, commercial exchanges and libraries as symbols of their wealth, power and importance in a struggle to establish themselves as the country's 'second city'. Fourth, landscapes of power engender a sense of loyalty to a place, an elite or a dominant creed. We have already touched in Chapter 5 on the role played by landscape in reproducing national identity, and most capital cities have monumental spaces that perform this function. Trafalgar Square in London, for example, is an unashamed exhibition of British imperialism and military might and serves as a focal point of patriotism. At a more personal level, public statues celebrate and venerate particular political leaders and dynasties, as does, more subtly, the naming of public buildings, squares or roads after local or national 'worthies' (Atkinson and Cosgrove 1998; Azaryahu 2012; Johnson 1995; Osborne 1998).

These functions are performed by landscapes of power through architecture and through the ordering of space. Architecturally, the size, shape and building materials of particular buildings and monuments can express power in terms of command over resources, wealth and property. The architectural style used may symbolise certain discourses of power and place. For example, classical architecture is often used for government and judicial buildings because it implicitly suggests a link to the classical ideals of justice and democracy (Cornog 1988). More explicitly, Napoleon copied the triumphal arches of ancient Rome in building the Arc de Triomphe in Paris in order to identify his empire with the power and longevity of the Roman Empire. Similarly, the use of sculpture, statues, murals, inscriptions and other symbols on and in monuments and buildings can explicitly convey political messages. In a more mundane, everyday context, Yanow (1995) describes how the size, design and building materials of a community centre in a working class neighbourhood in Israel contrasts with the surrounding houses, emphasising state power over the community.

Power is also expressed through the ordering of space. This can include the installation of physical barriers to movement across space and the use of surveillance technologies that enable a direct coercive exercise of power (see Box 6.1). It can also include the ordering of space through planning, as in apartheid South Africa where different racial groups were spatially segregated (Western 1996), or the displacement of subordinate populations. When the Earl of Dorchester wanted to create a lake to show off his new stately home at Milton Abbas in Dorset in 1770 he moved a whole village because it was in the way (Short 1991), whilst power is exerted when residents or businesses are unwillingly moved to make way for modern urban developments and mega-events such as the Olympics (Watt 2013).

More subtly and routinely, however, power is expressed through the placement and alignment of features in the landscape. The central location of royal palaces, government buildings, monuments and – at a more mundane level – factories and markets demonstrates the power of the institutions that they represent. Other monuments and buildings express power through their visibility – they are meant to be seen and to be constant reminders to the subordinate population of an elite's power. The location of the Basilica de Sacre-Coeur overlooking Paris from the Butte Montmartre, for example, was deliberately selected both for the significance of the site in events during the Paris Commune in 1871, and so that the monument would act as an 'ideological fortress' reinforcing Catholicism and monarchy (Harvey 1979).

Through these various techniques, landscapes can symbolise power, but their contribution to the actual exercise of power requires a response from the individuals that they are aimed at. As Hook (2005) argues, a political monument in particular is an unsettling object that suggests the disembodied presence of an absent political leader, and thus requires the physical presence of a human subject 'to complete the circuit of power it had initiated' (p. 700). Yet, this requirement also opens up a space of ambiguity – the opportunity for the individual to reject, ignore or contest the message that the monument is intended to convey.

Contesting landscapes of power

As landscapes can symbolise the power of leaders, institutions and ideologies, so they can also become the focal points for opposition and resistance.

BOX 6.1   LANDSCAPES OF CONTROL

The main text of this chapter discusses the symbolic power of landscape, but landscape can also enforce the exercise of power in more direct, physical ways by restricting and controlling movement, segregating groups and enabling the surveillance of individuals. National frontiers have historically been marked by landscapes of control, with walls, fences, minefields and militarised check-points designed to keep hostile forces out and citizens in. Peace and globalisation have led to the dismantling of physical barriers on some borders, such as between the United States and Canada, and within the European Union, but in other parts of the world new barricades have been erected, to restrict illegal immigration (such as the border fence between Mexico and the United States) or on the pretence of domestic 'security' (such as the separation barrier between Israel and the occupied West Bank). At a smaller scale, landscapes of control assist the function of prisons and detention camps, preventing escape and segregating internal space, with zones of differing degrees of isolation or connection with the outside world, as Giaccaria and Minca (2011) describe for the Auschwitz concentration camp.

Borders, prisons and camps are exceptional spaces, yet landscapes of control have become increasingly normalised in the urban environment. The discursive framing of the 'War on Terror' after 2001 reimaged cities as potential targets for terrorist attacks, the 'domestic front' (Graham 2006). Accordingly, 'urban political life is being saturated by "intelligent" surveillance systems, checkpoints, "defensive" urban design and planning strategies, and intensifying security' (Graham 2006: 259). Government buildings, embassies, state banks and stock exchanges, airports and other vulnerable sites are 'protected' by steel fences, concrete barriers, no-traffic zones and surveillance cameras, controlling access. Yet, newly securitised spaces include places that have traditionally functioned as symbolic public spaces and locations for political meetings and demonstrations, such as the Mall in Washington DC. As Benton-Short (2007) describes, security around public monuments in the Mall was increased after 9/11 through the deployment of temporary crowd barriers (known as 'Jersey barriers') and the introduction of 24-hour video surveillance, both of which were opposed by critics as diluting the Mall's symbolism as a 'beacon of freedom'. Plans for more permanent measures, such as installing vehicle barriers and underground entrances to monuments with screening facilities, were also subjected to intense political debate, with politicians, civil liberty groups and the media protesting that 'these changes would impair public access and the character of the public space' (Benton-Short 2007: 439). An editorial in the *Washington Post* newspaper commented, 'In the name of public safety, Washington is slowly disappearing. Concrete barriers stand as silent sentinels between people and their national shrines. What's happening in the city may serve some security interests, but the denial of public access is not in the national interest' (quoted by Benton-Short 2007: 441). In this way, the direct power of landscapes of control can come into conflict with the symbolic power of monumental landscapes, provoking debate on the meaning of a particular place.

Key readings: Benton-Short (2007); Graham (2006).

Monumental landscapes, in particular, are open to contestation as the most overt expressions of power. Proposals to erect new monuments are frequently subjected to debate over their symbolism, intended or otherwise, especially when commemorating historic events or persons. In the United States, meanwhile, the public display of religious iconography in the landscape has become a focus of political and legal debate in recent decades, as Howe (2008) discusses. Initiatives by conservative groups to display the Ten

Commandments in courthouses and on local government buildings as an overt expression of Christian ideology have stoked the conflict, but legal challenges to such icons as breaching the constitutional separation of church and state have also targeted established displays of crosses and nativity scenes. Accordingly, 'what was once a "normal" landscape – the crèche on the village green at Christmas time, the hilltop cross overlooking the city – is enmeshed in a new and highly unstable web of legal meaning' (Howe 2008: 441).

The contestability of landscapes of power means that they are targeted as sites for protests and demonstrations by both political supporters and opponents. To occupy a landscape of power – a symbolic space such as Parliament Square in London, or Ti'ananmen Square in Beijing, a council chamber or a corporate headquarters, for example – is to directly challenge power, as is discussed further in Chapter 7. Equally, enrolling a symbolic landscape can amplify a political cause. The surge of Serbian nationalism that led to the Balkan wars in the 1990s was infamously fuelled by a rally held by the hardline leader Slobodan Milosevic at a national monument to an historic battle. More recently, the staging of rallies by the Tea Party movement beneath the Washington Monument in Washington, DC, has consciously evoked an association with the founding fathers of the United States. Hook (2005) describes two contrasting appropriations of the Strijdom Monument in Pretoria – commemorating a former apartheid Prime Minister: the unprovoked shooting of black men and women by a right-wing extremist on the day that Nelson Mandela's release from prison was announced; and the anonymous contamination of the monument's fountain with red dye, symbolising the bloody legacy of the racist regime. In other contexts, landscapes of power have been subverted by graffiti, fly-posting and vandalism.

In totalitarian regimes, the capacity of the public to either oppose political monuments or openly challenge their meaning is limited, such that they become objects of resentment and commonly emerge as early targets of revenge during regime change. As communist regimes collapsed across Eastern Europe and the former Soviet Union in 1989, thousands of statues of Marx, Lenin and national communist leaders and monuments of 'heroic' workers and soldiers were removed from towns and cities. Some were violently demolished, others transported to museums such as Statue Park in Budapest (James 1999) (Plate 6.1). A similar cleansing of the monumental landscape occurred with decolonisation across Africa and Asia, removing statues of European monarchs such as Queen Victoria and King George V of Britain and colonial administrators such as Lord Delamere. As Larsen (2012) describes for Kenya, statues were often removed pre-emptively by out-going colonial regimes, though with the collusion of new post-colonial governments, 'while the colonial administration had wanted to remove colonial statues to prevent them from being subject to vandalism in post-independence Kenya, the new Kenyan Government sought their removal in an effort to divest the city of these symbols of colonial oppression, (Larsen 2012: 51). In some cases the statues were replaced by new national monuments, but in others the vacated spaces were absorbed back into the everyday city landscape:

> The *Daily Nation* reported how, 'Soon after the statue [of Lord Delamere] had been taken away into storage, sellers of wood carvings and curios had set up "pitch" at the foot of the empty plinth'. In this way, the space in which they statue was erected to create a European settler identity was quickly appropriated by members of Kenya's African population. Through their everyday activities, centred on the site, the space absorbed into Nairobi's 'ordinary landscape', while at the same time remaining part of its symbolic landscape through a change in the use of the space and in the meaning anchored in the site.
>
> (Larsen 2012: 49–50)

Indeed, the ability of urban societies to naturally neutralise the political symbolism of monumental landscapes should not be under-estimated. Iraqi dictator Saddam Hussein erected numerous monuments to himself, including the notorious Victory Arch. Standing forty metres tall, the monument comprised two massive hands, modelled

Plate 6.1   Monument to Hungarian Communist leader, Bela Kun, relocated to the Statue Park, Budapest

Photograph: Michael Woods

on the president's own, clasping inter-locking steel swords, each twenty-five metres long and cast from gun metal. Around the base of each hand were placed 5,000 Iranian helmets taken fresh from the battlefield, representing war spoils. The monument was a crass symbol of Saddam's despotic regime, yet, though other monuments were destroyed in the wake of the American invasion and the execution of Saddam, the Victory Arch survived. Prominent Iraqi dissident Kanan Makiya in an essay on the monument observed that 'cities collect objects like these, and then time transforms their meaning' (Makiya 2004: 132), and mused, 'what will future generations of Iraqis see in this monument: a symbol of the demonic machinations of one man . . . or an unforgettable testament to their country's years of shame?' (ibid.: 133–4). After several years of neglect, work started on restoring the monument in 2011 as an act of reconciliation, with a government spokesperson telling the *New York Times*: 'We are a civilized people and this monument is part of the memories of this country' (Myers 2011).

The question of how to deal with the monuments and landscapes of previous political regimes has been especially pronounced in Berlin, as discussed in an extended case study below.

Berlin: a contested landscape of power

Few urban landscapes have been inscribed with the marks of contrasting political regimes as that of Berlin during the twentieth century. The German capital has been a stage for the performance of power by successive imperial, Nazi and communist regimes, and following the collapse of communism in 1989 and the re-unification of the divided city contests have emerged around the redevelopment of symbolic sites and the commemoration of the past.

The landscape of power of Berlin at the start of the twentieth century recorded the rise and ambitions of the Prussian state over the preceding two centuries (Stangl 2006). Its centrepiece, the Brandenburg Gate, had been erected in 1795 and modelled on the entrance to the acropolis in classical Athens, reflecting the aspirations of the new kingdom (Plate 6.2). It had originally been conceived as a symbol of peace, surmounted by a sculpture of Eirene, the Greek goddess of peace, driving a chariot drawn by four horses (known as the Quadriga). However, in 1814 following the end of the Napoleonic wars, the gate was reinterpreted as a symbol of victory, Eirene was redesignated as Victoria, Roman goddess of victory, and an iron cross was added to the Quadriga to represent Prussian power. Further along the avenue of

Plate 6.2 The Brandenburg Gate, Berlin

Photograph: Michael Woods

Unter den Linden, towards the imperial residence at Berliner Schloss, a massive equestrian statue of Frederick the Great had been installed in 1851, symbolising Prussian military power and the authority of the imperial house.

The ambitions of the new German empire, which had been formally constituted only in 1870, were also represented in the size and ornate style of the Reichstag parliament building, which opened near the Brandenburg Gate in 1894. Yet, from the start the Reichstag had been a focus of contestation, not least over the inscription 'Dem Deutschen Volke' (To the German People), which had been opposed by the Emperor because of its democratic implication. In the turbulent years after German defeat in the First World War, the Reichstag became a stage for political acts by competing factions, culminating in an arson attack in 1933, which the Nazis falsely blamed on Communists as a pretext for seizing power.

The Nazis recognised the significance of landscape in communicating power, referring to architecture as 'words in stone', with 'the potential to both shape the present population and to leave a lasting legacy that would endure into the future' (MacDonald 2006: 108). Across Germany, Nazi redevelopment projects demolished 'inappropriate' buildings such as synagogues and restored more 'correct' Germanic architecture. In Berlin – as in Nuremburg in southern Germany, site of party rallies (see MacDonald 2006) – the emphasis was more on the spectacular exposition of power. Icons of Prussian imperialism were appropriated to symbolise patriotism and continuity, with the Brandenburg Gate adopted as a party symbol and the square in front of it used to stage parades and pageants (Stangl 2006). For new buildings, Hitler ordered that the architectural style would be, 'by necessity, classical – not only for the use of heroic forms, but because the Greeks and Germans were racially linked' (Balfour 1990, quoted by Till 2005: 40).

More crudely, the Nazi regime equated power with scale. Buildings such as the massive new Reich Chancellery were intended to intimidate and impress. Five-metre high doors led to a large reception decorated in mosaics and on to a 150-metre gallery, longer than the Hall of Mirrors in the Palace of Versailles. On seeing the plans, Hitler reportedly exclaimed that the walk would give visitors 'a taste of the power and grandeur of the German Reich'. Simply the construction of such buildings was an act of power, requiring the clearance of streets and houses, and the quarrying and transport of building materials by forced labour (Till 2005). Hitler and his architect, Albert Speer, planned to transform Berlin into a new capital, Germania, based on Rome but superscaled, including a domed 'people's hall' that would have been the largest enclosed space in the world. Till quotes Hitler's presentation of the plans:

> Nothing will be too good for the beautification of Berlin . . . . One will arrive there along wide avenues containing the Triumphal Arch, the Pantheon of the Army, the Square of the People – things to take your breath away! It's only thus that we shall succeed in eclipsing our only rival in the world, Rome . . . . Let [the center of the new Berlin] be built on such a scale that St Peter's and its Square will seem like toys in comparison!
>
> (Adolf Hitler, quoted by Till 2005: 40)

Germania was never completed. The Second World War left one-third of Berlin's housing destroyed, 75 million cubic metres of rubble, and an imperative both to rebuild and to 'rid the city of any remnants of the "Nazi disease" and to create a new slate to build the future' (Till 2005: 41). More immediately, however, the erection of first barriers and later the Berlin Wall between the Soviet-occupied sector in the east, and the American, British and French sectors in the west, imposed a landscape of control that prevented movement across the city and propped up the Communist regime of East Germany by restricting emigration and contact with the West. On the western side the power of the wall was subverted by graffiti, on the east it was heavily guarded.

Behind the wall, the communist East German government refashioned the city landscape according to its own ideological principles, as Till observes:

> In East Berlin, the construction of the neoclassical 2,300-meter Stalinallee (formerly Frankfurter

Allee and now Karl-Marx-Allee) in the 1950s was a symbol of the 'city of tomorrow'. Modelled after great Moscow's Gorki Street, the great boulevard of classical Greek proportions and mixed-use design was said to embrace local architectural traditions and adhere to socialist-realist tenets of urban design.

(Till 2005: 42)

The project involved the eradication of the Nazi and imperial landscapes of power, including the removal of Frederick the Great's statue and the demolition of the imperial palace, Berliner Schloss (Stangl 2006). A concrete and glass 'Palace of the Republic', housing

the East German parliament, was built in place of the latter, suggesting an illusion of transparency and democracy in an authoritarian state (Staiger 2009). As in other socialist cities, ideological dominance was reinforced by statues of Communist leaders, most notably of Karl Marx and Friedrich Engels emphasising the German origins of Marxism. Elsewhere, murals presented an idealised vision of life in East Germany, as on the House of the Teacher (Plate 6.3).

The rapid collapse of the East German regime in November 1989 was marked by a failure of the landscape of power, as crowds took sledgehammers to the Berlin Wall and reclaimed the Brandenburg Gate – which had been isolated behind barriers for

Plate 6.3 Communist-era mural on the 'House of the Teacher', Berlin

Photograph: Michael Woods

forty years – as a public space. The reunified city, however, was confronted by a number of conflicts over the redevelopment of the urban landscape and its engagement with the past (see Forest *et al.* 2004; Till 2005).

First, as the 'death zone' around the wall has been opened up for development, its colonisation by trasnational corporations and shopping malls has been controversial, with critics lamenting the commodification and neutralisation of history at Potsdamer Platz and Checkpoint Charlie. As Ladd (2004) observes, 'the new masters of Potsdamer Platz invoke the past, but not in the subversive and provocative ways it is done elsewhere in Berlin. This history is much closer to the Disney version' (p. 132). Moreover, Allen (2006) argues that the redeveloped Potsdamer Platz represents a more subtle exercise of ambient power, 'something about the character of an urban setting – a particular atmosphere, a specific mood, a certain feeling – that affects how we experience it and which, in turn seeks to include certain stances which we might otherwise have chosen not to adopt' (p. 445). The Sony Center in Potsdamer Platz works in this way, Allen suggests, not by exclusion but through seduction – drawing people into a private space that feels like a public space: 'the suggestive pull of the layout and design of the plaza, the feeling of openness inscribed in the space, have a seductive presence' (p. 451). Yet, within the space, 'choices are restricted, options are curtailed and possibilities are closed down by degree through the forum's ambient qualities' (p. 454), prompting individuals into consumption decisions and quietly exerting commercial power in a place once defined by overt political power (see also Box 6.2).

Second, debates have surrounded the renaming of streets in former East Berlin that had been named after Communist heroes by the East German regime as part of their landscape of power. In 1993 the city's Senate assumed powers to rename streets in areas associated with the 'capital city function' and adopted a programme of restoring historical names that brought it into conflict with the more moderate district councils. One particular flashpoint was *Otto-Grotewhol-Strasse*, previously known as *Wilhelmstrasse*. The problem, as Azaryahu (1997) observes, 'was that this traditional name was laden with historical associations and nationalistic meanings unequivocally linked to the German Reich. A restoration of the old name, therefore, could also be understood as an attempt to imply that German reunification also meant the restoration of the Reich' (p. 487). Instead the district council proposed the name *Toleranzstrasse* (Tolerance Street) as a symbol of a new, non-aggressive, Germany polity. This, however, proved unacceptable to more nationalist politicians who sought to recreate the previous global importance of Berlin, and who launched a court challenge. Eventually the Berlin Senate ruled that the street be renamed *Wilhelmstrasse*.

Third, efforts to reconcile past atrocities by creating memorials to the victims of the Nazi regime have stoked new arguments, as Till (2005) described for two critical sites. The first, the former location of the Gestapo headquarters, situated alongside the Berlin Wall, had been left derelict by the post-war West Berlin city authorities – which campaigners perceived as 'a metaphor for the postwar German desire to forget' (Till 2005: 65). The campaigners sought to reclaim the space, starting by naming it:

> They began calling this area the 'Gestapo Terrain' (Gestapo Gelände), a name that evoked dark personal memories for an older generation but for this second generation stood for those memories, social hauntings, of a traumatic past that could never be known to them personally. The name signified the second generation's emotional search for a past that would always be unknown; the landscape embodied what it meant for them to be German.
>
> (Till 2005: 80)

In 1982, the (West) Berlin Parliament agreed to a public art competition for a memorial at the site. However, divisions soon developed over the nature of the memorial. Whilst some survivors' and human rights groups called for a *Mahnmal*, literally 'a memorial of admonishment', others argued for a museum as a more active, critical engagement with the Gestapo atrocities. Frustrated by a lack of official progress, activists occupied the site in the 1980s, first

BOX 6.2 LANDSCAPE, POWER AND CONSUMPTION

Monumental spaces are clear demonstrations of the exercise of power through landscape, but in coining the term 'landscapes of power', Sharon Zukin (1991) was more concerned with expressions of economic power by global capital through the landscape. As such, she notes the uneven power relations exhibited in the contrast between the vernacular landscapes of working-class districts and the 'landscape of power' of corporations and economic capital, evident in the skyscrapers of Lower Manhattan, but also in the fantastical landscapes of Disneyworld or Times Square. These landscapes use the extraordinary to exert a subtle power in shaping lifestyles and encouraging consumption. This analysis resonates with a wider political-economy argument that in modern capitalist society, mass consumption is employed by the state to maintain social order in place of coercion. An interesting example of this is discussed by Thornton (2010) in a study of two neighbouring sites in Beijing. The first, Ti'ananmen Square, was constructed as a monumental landscape of power after the Communist revolution, dominated by the tomb of Chairman Mao, government buildings and political sculptures. The second, The Place, is a large shopping mall that reflects economic liberalisation and the rise of consumerism in China. Yet, as both spaces are ultimately controlled by the Chinese Communist Party, Thornton argues that The Place represents the fusing of global capital and Chinese communism to create a new vehicle for the projection of state power.

Key readings: Zukin (1991); Thornton (2010).

illegally digging to uncover remains of the complex and then establishing a temporary public exhibition, the 'Topography of Terror'. Continuing debates meant that work on a permanent museum did not start until 2009.

The second site is the Holocaust Memorial to the murdered Jews of Europe, opened in 2005 and situated between the Brandenburg Gate and the site of Hitler's bunker. The location, decided after some debate, was regarded as important by memorial advocates who believed that 'at such a historic location, the establishment of the memorial would purge Hitler's ghost, the lingering presence of evil, from the German body politic' (Till 2005: 175). More controversial, however, was the design. The memorial campaign's vision for the monument emphasised German guilt and shame, envisaged that 'German visitors should walk through a hall of horrors, enter in guilt and darkness, and reemerge redeemed' (ibid.: 178). This vision was criticised from several quarters and the design remained problematic and two proposals were rejected by the German Chancellor.

The eventual design, by Eisenman and Happold, avoids the use of overt symbols, consisting of 2,711 concrete slabs varying in height and arranged in a grid pattern on a sloping area of 19,000 square metres – intended to be disorientating and claustrophobic.

As Cochrane (2006) writes, 'Berlin is a city of paradox – a city of monuments and memorials and of absences' (p. 12). Its contested landscape of power is shaped by the legacy successive political regimes, but also by the problem of marking the murderous actions of these regimes. The problem of presence and absence in the landscape – what to preserve and what to exclude; what to remember and what to forget – is at the heart of its urban politics.

Contesting communities

Places are more than landscapes. They are also defined by the people who identify with them, and each other, as a community. The connection between community and place, however, can be complex. People may identify as a community through their shared experience of living in the same locality, however, residential proximity does not invariably translate into a community, especially if neighbours socialise in different networks, work in different workplaces and utilise different public facilities and services, or if they are divided by ethnicity, religion or class. Equally, people who identify with a particular place as part of a community may not necessarily be resident locally, but may include ex-residents who have moved away, tourists and other regular visitors, or individuals for whom the place has a symbolic significance.

Place and community are rather bound together by the shared understandings that members of a community project onto a particular geographical location, referred to by Martin (2003) as 'place-frames'. As place-frames are shared in a consensus by members of a community, they are normally latent. However, when the understandings of place that they represent are challenged or threatened by other actors, they can become the focal point for political mobilisation in what Pierce *et al.* (2011) characterise as 'the politics of place'. Pierce *et al.*, though, further argue that a focus primarily on 'place-frame' underplays the significance of external networks in the processes through which places are constituted and contested. As discussed at the start of this chapter, relational approaches in human geography have emphasised that places are not bounded territories, but are intersections of wider social, economic and political trajectories. Pierce *et al.* accordingly propose a model of 'relational place-making', that identifies and explores the key place-frames that shape perspectives on the bundles of relations that come together in place, as well as the key actors and institutions who construct competing place-frames, and proceeds to unpack and interrogate the connections that inform the positionalities of actors engaged in contesting place.

They illustrate this approach with an example of a land use conflict in Georgia, in the southern United States, where a hospital proposed to expand by acquiring and developing neighbouring land, involving the demolition of around fifty houses. The hospital authorities drew on a place-frame of the city as an economic centre, 'with the hospital as a key job/growth engine

serving a multi-county area' (Pierce *et al.* 2011: 65). The plans were opposed by local residents, whose place-frame 'emphasized a neighbourhood focused on single-family homes (owner and renter occupied), sidewalks, and friendly encounters between neighbours while walking with pets, children or to a local store' (ibid.), and positioned the hospital as just one of many sites in the area. However, whilst the two place-frames articulated contrasting discourses of place, they were both embedded in entwined social and economic relations, which also connected individual actors on either side:

> On the *economic* side of the conflict, some of the hospital administrators, doctors, other staff and patients also live in Normaltown where the hospital is located and which was the site of such conflicting frames. They have friends there, they shop at stores or eat at restaurants there, and they drive or walk to and from the hospital on the same streets and sidewalks that local residents use. On the *neighbourhood* side, Normaltown residents, too, have complicated and intersecting relations with the institution of the hospital (as a place where a child was born, a broken bone set, or cancer treatment provided) and with its employees.
>
> (Pierce *et al.* 2011: 65)

Consequently, Pierce *et al.* note that the two place-frames in fact depended on each other, and observed that the shared interests and social connections between actors both shaped the dynamics of the dispute and enabled a resolution to be found, in which neighbourhood residents were a given a more direct say in the development plans.

In the second part of this chapter we explore further the engagement of communities in contesting place as part of what following Pierce *et al.* we can call processes of relational place-making. We focus in turn on three particular aspects of a community's attachment to place: first, conflicts over changes to the landscape and land use; second, struggles around access to public space; and third, the contested claims to space made in performances of community identity. All three cases, we suggest, fundamentally concern a community's power over places with which it identifies.

Contesting everyday landscapes

Landscape and land use serve both practical and symbolic functions for place-based communities. The features of the built environment are the houses in which local residents live, the factories and offices where they work, the shops they use, the schools where their children are educated and so on, whilst the natural landscape offers environmental amenities. These sites are also the spaces in which members of a community meet each other and perform the social interactions that glue a community together, from the pub to the school-gate. At the same time, the appearance of a place can signify something of the identity, culture and aspirations of a community, from the presence or absence of different religious buildings to the maintenance and upkeep of property. Moreover, as a place is lived in, so sites within it accumulate associations with both collective and personal memories, as Hayden observes with respect to the American city:

> Identity is intimately tied to memory: both our personal memories and the collective or social memories interconnected with the histories of our families, neighbors, fellow workers and ethnic communities. Urban landscapes are storehouses for these memories, because natural features such as hills or harbors, as well as streets, buildings and patterns of settlement, frame the lives of many people and often outlast many lifetimes.
>
> (Hayden 1995: 9)

Changes to a landscape therefore resonate through a community in a multitude of ways and can become a catalyst for political mobilisation, especially if they are perceived to threaten elements of community identity.

Opposition is hence frequently mobilised to developments that disrupt the landscape, such as new roads, railways, airports, houses, industrial estates, windfarms, power stations and so on (Della Porta and Piazza 2007; Walton 2007; Woods 2005), or that introduce new people into the area that are perceived not to 'fit' with the local community, such as prisons, travellers' camps or asylum processing centres (Che 2005; Hubbard 2005). Protests of this kind are commonly dismissed as self-interest or 'NIMBY-ism', though as Box 6.3 discusses, such easy labels can disguise complex underlying processes. More accurately, conflicts over land use development tend to reflect competing place-frames that are embedded in different readings of wider social and economic power-geometries. This can be seen, for example, in the case of opposition to wind turbine power stations, or 'wind-farms', that have become commonplace in rural areas in many parts of Europe, Australia and North America (Phadke 2011; van der Horst and Vermeylen 2011; Zografos and Martinez-Alier 2009). As Woods (2003) demonstrates, proposals for new windfarms tend to frame their sites as functional spaces, resilient to change and as appropriate for wind power due to their environmental conditions; whilst opponents employ place-frames that portray the same sites as aesthetically attractive and ecologically sensitive places in which wind turbines would be 'out of place'.

Community mobilisations against new land use developments can also be motivation by frustration at a perceived lack of power over the landscape, by a belief that developments are being imposed on the community. Indeed, if proposals for large infrastructural developments such as windfarms are supported by place-frames that emphasise functional or strategic concerns, the significance of local interests can be marginalised and emotional objections excluded from the deliberation process (Ku 2012; Woods 2003). However, a relational analysis reveals that place-based conflicts are more than local–outsider struggles (Hogenstijn et al. 2008), but make wider associations both discursively and organisationally. The competing place-frames in windfarm conflicts both acknowledge the challenge of broader environmental change, but they differ in emphasis, contrasting local and global conceptualisations of nature (Woods 2003); whilst both pro- and anti-windfarm campaigns frequently enlist supporters outside the immediate locality. Indeed, the capacity of a local campaign to 'jump scales' and engage with centres of social power such as national and regional governments, transnational corporations and the national and international media

BOX 6.3   NIMBY OR NOT NIMBY?

Local opponents of land use developments or social projects are often described as 'NIMBYs', as articulating a 'Not In My Back Yard' argument, in which their opposition is not to a type of development in principle, but to its location. For example, campaigners may support renewable energy in principle, but oppose a windfarm in their locality as being in the 'wrong place', or support improvements to highways but oppose a new road past their property. The implication of this label is that opponents are duplicitous in their opinions and motivated by self-interest. The term is widely used in popular debate, but Wolsink (2006) has criticised the description of groups as 'NIMBY' in academic studies and the suggestion that 'NIMBYism' is an explanation for political action. He argues that 'NIMBYism' is a poorly defined and pejorative concept and that empirical studies of local conflicts show that only a small part of opposition is driven by NIMBY concerns. Wolsink observes that the tag of 'NIMBYism' can be used to dismiss local concerns, and proposes that the language of 'NIMBYism' needs to be interrogated as part of a wider analysis of the political and economic processes that shape siting controversies.

In a reply to Wolsink, Hubbard (2006) similarly calls for a more critical approach to the concept of 'NIMBYism', but argues that the term reflects the narratives deployed in many local campaigns and as such is a legitimate way of exploring ambivalent identity politics that inform struggles over place. In particular, he notes that 'NIMBY' opposition can be a way of translating fears about social otherness into what may be perceived as more 'legitimate' environmental concerns in protests against housing developments, asylum centres or actions against sex work (see also Hubbard 2005).

The debate between Wolsink and Hubbard did not resolve issues about the usefulness of otherwise of the 'NIMBY' label, but it does remind us of the need to be careful about the labels that we use in research and the imperative of critiquing and interrogating taken-for-granted ideas.

Key readings: Wolsink (2006); Hubbard (2006).

---

in an enlarged 'space of engagement' is identified as a critical attribute by Cox (1997).

Similar dynamics are evident in conflicts that revolve not around opposition to new developments, but campaigns to save threatened facilities and sites of community interaction. Protests against the closure of schools, post offices and libraries, or even stores and shopping malls, can be motivated not just by the loss of the service provided, but also by concerns over a dilution of community life (Parlette and Cowen 2011). In these cases, defensive protests can transmute over time into proactive initiatives that seek to reclaim spaces and reassert community power over place by taking over empty schools, stores or factories as community-run facilities. Sometimes this may involve providing the same service but as a voluntary or community enterprise, such as a community shop or

pub, in other cases it involves putting the building to a new use, but continuing to provide a space for community interaction.

The contestation of place can be particularly pronounced when land use changes do not only threaten an individual community facility or the appearance of the landscape, but alter the composition of the community itself by facilitating colonisation of a place by a new social group, as in the process of gentrification. Gentrification involves the redevelopment and renovation of a neighbourhood or district, increasing property values and displacing established, often working-class residents, in favour of higher-income newcomers (see Box 6.4). As social interaction between established and new residents is commonly limited (Mazer and Rankin 2011), gentrification in effect creates two communities

coexisting in the same place but with competing place-frames. Furthermore, the loss not only of affordable housing but also of shops, bars, cafes, churches and other social spaces associated with the established community means that the community becomes progressively diminished and weakened, with decreasing political power to influence the development of the neighbourhood (Martin 2007).

Gentrification and opposition to gentrification have been observed in both urban and rural areas across large parts of the world, but probably the best documented example is the Lower East Side of New York. Traditionally the first home to newly arrived immigrants, the area was first settled by German migrants in the mid-nineteenth century. As the industrious German settlers prospered and moved uptown, their place was taken by new groups of immigrants from eastern and southern Europe, including a large Jewish population. As immigration to the United States peaked in the early 1900s, the pressure on space in the Lower East Side became enormous. Tenement buildings of six or seven storeys would house up to two dozen families, in conditions that were unsanitary, vermin-infested,

severely overcrowded and deprived of natural sunlight. Amid the poverty a strong sense of community and self-help developed. Trade unions were organised, relief charities were established, and adult education classes flourished. Many of the beneficiaries of this education became politically active, campaigning for better housing and public services.

The first attempt to improve conditions on the Lower East Side was made in 1890 when the worst tenements were demolished to create parks and open spaces. However, the major phase of slum clearance came during the 1930s, 40s and 50s, with many tenements replaced by new public housing complexes. The district continued to attract newly arrived immigrants, notably the Chinese and Puerto Rican communities. This continuing mêlée of ethnic communities combined with an eastward drift of 'bohemians' from Greenwich Village following the dismantling of the Third Avenue Elevated Railroad in 1955, to sustain a vibrant alternative cultural and political scene in the Lower East Side, and an enduring sense of community.

However, the combination of low property prices and the newly fashionable bohemian atmosphere

BOX 6.4 GENTRIFICATION

Gentrification involves the redevelopment of property in a neighbourhood by and for affluent incomers leading to the displacement of lower-income groups who are unable to afford the inflated property prices. The term was first coined by a sociologist, Ruth Glass, in 1964 to describe the renovation of working-class districts in London. Similar processes have since been observed in most Western cities as well as in many rural areas. Urban gentrification is often associated with neighbourhoods with large older properties, such as tenements, that can be easily converted into apartments but which have become run-down due to out-migration. At the start of the process property prices are cheap compared to other parts of the city thus allow significant profits to be made. Classic examples include Islington in London, Society Hill in Philadelphia and Waterlooplein in Amsterdam, as well as the Lower East Side of New York. As Neil Smith (1996) discusses, there are a number of different explanations that have been proposed for gentrification, including cultural theories linked to changing consumption patterns and economic arguments about the benefits of urban living. Smith also examines the role of property developers and speculators in fuelling gentrification, for example by raising rents to force out lower-income residents.

Key reading: N. Smith (1996).

also appealed to property developers who began moving in during the 1970s, buying old run-down tenement buildings and refurbishing them as luxury upmarket apartments. To accelerate the process, tenants were evicted or forced out by rent increases so that landlords could redevelop property for sale or rent at much higher prices (N. Smith 1996). The developers and middle-class newcomers articulated a place-framing of the Lower East Side as a fashionable but edgy frontier neighbourhood, captured in its rebranding as the 'East Village' to suggest an extension of adjacent Greenwich Village. This place-frame was underpinned by private and corporate capital and by the collusion of the city authorities and media, and further demanded the creation of spaces for middle-class consumption in the neighbourhood, displacing established businesses and prompting the enforced sanitisation of Tompkins Square Park by evicting perceived 'undesirable' users such as rough sleepers (N. Smith 1996).

Resistance to gentrification has been organised since the 1980s through grassroots' movements such as the Lower East Side Collective (Newman and Wyly 2006; Sites 2003; N. Smith 1996). These have articulated a counterveiling place-frame of the Lower East Side as a culturally rich and diverse community, embedded in history. The struggle has ranged over different focal points during the ensuing three decades, including the demolition or redevelopment of churches and public buildings (Plate 6.4), the

Plate 6.4   Protest against gentrification in East Village, New York

Photograph: Michael Woods

commodification of tenant blocks as an 'historic district' for tourist consumption, and proposals by the city council to sell for development plots of one-time derelict land that had been cultivated by neighbourhood groups as 'community gardens' (Schmelzkopf 2002; Smith and Kurtz 2003).

The success of neighbourhood movements in resisting gentrification has been mixed. Ley and Dobson (2007) suggest that neighbourhood political mobilisation has combined with other factors to block or stall gentrification in some districts of Vancouver, whilst research in Atlanta has suggested that neighbourhood organisations with settled structures and a track record of delivery have some capacity to protect the political participation of long-term residents in gentrifying areas (Martin 2007). In the Lower East Side, protesters succeeded in halting the development of community gardens, though partially through the help of external philanthropists who stepped in to buy and protect plots. Overall, though, the clear trend is with the gentrifiers.

Struggles over public space

Public spaces are important for the practice of community. They are the places in which neighbours see each other and meet together, hold conversations and take part in community events. They are also spaces in which community members encounter people from outside the community, and thus through which they represent the community to the outside world. Yet, public spaces can also be objects of conflict and struggle within a community. In theory, a public space is open to anyone, but in practice the use of public space is commonly controlled and regulated. This was done overtly and in a racially motivated way in apartheid South Africa and during segregation in the southern United States, in which access to public space was restricted by race. Such racist policies are now illegal in most democratic states, as are other forms of explicit discrimination, but certain social groups can still be excluded by more subtle forms of exclusion.

The use of public spaces by homeless people, for instance, is a widespread focus of contestation.

Homeless people gather and sleep in public spaces such as parks because they do not have private properties of their own. In doing so they are arguably asserting a right to public space as citizens. However, their presence may be intimidating to other local residents, who feel excluded from the public space by fear. Accordingly, initiatives to discourage rough sleeping in public spaces are frequently legitimised as actions to 'reclaim' public space for the community. This may be achieved through policing, surveillance cameras, locking spaces at night or changing the landscape. In Horton Plaza Park in San Diego, California, for example, a mid-1990s redevelopment that involved closing public toilets, removing benches and replacing lawns with prickly plants and flowers was described by the city council as an initiative to 'return the space to full public usage by all segments of the community . . . [and to] dilute the influence of the so-called undesirables' (quoted by Staeheli and Mitchell 2008: 60).

While homeless people may not be perceived by other local residents as 'part of the community', struggles over the use of public spaces by young people reveal a more complex politics around differential rights to public space within a community. As Staeheli and Mitchell (2008) observe:

> Teenagers and young adults occupy an ambiguous position within the public, with some rights accorded to them, but other rights withheld. What does seem unambiguous, however, is the wariness that many older adults feel around youth, and in particular, young males or young people of color. . . . By their very presence, youth can be challenging and threatening, disrupting the feeling of safety
>
> (Staeheli and Mitchell 2008: 85)

Strategies for regulating the presence of young people in public spaces often involve prohibiting certain activities that are associated with youth but are represented as disruptive or threatening to other users of the space, such as skateboarding, ball games, cycling, playing music or even 'loitering'. These may be banned altogether, or restricted to particular areas or to designated spaces elsewhere, such as 'skateboard parks'. The consequence is to define the right to public space in terms of appropriate forms of activity or behaviour, which are subjective judgements often reflecting commercial interests. As Staeheli and Mitchell (2008) note of a youth curfew in the semi-public space of the Carousel Center Mall in Syracuse, New York, 'youth are apparently part of the public when consuming, but not when socializing – or at least not on Friday or Saturday evening' (p. 86) (see also Collins and Kearns 2001). In turn, young people who are discouraged from gathering in public spaces create their own communal places in more marginal spaces of a town or city, often on private property thus provoking new conflicts (Panelli *et al.* 2002).

Questions of access to public space have become more complex as distinctions between public and private space have been increasingly blurred. In many new developments there may be no municipally owned public space, with public spaces limited to those provided, owned and managed by developers, often with access restricted to the residents of a 'gated' community (McGuirk and Dowling 2009). More informally, privately owned and managed shopping malls have become the de facto public spaces of many communities (Kohn 2004), whilst urban spaces have also been 'privatised' by transfers of ownership as part of redevelopment deals or the franchising of management and commercial concessions (Staeheli and Mitchell 2008).

Parades, pageantry and politics

A further arena for the contestation of connections between place and community is the performance of parades, marches, carnivals and other public events. As communal events, such performances can convey representations of place that identify them with particular social groups or certain cultural, political or economic interests. These may be acts of resistance towards external powers, as in celebrations of indigenous cultures, or they may be pageants aimed at reinforcing elite narratives of place within a community (Hagen 2008; Woods 1999).

Where such events take place is also important. For a festival or rally to occupy a central park or plaza, or for a parade to follow a route through the centre of a town or city, or around a particular neighbourhood, is to make a claim to space. O'Reilly and Crutcher (2006), for example, describe the spatial politics of the Black Men of Labour and gay Southern Decadence parades in New Orleans, both of which happen on Labor Day. The two parades briefly pass each other on one street corner, but trace routes through different neighbourhoods that 'demonstrate concerns with community memory, respectability, tradition, social ties and local economies' (p. 246). The parades are therefore 'acts of territoriality', through which 'each community lays claim to New Orleans' streets and neighbourhoods to the temporary exclusion of other actors, and makes explicit their community's presence in these areas' (O'Reilly and Crutcher 2006: 247).

However, the claiming of public space for parades and events by extremist groups can be designed as acts of intimidation against other social groups within the locality. Accordingly, parades and public events can be contested as part of conflicts over the positioning of place within wider social and political struggles. Annual parades by Protestant groups in Northern Ireland supporting the union with Great Britain, for example, are contentious demonstrations of power and claims to territory, following routes that occupy city centre streets, but which in some cases also pass through Catholic neighbourhoods sympathetic to Irish nationalism (Cohen 2007). As part of the 'peace process', attempts have been made to defuse the inflammatory potential of the parades by rerouting them away from Catholic areas, yet this has been resisted by participants who perceive it as an attack on their community traditions.

Indeed, 'tradition' can be a polarising idea in the context of place and performance. The 'traditions' embedded in some overtly non-political rituals and events such as carnivals and festivals have been challenged in some cases as reproducing racist, sexist or bigoted conventions. A classic paper by Susan Smith (1993), for example, describes a complaint against the 'traditional' participation of blacked-up characters in an annual children's fancy-dress parade in a small town in Scotland. However, as in the case described by

Smith, such challenges can serve to galvanise local communities to mobilise in opposition to perceived outside threats to their local traditions, and thus to place identity.

Summary

This chapter has demonstrated that the contestation of place is often a central element in political conflict. This arises because the meaning of place is not value-neutral. Different actors – who might be individuals or organisations – socially construct different places co-existing over the same territory and tensions are generated when elements of the different 'imagined places' prove to be incompatible. As actors then move to promote or protect their particular 'discourse of place', political tensions can become political conflict. This can take a range of different forms and can be focused on a whole range of different expressions of place. In some cases it is the interpretation of certain features in the landscape that is at issue; in others how a place is represented through pageantry; in yet other cases the conflict may revolve around the impact of development on the character and identity of the local community. Usually these kinds of conflict are not just about place. They are also about class, or race or gender or other social divisions. But at the same time they are not entirely reducible to class or race or gender because of the significance of place in framing the dispute. It is by recognising and exploring the role of place in political conflicts of this kind that geographers can make a distinctive contribution to understanding such processes.

Further reading

The case study of Berlin is a good place to start exploring the construction and contestation of landscapes of power. Stangl (2006), 'Restoring Berlin's unter den linden: ideology, word view, place and space', in *Journal of Historical Geography*, 32, 352–376, is an excellent account of the changing political symbolism of the city's central avenue and its surrounding monuments and buildings. Karen Till's book *The New Berlin: Memory, Politics Place* (2005), and Cochrane (2006) 'Making up meanings in a capital city: power, memory and monuments in Berlin', in *European Urban and Regional Studies*, 13, 5–24, are both good studies of the post-Cold War politics of the Berlin landscape and the problem of remembering the past.

The concept of 'relational place making' is introduced by Pierce, Martin and Murphy (2011), 'Relational place-making: the networked politics of place' in *Transactions of the Institute of British Geographers*, 36, 54–70. For a good example of community opposition to redevelopment, see Ku (2012) 'Remaking places and fashioning an opposition discourse: struggle over the Star Ferry pier and the Queen's pier in Hong Kong', in *Environment and Planning D: Society and Space*, 30, 5–22. Neil Smith's classic book, *The New Urban Frontier* (1996) is still a key reading on gentrification, and particularly the Lower East Side, whilst Ley and Dobson (2008) 'Are there limits to gentrification? The contexts of impeded gentrification in Vancouver' in *Urban Studies*, 45, 2471–2498, is an interesting discussion on resistance to gentrification.

For more on struggles over public space see Staeheli and Mitchell (2008) *The People's Property?*, and for discussion of parades, politics and territory see O'Reilly and Crutcher (2006) 'Parallel politics: the spatial power of New Orleans' Labor Day parades' in *Social and Cultural Geography*, 7, 245–265; and Cohen (2007) 'Winning while losing: the apprentice boys of Derry walk their beat', in *Political Geography*, 26, 951–967.

Web resources

Several of the 'landscapes of power' discussed in this chapter can be explored using Google Earth and Streetview at the following coordinates:

Berlin – Brandenburg Gate	52° 30' 58" N 13° 22' 39" E
Berlin – Checkpoint Charlie	52° 30' 30" N 13° 23' 24" E
Berlin – Holocaust Memorial	52° 30' 51" N 13° 22' 41" E
Berlin – Postdamer Platz	52° 30' 35" N 13° 22' 30" E
Berlin – Topography of Terror	52° 30' 24" N 13° 22' 59" E
Baghdad – Victory Arch	33° 18' 20" N 44° 32' 1" E
Beijing – Ti'ananmen Square	39° 54' 10" N 116° 23' 32" E
London – Trafalgar Square	51° 30' 27" N 0° 7' 40" W
Milton Abbas	50° 49' 0" N 0° 16' 50" W
Paris – Basilica de Sacre-Coeur	48° 53' 13" N 2° 20' 35" E
Paris – Arc de Triomphe	48° 52' 20" N 2° 18' 4" E
Saltaire	53° 50' 13" N 1° 47' 25" W
Washington DC – The Mall	52° 30' 58" N 13° 22' 39" E

Google Earth can also be used to explore evidence of gentrification and sites of resistance in the Lower East Side of New York. Tompkins Square Park (40° 43' 35" N 73° 58' 53" W) is at the heart of the East Village and on its south-east corner is St Brigid's Church, rescued from demolition by protests in 2006. To the south, 6th Street and Avenue B Community Garden (40° 43' 27" N 73° 58' 55" W) is one of the larger gardens saved from sale and development. The commodified historical district is centred on the Tenement Museum on Orchard Street (40° 43' 5" N 73° 59' 24" W).

Geographies of representation and mobilisation

Introduction

This chapter will focus on the way in which political formations shape and are energised by processes and possibilities of democracy. We will argue that democracy in all its forms is thoroughly geographical. This includes not only the outcomes of electoral democracy – where voting patterns are geographically differentiated, and where the territorial structure of electoral systems can change results – as discussed in the first half of the chapter, but also more participatory forms of democratic mobilisation – for example, through social movements – which the second half of the chapter will discuss.

'Democracy' (the word was coined in Ancient Greece to denote the rule or power [*kratia*] of the people [*demos*]) refers in the broadest sense to any process in which many, most or all of the people affected by decisions are able to participate in making or influencing those decisions. Democracy can thus play a role in all spheres of social life, from the family to the workplace or economy to international relations. This chapter will focus primarily on political democracy, the sphere of public participation in or influence over the policies and decision-making processes of states. However, it is important to bear in mind that no strict boundary can be drawn between political democracy and cultural or economic democracy.

Political democracy is now recognised by many people around the world as the chief source of the legitimacy of states, though the expansion of democratic institutions and practices among states has not been constant or geographically uniform

(see Box 7.1). Because modern societies are so large and complex, the most typical form taken by democracy at the national level is representative, rather than direct. In a representative democracy, eligible voters periodically elect a small number of representatives, for example MPs in the UK or Members of Congress in the US, who are engaged full-time in political decision-making on behalf of the constituents who elected them. If these elected representatives don't govern the way most of the voters in their districts want them to, they can be replaced at the next election by someone else. This system allows for fairly efficient political deliberation and decision-making, but it distances the average voter from direct involvement in day-to-day government. This can dilute the perceived legitimacy of government decisions. Thus many modern democracies also have plebiscitary procedures (referenda, initiatives or petitions whereby individual citizens can directly express their preferences on particular issues) in place for special occasions.

This basic sketch of modern electoral democracy does not, however, give a complete picture. For one thing, and quite obviously, not every nation-state or political entity has democratic institutions in place. There are many parts of the world ruled by monarchs, despots or military *juntas* who explicitly reject and try actively to suppress any moves toward democracy. Less obvious, though, but equally important, is the fact that even long-established, iconic democracies such as the United Kingdom or the United States only 'work' as democracies to the extent that their official electoral systems are charged, animated and sometimes challenged by the energies of *civil society*

(Cohen and Arato 1992). 'Civil society' is a somewhat elusive term usually used to refer to the range of activities undertaken by people in the sphere 'between' the private, domestic realm of the household, the closed, privately regulated sphere of the workplace, and the official, institutional realm of the state. Core features of civil society include participation in market exchanges and unofficial forms of political association and expression. What these two kinds of activity have in common is that they bring together strangers, people otherwise unknown to each other, in order to negotiate and pursue mutually compatible goals.

In political terms, civil society can be seen as the realm of 'the public', in which matters of public relevance are debated, public opinion is formed and public pressure put on (especially state) institutions to address specific issues. Richard Sennett and Jürgen Habermas have famously argued that 'the public' in this sense, or the bourgeois public sphere, emerged in major European cities in the later eighteenth century (Sennett 1977; Habermas 1991). In the coffee houses, pubs and theatres of Paris and London, mutual strangers began, in unprecedented numbers, to hold informal discussions and debates on matters of the day, share information and interpretations of the latest news, and formulate demands addressed to their governments. This kind of activity seems fairly unremarkable 250 years later. Yet at the time it was a new experience. The modern idea of 'citizenship' born at that time involves having an equal right to participate in decision-making *regardless of exactly who one is*, from what profession, religious faith, gender or ethnic group (see Chapter 4, and Box 4.1). The public debates in coffee houses and parlours led to revolutionary

BOX 7.1   GEOGRAPHIES OF DEMOCRATISATION

The discussion in this chapter relates primarily to advanced liberal democracies in which citizens enjoy wide-ranging social and political rights, including the ability to choose (and remove) governments through fair and free elections. However, much of the world's population does not have these freedoms. In over sixty states, either power is exercised by unelected totalitarian regimes, or a superficially 'democratic' system is restricted by the suppression of opposition parties, vote rigging, voter intimidation and controls on the freedom of speech. Since the 1980s there have been a number of high-profile instances of 'democratisation', notably in central and Eastern Europe, South Africa, and parts of Asia and the Middle East (the so-called 'Arab Spring'). These events have been positioned by some commentators as forming part of a 'third wave of democratisation' (Huntington 1991). According to Huntington's model, the 'first wave of democratisation' began in the United States in the early nineteenth century and continued to 1922, embracing the establishment of parliamentary democracies and the universal franchise in Europe, North America, Australia, New Zealand and parts of Latin America. The 'second wave' followed the end of the Second World War and lasted only until around 1962, during which time democracy was reintroduced in parts of Europe and confirmed in many newly independent post-colonial states such as India. The 'third wave of democratisation', it is argued, began with the overthrow of the Salazar dictatorship in Portugal in 1974 and continues to the present day, having reached its crest with the democratisation of central and eastern Europe in the early 1990s.

The democratisation of states is of interest to political geographers because, as Bell and Staeheli (2001) describe, democratisation is often conceived of not just as a historical shift, but also as a geographical process of diffusion. The task of mapping the tide of democratisation has become an industry in its own right, involving government agencies, policy think-tanks and academic researchers. Bell and Staeheli (2001) argue that in order to measure the diffusion of democracy on a global scale, mechanisms have needed to be constructed to facilitate cross-national comparisons. In particular this has been done through the use of a 'democratic audit', surveying states against a checklist of political rights and civil liberties, such as fair elections and a free press. However, Bell and Staeheli contend that this approach reduces democracy to a set of procedures and institutions. For example, they note in the case of the audit used by the Freedom House think-tank that 'in attempting to evaluate the openness of a society to dissenting opinion, the survey team examines the kind of laws and institutional protections for speech, but pays limited attention to actual speech within the country' (p. 186).

Thus, the evaluation of democratisation is a subjective process that is inevitably informed by strategic, geopolitical, considerations. With this in mind, four further observations can be made. First, the promotion of Western-style parliamentary democracy can involve the imposition of inappropriate institutions and procedures to replace traditional forms of political organisation with strong democratic elements, such as tribal councils. Second, strategic considerations can mean that states with poor human rights records are defended as 'democracies', whilst states in which free elections lead to the rise of parties that are Islamist (e.g. Turkey and Algeria in the 1990s), xenophobic right-wing (e.g. Austria in the 1990s) or leftist anti-American (e.g. Nicaragua in the 1980s, or Venezuela in the 2000s), are subject to international condemnation. Third, the transition to democracy is frequently contested, with struggles over the nature of emergent democracy, and in some instances popular protests and even military intervention to remove elected governments in the name of 'protecting democracy', as in Egypt in 2013. Fourth, the 'democracies' of Western states such as the USA and Britain contain flaws that can produce outcomes that might be condemned as 'undemocratic' if they were to occur elsewhere – as the discussion of the 2000 US Presidential election in this chapter illustrates

Key readings: Bell and Staeheli (2001); O'Loughlin (2004).

claims for political rights that replaced absolute monarchies with liberal democracies. However, political rights such as the right to vote, freedom of speech, freedom of assembly and freedom of association, were initially limited to elite groups, and such rights have only slowly become available to most adults in practice over the past 200 years. The expansion of citizenship rights over this period effectively transformed the definition of 'the public'. It did not occur 'naturally' but was driven by struggles taking place in and over civil society, most notably in the form of the multinational women's suffrage and women's rights movements of the nineteenth and twentieth centuries, the global anti-colonial movements over roughly the same time period, and the Civil Rights movement of the 1950s–1970s.

Such movements illustrate very well why official democratic institutions are not the whole story of democracy, and why this chapter examines both electoral geographies and the geographies of social movements.

Electoral geographies

In a democratic society, the right to vote – and consequently the right to select and remove governments – is perhaps the most fundamental right of the citizen. However, it should also be noted that the outcome of elections rarely reflects the pure, rational decision of the electorate. Geography keeps getting in the way. In this part of the chapter we explore the two main ways in which this happens – first, when local factors influence voting decisions, and second, when the geographical structure of the voting system distorts the result. The significance of these mediations is then illustrated with a detailed discussion of American elections between 2000 and 2012.

Geographical influences on voting behaviour

The mapping of voting behaviour is one of the oldest elements of political geography, dating back to 1913 when French geographer André Siegfried produced maps comparing party support in the *département* of Ardèche to the region's physical, social and economic geography. Siegfried's work was simplistic, descriptive and tended towards environmental determinism, but revealed an essential truism – that voting patterns vary spatially and that there is a relationship between these and the spatial distribution of other social and economic entities. This is not entirely surprising. In most advanced democracies the party system is based on historical social, cultural or economic 'cleavages' – for example, between classes, or between religious or ethnic groups (Rokkan 1970) (see Box 7.2 on electoral geographies of emerging democracies).

As these social groups tend to be geographically concentrated, the parties associated with them will also find their support varying spatially. So, social democratic parties built on working-class mobilisation have historically secured more support from industrial areas with a higher working-class population, whilst pro-employer conservative parties have attracted support from more middle-class suburban and rural districts. Similarly, the tendency of black Americans to vote Democrat is reflected in the correspondence between voting patterns and the racial composition of neighbourhoods in cities such as New York and Los Angeles.

When the spatial distribution of classes or ethnic groups shifts over time, so the associated geography of voting evolves. For example, middle-class migration from British cities to suburbs and rural areas in the 1960s and 1970s resulted in an increasing polarisation of rural–urban voting patterns – which weakened

---

BOX 7.2 ELECTORAL GEOGRAPHIES IN EMERGING DEMOCRACIES

The study of electoral geography has tended to focus on stable, established democracies with party systems that are embedded in political cleavages with deep historical roots. In emerging democracies, electoral geographies reflect much more fluid and unstable party groupings and voting patterns, and are entwined with a wider array of political struggles, including struggles over the process of democratisation. In Nicaragua, for example, democracy was introduced after the overthrow of the Somoza dictatorship in 1979, with the left-wing Sandinista Front for National Liberation (FSLN) governing until 1990, when it was defeated in elections by a liberal coalition. However, electoral contest between the left- and right-wing blocs in Nicaragua is supplemented by struggles in other political arenas, such that 'after their first electoral defeat in 1990, the FSLN retained significant political control of key institutions including the police, the armed forces and the media as well as the capacity to mobilise large numbers of supporters to take to the streets and protest when necessary, a situation described as the FSLB "governing from below"' (Cupples 2009: 113). Both blocs have tried to influence electoral law and administration to their advantage, including manipulating geography, for example, by revoking the right of citizens with a voter identity document to vote at the corresponding polling station even if their name did not appear on the electoral roll. As Cupples notes, opponents argued that this reform would create 'a form of geographic striation in which people that the FSLN wanted to deter from voting would be conned into going from one polling station to another in the hope of finding their name on the roll and would eventually give up' (ibid.: 115). Even at the micro-scale, Cupples shows that the geography of the polling station became significant, with party observers positioned strategically to watch voters and officials.

In Thailand, the process of democratisation has been similarly framed by a class struggle that has a clear geographical expression. As Glassman (2010) documents, tensions reached an apex in September 2006 when a military coup overthrew the populist elected prime minister, Thaksin Shinawatra, who drew support primarily from rural areas. The coup was implicitly supported by the opposition neoliberal Democrat Party, with its electoral base in middle-class voters in the capital, Bangkok, alarmed by Thaksin's attempts to consolidate his power, including by challenging the influence of Democrat-supporters in the military. As such, claims were made that the coup was 'a necessary – if unfortunate – step in the restoration of non-authoritarian political rule' (Glassman 2010: 1317) (Interestingly, similar rhetoric was used to justify the removal of elected president Mohammed Morsi by the military in Egypt in 2013, with a similar cleavage in support between the capital and provinces). Democracy was restored following a referendum on a new constitution in 2007, the results of which were polarised between Bangkok and the capital, and a similar geographical cleavage in subsequent elections returned pro-Thaksin supporters to government.

As the above cases have demonstrated, the line between electoral politics and civil society in emerging democracies can be vague and regularly transgressed. This is especially the case where parties have evolved from national liberation movements and enjoy near-monopoly support. Bénit-Gbaffou's (2012) analysis of the African National Congress (ANC) in Johannesburg, for example, shows that the local branches of the ANC act as the main platform for political mobilisation, debate and securing resources, more so than elected ward councils or civil society groups. Indeed, the strength of the ANC is such that it has restricted the development of an independent civil society, in part because of a mentality that sees rivals as 'enemies', as Bénit-Gbaffou notes from observation at public meetings:

Not knowing to what extent one is being watched by the others in those public spaces, and if one's discourse will be reported to the ANC, participants in public meetings are generally very careful not to associate themselves, or even be perceived as associating themselves, with the political "enemy" – which varies according to local contexts. In some places, the enemy is COPE [a rival party]; in others, it is radical social movements like the Anti Privatisation Forum. Inviting them to public meetings or to forums, engaging with them publically is seen as a betrayal, and their members are fled and ostracized like contagious disease carrying people, even if other organizations might be sympathetic to their struggle.

(ibid.: 186)

Key readings: Bénit-Gbaffou (2012); Cupples (2009); Glassman (2010).

again when urban working-class solidarity was undermined by economic change in the 1980s (Johnston and Pattie 2006). However, if voting patterns simply reflected the political preferences of socio-economic groups then the geography would be purely coincidental. On the contrary, electoral geographers have argued that geographical factors can amplify social biases in voting.

First, people tend to vote in a similar way to their neighbours, even if their own socio-economic status suggests that their loyalties should lie elsewhere. This

'neighbourhood effect' operates because individuals' interpretation of political news and issues is mediated through local discussion creating a predisposition for people of all backgrounds to adhere to the dominant political narratives of their locality. Early studies focused on the importance of the class composition of a locality over-riding individual class identity, and whilst these arguments have been critiqued (see Johnston and Pattie 2006), more recent work has assembled evidence of more complex dynamics at a variety of scales. For example, Johnston et al. (2005a) show that non-Conservative voters in rural areas of Britain were more likely to vote Liberal Democrat rather than Labour than similar voters in urban areas, whilst Walks (2010) found that living in a gated community amplified the Conservative voting behaviour of residents in Canada. Moreover, analysis has confirmed that the neighbourhood effect can be attributed to residents talking to each other, and thus is stronger in tightly knit communities: 'voters who are deeply embedded in their local milieux . . . are much more likely to vote for the most-preferred party locally than those who are much more isolated and lack such well-developed local social networks' (Johnston *et al.* 2005b: 1457; though see Walks (2006) for a contrasting conclusion from Canada).

Second, party loyalties can be disrupted by personality politics and issue voting. A 'friends and neighbours' effect means that candidates can generally expect to poll more strongly in their home areas (Arzheimer and Evans 2012); whilst anomalous results can be produced when specific local issues overshadow issues in the national campaign. For example, in the 2001 British General Election the victorious Labour party lost a seat, Wyre Forest, to an independent candidate campaigning on the single issue of saving a local hospital from closure. Third, variations in the level of campaigning by candidates and parties between different electoral districts or constituencies can influence both the voter turnout and party support. This is becoming increasingly significant as voters behave more like consumers, selecting between different competing party 'brands' rather than following traditional class or ethnic loyalties (Johnston and Pattie 2006).

Gerrymandering and malapportionment

Few governments or political leaders are elected simply on the basis of the number of votes cast. In most electoral systems, the vote is effectively filtered, either by the election of representatives from geographical constituencies to a legislature, or by the operation of an electoral college. In 'first-past-the-post' electoral systems, such as those used in Britain, Canada and the United States, where the winning candidate 'takes all', this means that votes cast for a losing candidate in any constituency are effectively 'wasted'. A party that loses in every constituency by just one vote will not be represented in the legislature, whilst a party that wins in every constituency by just one vote will hold all the seats. As such, parties are discriminated against if their support is geographically dispersed and advantaged if their vote is concentrated in particular localities. Historically, this bias has meant that small parties with nationwide support but no 'strongholds', such as the British Liberal party, have been under-represented. It can also produce dramatic outcomes such as in the 1993 Canadian election when the governing Progressive Conservative Party collapsed from 169 seats to just two (see also Johnston 2002a).

Which party benefits most from biases in the electoral system will depend in part on how the boundaries of electoral districts or constituencies are drawn. For example, Figure 7.1 depicts four towns with equal populations but different divisions of support between two parties, which must be organised into two constituencies. If town A and town B are paired together in one constituency and towns C and D in the other then the Red party and the Blue party would each win one constituency. However, if town A was paired with town C, and town B with town D, both constituencies would be won by the Red party.

The deliberate manipulation of electoral district boundaries for political gain is known as a 'gerrymander' after a nineteenth-century Governor of Massachusetts, Eldridge Gerry, who authorised a

Town A Red party: 700 votes Blue party: 300 votes	Town B Red party: 500 votes Blue party: 500 votes
Town C Red party: 400 votes Blue party: 600 votes	Town D Red party: 550 votes Blue party: 450 votes

Figure 7.1 Different ways of drawing constituency boundaries can produce different election outcomes

peculiarly shaped electoral district in Essex County. When a local newspaper's cartoonist saw a map of the new district he accentuated its resemblance to a salamander, or as his editor named it, a gerrymander. There is some debate about the correct application of the term today (Johnston 2002a; Johnston 2002b; Moore 2002), but no shortage of examples of fragmented, sinuous or otherwise misshapen electoral districts designed, for instance, with the distribution of racial groups in mind (Figure 7.2), especially in the United States, where districting is controlled by partisan legislatures (Yoshinaka and Murphy 2009). In Britain, responsibility for delineating parliamentary constituencies lies with an independent Boundary Commission. However, party biases can still arise. *Malapportionment* can occur between reviews as urban-to-rural migration steadily reduces the electorate of urban constituencies whilst increasing that of rural constituencies. This benefits the Labour party over time, but boosts the Conservatives immediately after a review (Johnston 2002a).

Constituency geography can also have other, unintended, effects on election result. For example, voter turnout decreases with distance to the polling station, which means that as parties will have differing strength in different parts of a constituency the location of polling stations can influence election results (Orford *et al.* 2009).

US electoral geographies, 2000–2012

An example of how geographical influences on voting behaviour and the way in which votes are translated into a national result by the geographical architecture of the electoral system can affect the outcomes of elections can be seen in the United States at the start of the twenty-first century. The presidential elections of 2000, 2004, 2008 and 2012 were all close, with the margin of difference in the popular vote between the Democrat and Republican parties ranging between 0.5 per cent (in 2000) and 7.2 per cent (in 2008) (see Table 7.1). The closeness of the vote also indicated a polarised electoral geography, with only nine states switching between the parties during these four elections. This geographical cleavage – popularly represented as 'red states' (Republican) versus 'blue states' (Democrat) – reflects a complex cultural, ethnic and religious cleavage. Older, white, evangelical

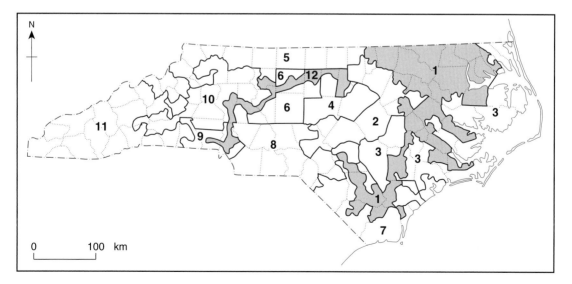

Figure 7.2 US House of Representatives electoral districts in North Carolina. The shading illustrates how misshapen districts 1 and 12 have become for partisan purposes

Table 7.1 United States presidential elections, 2000–2012

	2000		2004		2008		2012	
	Democrat (Gore)	Republican (Bush)	Democrat (Kerry)	Republican (Bush)	Democrat (Obama)	Republican (McCain)	Democrat (Obama)	Republican (Romney)
Popular vote (%)	48.4	47.9	48.3	50.7	52.9	45.7	51.1	47.2
Popular vote (votes)	50,999,897	50,546,002	59,028,494	62,040,610	69,498,516	59,948,313	65,899,660	60,932,152
Electoral college votes	266	271	251	286	365	173	332	206
States won	21	30	20	31	29	22	27	24
Average electoral college votes per state won	12.7	9.0	12.6	9.2	12.6	7.9	12.3	8.6
Popular votes per electoral college vote won	191,729	186,517	235,173	216,925	190,407	346,522	198,492	295,787
States gained from other party from previous election		Arizona Arkansas Florida Kentucky Louisiana Maine Missouri Ohio Nevada Tennessee West Virginia	Maine	Iowa New Mexico	Colorado Florida Indiana Iowa Nevada New Mexico N Carolina Ohio			N Carolina Indiana

Christian and culturally conservative people are more likely to vote Republican, and younger, black and culturally liberal people are more likely to vote Democrat. As the geographical distribution of these groups is uneven, so the geography of party support varies, with the Democrats dominant in urban areas and in north-east and west coast states, and the Republicans dominant in small towns and rural areas, and states in the Mid West and South. However, there is also a neighbourhood effect, which exaggerates this polarised electoral geography (McKee and Teigen 2009; Morrill *et al.* 2011).

In close elections, where votes are cast can make a big difference. This was most clearly demonstrated in the 2000 presidential election, in which Republican George W. Bush was elected with a majority of electoral college votes, despite polling nearly 540,000 *fewer* actual votes than Democrat Al Gore. Geography helped to produce this outcome in three critical ways.

First, Bush polled popular votes where they counted for more in the electoral college. The American President is elected not by a popular vote but via an electoral college in which each state has a designated number of electors roughly proportionate to its population. The candidate who polls most votes in a state gets all of that state's electoral college votes (except in Maine and Nebraska, which have slightly different systems). Although the allocation of electoral college votes is roughly proportional, there are discrepancies. California, as the largest state, has fifty-four votes, or the equivalent of one electoral college vote for every 550,000 residents. Wyoming, in contrast, has three votes, or the equivalent of one

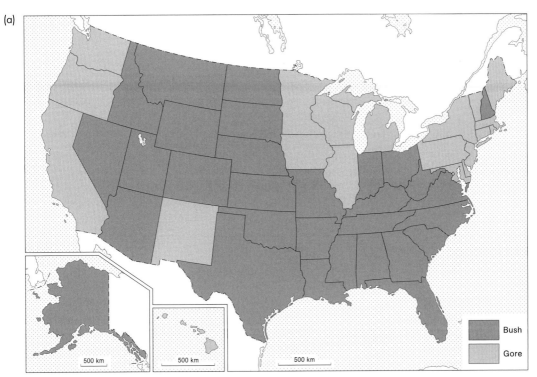

Figure 7.3 States won by George W. Bush and Al Gore in the 2000 US presidential election. (*a*) A conventional projection of the United States. (*b*) States represented proportionally to the vote in their electoral college. (*c*) States represented proportionally to the relative weight of each resident's vote

(b)

Bush

Gore

(c)

Bush

Gore

electoral college vote for every 150,000 residents. In other words, a vote cast in Wyoming has three times the value of one cast in California (see also Warf 2009). In 2000, Gore won large states including California, New York, Pennsylvania and Illinois, but Bush led in more small states where each individual vote counted for more (Figure 7.3). Thus, Bush's 4.5 million votes in the thirteen small states of Alaska, Arkansas, Idaho, Kansas, Mississippi, Montana, Nebraska, Nevada, New Hampshire, North and South Dakota, West Virginia and Wyoming, netted him fifty-six electoral

college votes, whereas Gore's 5.7 million votes in California gave him only fifty-four electoral college votes (Archer 2002; Johnston *et al*. 2001).

Second, Bush also gained more from the efficiency of his vote spread. Both Bush and Gore won a number of states by margins of less than 2 per cent, but whereas Gore piled up huge majorities of more than a million votes in California and New York, Bush was crucially declared the winner in Florida by 537 votes. This narrow margin brought Bush twenty-five electoral college votes – equivalent to the combined college votes of the eight smallest states – making the vote of each one of the 537 electors 1,584 times more influential that of the average voter (Warf and Waddell 2002).

Third, Bush benefited more than Gore from geographical variations in the administration of the election. His narrow lead in Florida was assisted by a controversial state law restricting the ability of convicted felons to vote, which disproportionately discriminated against Democrat-leaning black voters, and which did not apply in other states. At a more local level, the ability of wealthier (and Republican-leaning) counties to purchase modern technology for casting and counting votes, whilst poorer (Democrat) counties made do with antiquated equipment, made a small yet crucial difference in reducing errors and completing recounts in Florida before the deadline imposed by the Supreme Court.

Although the electoral geography of the United States appears to be entrenched, it is in fact susceptible to demographic trends, especially the prevailing flow of migration from the north-east and mid-west to states in the south and west. This has had two effects. First, growth states in the south and west have gained additional congressional seats, and this has arguably benefited the Republicans in elections to the House of Representatives. Second, however, analysis has shown that migrants to states such as Colorado, Florida and Nevada are more likely to be Democrat-leaning than existing residents. As these states have been finely balanced politically, migration has had a significant impact in making these states more likely to vote Democrat, with a consequence for national results, especially as the source states for migrants tend to be more firmly dominated by one party (Jurjevich and Plane 2012; Robinson and Noriega 2010). Accordingly, migration has helped to tip the balance of advantage in the electoral college from the Republicans in 2000 and 2004 to the Democrats in 2008 and 2012.

Geographies of social movements

Elections are only one form of democratic practice, and whilst the spread of electoral democracy across the world still continues, in many established democracies there is evidence of some degree of public disengagement from the electoral process: in declining voter turnout and disillusionment about the ability of parties and elected representatives to reflect public interests. Popular demonstrations against austerity policies adopted by governments in Europe, commonly with broad support from major political parties, have been one expression of this, and can be considered to be part of a broader mobilisation of social resistance, labelled by some as 'contentious politics' (Leitner et al. 2008), in which citizens have sought a more direct engagement in the political process through protests, direct actions and the construction of alternatives to hegemonic state policies. These public mobilisations are also referred to as 'social movements' (see Box 7.3), and have particularly gained prominence since the 1960s, although they have much older roots. The labour movement, for example, is a classic social movement that has organised workers into trades unions and employed a range of tactics to further its cause, including involvement in electoral politics. The late twentieth century, however, saw the emergence of 'new social movements', focused not on the class struggle but on a plethora of non-material issues (Fainstein and Hirst 1995; della Porta and Diani 1999). Examples include the movement for gay rights, opposition to the Vietnam War, the environmental movement, as well as mobilisations against the South African *Apartheid* regime, against Soviet-controlled authoritarian communist rule in the former Soviet Bloc and, currently, against the injustices of globalised neoliberal capitalism.

In addition to changing opinion and policy, successful social movements often also result in the formation of more permanent organisations (nonprofits and NGOs) that formalise advocacy on particular issues, and may grow to be global in their field of operations (as in the case of Amnesty International). Social movements can likewise lead to the establishment of new state agencies, such as the US Environmental Protection Agency. These highly visible movements, though, are in one sense only the tip of the iceberg. Below the radar of major media reporting on issues of national or international

BOX 7.3   SOCIAL MOVEMENTS

Diani (1992) defines a social movement as 'a network of informal interactions between a plurality of individuals, groups and/or organisations, engaged in political or cultural conflict on the basis of a shared collective identity' (p. 13). There are four key components to this definition. First, a social movement comprises a number of different, independent, groups who share a common purpose, but may also on occasion adopt contradictory policies or strategies and be drawn into conflict with each other. Second, the links between the component groups are informal – there is no centralised leadership or command. Third, social movements are engaged in political activity. This distinguishes them from social clubs, voluntary groups and religious organisations. Fourth, the uniting force for social movements is a shared identity, not just shared interests. Put together, these components allow social movements to be distinguished 'from various forms of collective action which are more structured and which take on the form of parties, interest groups or religious sects, as well as single protest events or ad hoc political coalitions' (della Porta and Diani 1999: 16).

Key readings: della Porta and Diani (1999); Diani (1992).

significance, millions of more local and community-based political initiatives and mobilisations do their work all over the world, bringing concrete improvements to the lives of countless people. A complete view of political systems, then, requires that we pay attention not only to institutional frameworks (whether 'democratic' or not) but also to the underlying pressures for democratisation at work within the society, and to the ways in which relatively static institutional frameworks interact with these more fluid and unpredictable movements.

Once again, geography is critical to understanding how social movements are formed, organised and achieve their aims. Leitner *et al.* (2008) identify a number of ways in which social movements can be analysed geographically, but we will focus on three aspects here.

First, place is frequently important in the constitution and mobilisation of social movements. Social movements can coalesce around many different shared concerns and identities, and some of these have a geographical dimension. At a local scale, social movements can emerge as community mobilisations to represent the interests of a neighbourhood, to fight closure plans for schools or hospitals, to campaign against new developments, or for environmental improvements. At a higher scale, there are regional

and nationalist movements (see also Chapter 5), as well as movements tied to more amorphous places, such as 'the countryside'. Around the turn of the century, a rural movement briefly gained prominence in Britain with several large rallies and marches and a plethora of smaller direct actions. The movement brought together individuals with specific concerns such as the prohibition of hunting or agricultural incomes under the discursive frame of standing up for rural interests against urban power (Woods 2003, 2005). Although participants represented only a small proportion of the rural population, they were galvanised by a strong sense of rural identity and an emotional attachment to the British countryside that they perceived to be under threat (Woods *et al.* 2012).

Places are also important to social movements as sites of action. Protest, demonstrations and occupations commonly take place in plazas, squares and other public spaces selected not only for their logistical characteristics but also for their political symbolism – places such as Trafalgar Square in London, or the Mall in Washington DC (see also Chapter 6). As particular places of protest become identified with particular social movements, they also gain symbolism for the movement and contribute to its identity. Bosco (2006), for example, describes the case of the Madres de Plaza de Mayo in Argentina, who originated as an informal group of mothers who met weekly in the Plaza de Mayo in Buenos Aries for solidarity in their struggle for information on their children who had 'disappeared' during the military dictatorship. The movement grew into a human rights network with groups across Argentina, but retained the symbolic name of 'Madres de Plaza de Mayo'. Similarly when revolutionary protesters occupied an empty space in front of the Tianan Gate in Beijing on 4 May 1919, they gave definition to the emergent communist movement in China (Lee 2009). The site of the protest was later reconstructed as Tiananmen Square to form the most prominent public space in Communist China, and the revolutionary symbolism added potency to the occupation of the square by anti-Communist demonstrators in 1989.

Second, as the above examples have hinted, social movements do not only occupy public space, they also subvert and transform it. Public space is critical for social movements as it allows individual grievances to be converted into collective action. The ability of people to come together, organise and protest in parks, squares and streets, is what positions cities as drivers of social movements, as Low (2004) argues:

> The privilege accorded to cities as sites for democracy often rests on a set of micro-sociological foundations. This involves the simplification of the city into a political space whose occupants can face each other, encounter each other on the street, gather together, use their senses to negotiate their relationships with one another, struggle with one another, and so on.
>
> (Low 2004: 131)

Yet, public spaces are always contested spaces. State authorities may seek to control their use for political purposes through laws, surveillance, policing and physical barriers (see Chapter 6). Social movements may in turn attempt to transform public spaces into sites for radical political experiments (as discussed further in the next section). As Salmenkari (2009) argues in a comparison of protests in Seoul and Buenos Aries, such struggles over public space are central to the democratic process, and the performance of protest in public spaces is an important element in maintaining democracy:

> The manner in which demonstrations construct a public sphere must be viewed not only at the demonstration site but also in terms of the demonstration's contribution to the public sphere as a whole. In this sense demonstrations have an important equalizing effect. Although public space ideally is accessible to all, access for some social groups in fact is limited. Demonstrations provide a chance to address the public and the elites without having to pay for a forum and in ways that make short-term engagement possible. This opens the way not only for disadvantaged groups like workers or opponents of the free trade agreement when the Korean government professes neoliberal labor-hostile and market-friendly principles, but also for socially marginalized groups. Movements of the unemployed in Argentina and the disabled in Korea not only articulate group demands but also literally make these groups socially and politically visible. Therefore, demonstrations are an important means for maintaining a more inclusive public sphere.
>
> (Salmenkari 2009: 256)

Furthermore, social movements by their nature have a propensity to blur the boundaries between public and private space. Feminist geographers have argued that the traditional association of politics with public space and the public sphere has contributed to the political marginalisation of women. Bondi and Domosh (1998), for instance, note that even today the freedom of women to use public space remains qualified by class-specific, gendered notions of production and consumption. Women's presence in *unregulated* public spaces *unrelated to consumption*, they suggest, is still often not seen to be as 'natural' or unproblematic as is men's. Marston and Mitchell (2004) have accordingly challenged this distinction between the purportedly 'political' public spaces of citizenship and the private realm, arguing that a range of other locations not traditionally considered public spaces serve as settings for much of the important work of citizenship.

They highlight as an example the role of the late nineteenth-century 'domestic feminism' movement in the US, which was about 'extending the rules that "governed" middle-class households – rules about personal hygiene, food preparation and eating, marriage, childrearing and health care – beyond the bounds of the private domicile into the city', in order to 'rationalize the city' (p. 103). They show that this movement was to become a key source of basic policies of the twentieth-century welfare state.

Many contemporary social movements operate across both public and private space. Social movements concerned with gender equality, gay rights, combating domestic violence, child poverty and so on, use interventions in the public sphere to affect changes in private spaces. The squatter movement in Berlin from the 1960s onwards, has used the occupation of (empty) private spaces to make public political statements (Vasudevan 2011), whilst the occupation of public spaces by social movements can involve the creation of temporary private spaces. The BBC's interactive photo-map of the occupation of Cairo's Tahrir Square during the revolution of 2011, for example, shows how pointless it is to try to separate the public and the private (the image can be accessed at http://www.bbc.co.uk/news/world-12434787). The public politics represented by the protests were so effective because the occupation of the square continued for weeks on end. Yet to last long enough to give its 'public' message any staying power at all, the occupation needed to incorporate core elements of 'private' life. Thus, while the BBC map shows activities that would fall under a more traditional understanding of 'publicness', for example, the main stage and of course a wired area for bloggers near the centre, a newspaper wall near the upper right, an area for public artwork off the map to the lower right, these stations share space with a camping area on the left centre, and toward the top, a kindergarten, a water point, food stalls and toilets. At the upper right corner is a medical clinic. 'Occupation' (and this goes as well for the 'Occupy' movement) cannot be simply 'public'.

Third, in order to achieve their aims, social movements need to overcome the problems of space and scale (Nicholls 2009). Social movements often start with specific local concerns, but deal with issues that cannot be resolved locally and therefore need to extend their activities to other places and engage with higher-scale actors. Moreover, the potency of social movements as a form of *democratic politics* rests on their capacity to mobilise large numbers of people, which also requires building coalitions or networks of like-minded groups across multiple places. The immigration rights movement in France, for example, emerged from specific urban struggles in Paris, but developed as a national movement, engaging with national government policies, by reaching out to connect the Paris hub with localised campaigns across the country (Nicholls 2011).

The networks developed by social movements have been described by some commentators as 'rhizomic', employing a metaphor proposed by post-structuralist writers Gilles Deleuze and Felix Guattari to describe systems that are diffuse, non-hierarchical and heterogeneous without an organising centre, recalling the characteristics of biological rhizomes such as tubers or bracken. As Woods *et al.* (2013) observe, the metaphor of the rhizome is useful for capturing the agility of social movements, their linking together of sometimes diverse groups and campaigns, and the potential for protests to spread to different localities without any overarching coordination. However, Woods *et al.* also warn against over-stating the rhizomic nature of social movements, arguing that there is a tendency over time for social movements to move toward more rigid, hierarchical and formalised modes of organisation: social movements that begin as loose collections of individuals become formal NGOs with elected leaders, paid-up members and local branches. This transition can occur because a hierarchical structure of local branches linked to a coordinating centre is the most effective way of organising across space in a disciplined and stable way, and because it mirrors the hierarchical political geography of the state, permitting interaction with the state at multiple scales.

The networks of social networks are not just about linking up activists, but also about connecting with media outlets that can help to spread their message. The success or failure of a protest event, and of a social movement more generally, depends crucially on whether they garner enough media attention to be brought before a national audience capable of putting pressure on policy makers or other actors. This 'scale jumping' (Herod 2010), making the leap from a small, usually localised audience to a larger, perhaps regional or national public, has become absolutely central to engagement in the public sphere in the age of electronic broadcast media. Bringing regionally and place-specific racial discrimination to the attention of a national public who found it morally wrong was key to the success of the US Civil Rights movement in the 1960s, and similar kinds of scale-jumping remain crucial for the Arab Spring activists or the anti-corporate globalisation networks of the twenty-first century.

The difficulty is that scale-jumping puts the message intended by a political group at the mercy of the media who send it out to a larger public: the media often select aspects of public events that have little to do with the message. The one-way nature of television broadcasting also means that answering back directly is impossible. In recent years, the development of 'IndyMedia', a network of alternative broadcasters able to accompany and document major demonstrations, and the growth of independent blogging, has sought to counteract the selectivity of mainstream media (Castells 2012). These practices are aimed at ensuring that the process of jumping scales to reach larger publics does not completely sacrifice control over the message.

IndyMedia and blogging can also be positioned as part of a further form of networking through the internet as a kind of 'virtual public space' for democratic political discussion and action, in which social movements are at the fore. In this way, electronic technology has arguably overcome limitations of face-to-face interactions and has brought the world closer to an ideal of universal access to discussions. However, as Barnett points out,

> [t]here is nothing 'virtual' about the publics brought into existence through these networks . . . if 'virtual' is meant to imply that they are somehow immaterial. These networks are material in a double sense: they are embedded in a tangible infrastructure of institutions, organizations, technologies, and social configurations that are every bit as produced as roads, railways, and buildings . . . and they are material in the sense of being effective in shaping opinions, decisions and outcomes.
>
> (Barnett 2004: 195)

Many scholars and activists have pointed out that billions of people worldwide still lack reliable (or indeed any) connection to the internet (Crampton 2004), in some cases because of government controls on internet access (as in China and Iran). Neither is the internet an entirely free and egalitarian democratic space: the ability to post opinions does not mean that they will be taken seriously or evaluated on their merits; the proliferation of contributions can debilitate as well as facilitate effective democratic discussion; and governments have developed sophisticated means of monitoring internet communications. Nevertheless, as Crampton (2004) makes plain, the techno-material configurations of the internet, and of electronic communication more generally, have made some new forms of political activity possible. The more fluid politics of identity enabled by social networking sites or shared virtual worlds, for example, has opened up new possibilities for separating (or altering) links between 'who one is' and the positions one takes on public issues. New technologies have likewise changed the ways in which demonstrations are planned and carried out, improving the ability of protest organisers to adapt the course of demonstrations in real time to changing conditions (e.g. police tactics) (Pickerell 2006).

The rhizomic networks of social movements, the rise of independent media and the virtual public space of the internet and social media are all important in facilitating the mobility of ideas, images and messages. In this they have contributed to the phenomenon of nomadic protests, moving around between different sites of struggle, such as transnational political summits (see Chapter 9). However, social movements are also associated with a contrasting strategy that involves claiming space, 'staying put' and constructing territories of alternative political expression, as the next section explores.

Spaces and territories of alternative politics

In the years that followed the global financial crisis in 2008, a new wave of political movements mobilised around the world, including anti-austerity protests in Europe, the anti-capitalist 'Occupy' movement, and the revolutionary movements of the 'Arab Spring'. Whilst these movements emerged from different contexts and varied in their aims and targets, they shared a common tactic: to occupy public space and refuse to move. From the *Indignados* in Madrid's Puerta del Sol to anti-government protesters in Istanbul's Taksim Square, these actions demonstrated a fixidity that contrasted with the mobility of counter-globalisation protests in the previous decade (see Chapter 9 for more on these), as radical writers Hardt and Negri observed:

> [a] decade ago the alterglobalization movements were nomadic. They migrated from one summit meeting to the next, illuminating the injustices and antidemocratic nature of a series of key institutions of the global power system: the World Trade Organization, the International Monetary Fund, the World Bank, and the G8 national leaders, among others. The cycle of struggles that began in 2011, by contrast, is sedentary. Instead of roaming according to the calendar of summit meetings, these movements stay put and, in fact, refuse to move.
>
> (Hardt and Negri 2012: 7)

The claiming and occupation of space as protest has been exemplified in recent years by the Occupy movement (see Box 7.4), but has a long heritage – from revolutionary occupations of public squares, through factory sit-ins during industrial disputes, squatting of empty buildings, the transgression of 'whites-only' spaces in struggles against racial segregation, protest camps outside nuclear bases during the Cold War, to environmental encampments against the construction of new roads or airports. In some cases, occupations are intended as physical obstructions – preventing the building of a new road, for example – but they are also symbolic.

Physical places are still important to the public sphere, and so major city squares still in some sense serve as iconic public spaces. But, first of all, they have to made and re-made, claimed and re-claimed, often against forces of order seeking to shut down debate and dissent. This process requires the contestation of the rules of territory and the inherited cultural constructions of public space as a place where only some people belong. Spaces need to be claimed in particular as 'spaces of appearance' in which groups reject their exclusion or anonymity (Springer 2011: 538). Paradoxically, this often means overcoming the public-vs.-private distinction, so to speak, *in public*. Furthermore, the hybrid political public spaces created in this way can only be sustained and animated through establishing links between different places and by constructing geographical scales on which to address publics, that is, through the building and re-building of networks. They are rooted, geographically circumscribed, but they depend on the diffusion of support and the mobility of their message.

To achieve these aims, protest movements can adopt a number of tactics, including the discursive and material demarcation of the occupied space as a territory separate to the surrounding polity; the staging of symbolic challenges to established political and economic orders; and the articulation and practice within the occupied territory of alternative forms of social, political and economic organisation. Collectively, the tactics can be seen as part of what Routledge (1997) calls a postmodern politics of resistance, 'characterized by heterogeneous affinities that coalesce in particular times and places as activist assemblages. Eschewing the capture of state power, they nevertheless pose challenges to the state' (ibid.: 372).

Routledge illustrates this model through the example of a protest against the construction of a new motorway, the M77, through Pollock Park on the edge of Glasgow, Scotland. The protest campaign was organised by Glasgow Earth First! and involved both local people and participants who came specifically to join the protesters from elsewhere. It involved the occupation of the construction site with a semi-permanent protest camp as well as other tactics such as protest marches and sabotaging machinery. However, as well as forming a physical obstacle to the road building, the protest also mounted a series of symbolic challenges to the planning of space that had led to road proposals, to the cultural norms of a car-dependent society and to the power of the state. These symbolic challenges involved both the subversion of spatial order and the subversion of citizenship – the latter being articulated by the declaration of the protest camp to be a 'Free State' and the issuing of its own 'passports':

> The Free State represented the 'homeplace' and the focus of the resistance against the M77, articulating an alternative space that occupied symbolic and literal locations. It acted as a place where people who were interested in the M77 campaign could learn more and get involved. . . . The Free State stood as a critique of the environmental damage caused by road building and an example of how people might live their lives differently. Its politics of articulation interwove ecological, cultural and political dimensions.
>
> (Routledge 1997: 366)

These messages were reinforced by the mixture of symbols, icons and images created and employed within the camp:

> In addition to the totems and tree houses – themselves hybrid sites of habitation and tactical forms of protection for trees – the Free State comprised a mixture of symbols. Abandoned cars were used to create dramatic sculptures such as 'Carhenge'. A flag of the Lion Rampant girded the trunk of a tree near to the entrance of the Free State, next to which was an Australian aboriginal land rights flag. A wind-powered generator supplied power to a portable television and stood above a mobile phone. Next to images of celtic knots flew Buddhist-style prayer flags strung from the trees, on which the phrase 'Save our dear green place' was block-printed.
>
> (Routledge 1997: 367)

In these ways the Pollock protest camp communicated a message about a global issue – the environment – through the manipulation of a specific site. To do so it drew together in a unique, place-specific,

BOX 7.4 THE 'OCCUPY' MOVEMENT

The 'Occupy' movement refers to a wave of protests held in numerous cities around the world in 2011 that involved the physical occupation of space through encampments in prominent public locations. The movement was inspired by, and followed from, demonstrators known as *'los Indignados'* (the indignant) who set up an encampment in Madrid's Puerta del Sol during the Spanish election campaign in May 2011, and stayed beyond the election as a protest against the new government's austerity measures. The global Occupy movement broadened the focus to campaign for greater social and economic equality, challenging the concentration of wealth within capitalist society. It claimed democratic legitimacy with its slogan, 'We are the 99%', positioning itself against the 1 per cent of society with which wealth is concentrated.

The first camp to use the 'occupy' descriptor was established in Zuccotti Park in New York on 17 September 2011. By the end of October 2011, there were reported to be over 2,300 Occupy camps in over 2,000 cities worldwide. The numbers involved in the protests fluctuated – the Occupy Wall Street demonstration in Zuccotti Park had a core of around 100–200 people living in the camp, but was bolstered in protests and marches by crowds of up to 15,000 people. The camps also varied in their duration, as governments employed a mixture of legal and coercive methods to evict protesters. The Occupy Wall Street camp was forcibly cleared on 14 November 2011, whilst camps in London and Washington DC resisted efforts at eviction until February 2012.

The physical and symbolic occupation of public space was a core part of the movement's strategy. The camp in Zuccotti Park was deliberately set up to claim space in the heart of New York's financial district, responding to a call by the internet hacker group Anonymous to 'flood lower Manhattan, set up tents, kitchens, peaceful barricades and Occupy Wall Street' (the location of the camp in Zuccotti Park can be found on Google Earth at coordinates 40° 42' 33" N 74° 0' 40" W). Encampments in other cities were similarly commonly located in financial districts, or in prominent public plazas, spatially juxtapositioning their anti-capitalist message with the offices of financial institutions. In London, 'Occupy LSX' were prevented by police from setting up a camp outside the Stock Exchange, but instead occupied the precinct of St Paul's Cathedral in the City of London, with around 150 tents. Efforts by church authorities to evict the protesters through legal action helped to spark debates around the morality and ethics of global capitalism.

The Occupy camps not only presented physical challenges to the state and to private landowners, but also created spaces within which alternative forms of radical democracy and social organisation could be practised. Many Occupy camps placed a strong emphasis on education, inviting speakers to give lectures and run workshops on a wide range of topics, and adopted participatory decision-making processes including general assemblies, the use of hand signals to increase participation, and a 'progressive stack' system allowing participants from marginalised groups to speak first. In this way, Hardt and Negri (2012: 55) argued that 'the encampments are a great factory for the production of social and democratic affects'.

As such, the Occupy movement was rhizomic in its character, with no leaders or coordinating centre and spread by imitation rather than organisation. This in itself presented challenges to the police and city authorities, who were frustrated by an absence of accepted leaders with whom they could negotiate. However, Occupy also demonstrated the difficulty of successfully mounting entirely spontaneous and autonomous political mobilisations, with Uitermark and Nicholls (2012) showing that the most successful Occupy encampments were closely linked to existing local activist networks.

Key readings: Pickerell and Krinsky (2012); Uitermark and Nicholls (2012).

combination cultural symbols and signifiers from around the world that associated the site with a plethora of struggles from renewable energy to Aboriginal land rights.

Summary

This chapter has explored two different forms of political mobilisation in a democratic society and shown that geography is intrinsic to both. Participation in democratic politics is most commonly associated with voting. Yet, the conversion of votes into power is mediated by geography. Patterns of support for particular candidates or political parties tends to be

geographically concentrated – reflecting the spatial expression of social, class, ethnic and religious cleavages that frame party programmes – and therefore the geographical structure of the electoral system can make a critical difference in determining the result – as witnessed in the recent electoral history of the United States. Imperfections in the electoral system have contributed to an apparent growing disillusionment with electoral politics, with citizens searching for alternative ways of expressing their views, including through the direct actions of social movements. Yet, again, the mobilisation and operation of social movements is highly geographical. Social movements are frequently tied to place-based identities and associated with particular sites of struggle, but most overcome the constraints of space and scale if they are to be affective in achieving their aims. The claiming and occupation of public space is one key tactic, presenting both a physical and a symbolic challenge to state power and established social, political and economic orders.

Further reading

There are few books or papers that take a broad perspective towards discussing geography and democracy. An exception is a volume edited by Barnett and Low, *Spaces of Democracy* (2004), which includes chapters on democratisation, electoral geography and social movements. Chapters in *Spaces of Democracy* by Ron Johnston and Charles Pattie and by Richard L. Morrill outline some of the principles of electoral geography. For more specific examples, the spatial biases in the US electoral college are examined by Warf (2009), 'The U.S. electoral college and spatial biases in voter power' in *Annals of the Association of American Geographers*, volume 99, pages 184–204, whilst the impact of migration on the electoral geography of the western United States is investigated by Robinson and Noriega (2010) 'Voter migration as a source of electoral change in the Rocky Mountain West' in *Political Geography*, volume 29, pages 28–39. From a British perspective, the neighbourhood effect in UK general elections is analysed by Johnston et al. (2005b) 'Neighbourhood social capital and neighbourhood effects' in *Environment and Planning A*, volume 37, pages 1443–1459.

Two good overviews of the geographical aspects of social movements are provided by Leitner *et al.* (2008), 'The spatialities of contentious politics' in *Transactions of the Institute of British Geographies*, volume 33, pages 157–172, and by Nicholls (2009) 'Place, networks, space: theorizing the geographies of social movements', also in *Transactions of the Institute of British Geographies*, volume 34, pages 78–93. With a more empirical focus, the significance of places as sites of protest are discussed by Salmenkari

(2009) 'Geography of protest: Places of demonstration in Buenos Aires and Seoul' in *Urban Geography*, volume 30, pages 239–260. The occupation of space as a symbolic challenge to authority is explored further in Routledge (1997) 'The imagineering of resistance: Pollock Free State and the practice of postmodern politics', in *Transactions of the Institute of British Geographers*, volume 22, pages 359–376.

For more on the Occupy movement see a special issue of *Social Movement Studies*, volume 11, issues 3–4 (2012).

Web resources

There are numerous resources available on the internet with detailed information about both electoral geographies and social movements and protests. For American elections, Dave Leip's on-line Atlas of US Presidential Elections, http://uselectionatlas.org/, is an excellent source of information for elections back to 1824, with detailed results, analysis and interactive maps, whilst the University of Michigan has a website showing different cartographic representations of the 2012 presidential election result at www-personal.umich.edu/~mejn/election/2012/. For British elections, extensive but accessible election statistics, maps and information are archived on the United Kingdom Election Results site, http://www.election.demon.co.uk/.

Websites relating to the Occupy movement include occupytogether.org and occupywallst.org. A good collection of news reports and commentary on the Occupy movement, including a map of Occupy protests worldwide, can be found on *The Guardian* website at www.guardian.co.uk/world/occupy-movement. For more geographical perspectives, see contributions to a forum on Occupy on the *Environment and Planning D: Society and Space* blog, societyandspace.com (18 November 2011). The interactive map of protests in Tahrir Square, Cairo, referred to in this chapter can be found at http://www.bbc.co.uk/news/world-12434787.

Geographies of empire

Introduction

Our aim in this chapter is to explore the geographies of empire, imperialism and colonialism. Empires, especially in the form that they took from the nineteenth century onwards, have been described as 'an extensive group of states, whether formed by colonization or conquest, subject to the authority of a metropolitan or imperial state' (Jones 1996: 155). Similarly, the political relationship of imperialism, which is often based on the existence of empires, has been described as follows:

> The creation and maintenance of an unequal economic, cultural and territorial relationship, usually between states and often in the form of an empire, based on domination and subordination.
>
> (Clayton 2000a: 375).

There is a close relationship, therefore, between empires and imperialism. Empires are usually constituted on the basis of processes of imperialism. At the same time, the existence of empires enables imperial relationships of domination to come into existence and be reproduced. The meaning of colonialism is close to that of imperialism since it also refers to a relationship of domination between a state and other, usually distant, territories (Ferro 1997). Colonialism, however, is usually viewed as a specific form of imperialism since it involves more tangible and concrete forms of rule, including the settlement of people and/or the displacement or subordination of others (Said 1993).

And yet, while the above discussion is useful as a way of developing an understanding of the official definitions of the key concepts discussed in this chapter, it does little to give a flavour of the different meanings ascribed to the same term by various actors. Listen, for instance, to these different voices and note their different interpretations of the broader significance of empires and imperialism. In 1872, the prominent British politician Benjamin Disraeli spoke of the British empire as something that has enabled and reflected the greatness of Britain as an imperial country: 'a country where your sons, when they rise, rise to paramount positions, and obtain not merely the esteem of their countrymen, but command the respect of the world' (Disraeli in Kebbel 1882 vol. 2: 534). Jump forward to the early 1960s and note how the author and anti-colonialist Frantz Fanon refers to those subjugated to French colonial rule in Algeria as the *Wretched of the Earth* in his 1961 book (Fanon 2001). Take heed also of Ronald Reagan's reference to the 'evil empire' of the Soviet Union in his speech to the National Association of Evangelicals in Orlando, Florida in March 1983. The use of such language would become increasingly apparent over the 1980s as a means of countering Soviet influence in global politics. Note finally the way in which anti-war campaigners (Phillips and Jones 2008), political commentators and some academics in recent years have drawn attention to the creation of a *New Imperialism* (Harvey 2003) or *The Colonial Present* (Gregory 2004).

What unites these different uses of the language of empire? There is clearly a common understanding among these authors of empires as large-scale political units that are constituted on the basis of unequal power relations: namely those that exist between a metropolitan state and its satellite territories. At the same time, there are different understandings of the significance and nature of empires within these various statements. These different statements about empire and imperialism are certainly separated by time and by different attitudes towards empire. Attitudes towards empire in the nineteenth century, for instance, were generally far more supportive than they have been since the second half of the twentieth century. These different takes on empire and imperialism also reflect a different relationship between these authors and the empires about which they were writing. Disraeli was a key political figure in a state that benefited greatly from its position at the head of one of the largest empires ever created while Fanon identified with the 'wretched' people upon whom imperial violence was directed. There is finally a clear sense in which these authors are writing about empires that are constituted in slightly different ways to each other. The kinds of formal political relationships of domination that characterised the British Empire of the nineteenth century, for instance, are very different to the more informal – yet equally significant – power relationships that exist between different states in the contemporary world. These inconsistencies and contradictions begin to illustrate some of the difficulties in grappling with the notion of empire; it is a term that means different things to different people living in different places and different times. It is a value-laden term, which is socially constructed in different ways by different authors. And yet, it is also a term that demonstrates a high level of resilience: 'empire' and other related terms such as 'imperialism' and 'colonialism' have been used prominently in academic, political and public debate over a number of centuries.

Our aim in this chapter is to discuss the geographical constitution of empires. We seek to show here not just how empires possess geographical manifestations – for instance, in terms of their territorial extent – but also how Geography and geographical concepts have influenced the formation, reproduction and dissolution of empires. We begin in the next section by discussing the territorial extent of empires. Using a series of maps and tables, we provide an overview of the geographic extent of a European imperial project, as well as seeking to complicate this well-worn picture by

discussing the growing importance of other non-European empires during the twentieth century. We complement this discussion in the following section by elaborating on some of the geographical and other techniques used to create and reproduce empires. We focus in particular on the ecological, technological, geographical techniques that have been used to underpin imperial projects throughout history. The final substantive section explores the break-up of empires and the associated notion of post-colonialism. We explore in this section the significance of more informal imperial relations that help to constitute the political geographies of the contemporary world. We draw the chapter to a close with brief conclusions.

The territorial extent of empires

It is instructive at the beginning of this chapter to appreciate the extent to which imperialism has existed as a guiding principle of world political geographies. One key way in which we can think about the significance of empires in an historic context is by examining their territorial extent. Look at the map in Figure 8.1, which depicts the extent of the territories that have at some time or other come under the imperial control of European states. The significance of this map, of course, is the fact that it demonstrates the staggering fact that few people or territories have succeeded in maintaining their independence from the formal empires created by European powers; China and parts of the Middle East being notable exceptions. Even these latter cases have been come under the influence of the informal empires created by European powers. Think, for instance, of the unequal trade relationships, which have helped to structure the political and economic relationship between the UK and China (with regard to tea and opium) and the UK and the Middle East (with regard to oil). Whilst these regions may have maintained their formal independence from European states, they have been heavily influenced by the latter's political and economic needs and priorities. The significance of this relationship between European states and their distant colonies should not be under-estimated. Indeed, the geographer Blaut (1992, 1993, 2000) consistently argued the growth of

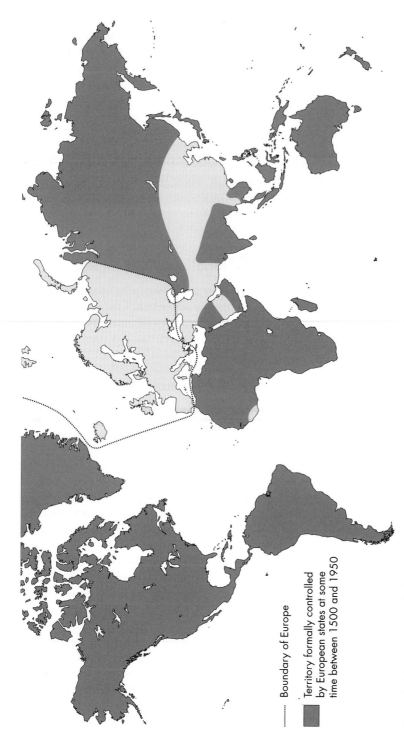

Figure 8.1 European control of non-European lands

Source: Taylor and Flint (2000)

Europe during the modern period can be linked to the great riches gained by them from their colonial ventures in other parts of the world. Crucially, this growth in the economic might of European states helped to sustain their continuing influence in other colonial ventures.

Perhaps because of their immense scale or perhaps because of a sense of collective 'guilt' felt by European academics about the imperial practices conducted by their own states in the past, it is noticeable that the main focus of academic attention for scholars in general – and specifically for Geographers – traditionally have been the empires created by various European states in the period after Columbus' so-called 'discovery' of America in 1492. Ashcroft *et al.* (1998: 189), for instance, have argued that studies of colonialism have traditionally been geographically limited, being grounded in 'European colonialist histories and institutional practices, and the responses (resistant or otherwise) to these practices on the part of all colonized peoples'. We do not wish to decry the significance of the work that has been conducted on modern European empires. Nor do we wish to somehow underplay the significance of European instances of imperialism during the modern period. At the same time, we think it important to position this European imperial project within a broader context. After all, while the French historian Marc Ferro (1997: viii) has argued that 'historical analysis [assumes] a vision of the past which Europeanizes the colonial phenomenon', he also maintains that 'other colonizations have also contributed to fashioning the present image of the planet'.

We can complicate these 'Europeanised' interpretations of imperialism in a number of ways. First of all, we can note the empires that characterised the pre-modern period (Jones and Phillips 2005). Consider, for instance, the territorial extent of the Roman Empire. Under Emperor Hadrian during the second century AD, the empire extended from England and Wales in the far west of Europe to Palestine in the Middle East and from North Africa to lands located in modern day Netherlands. Even more impressive was the Mongol Empire of the late thirteenth century, which extended from the Sea of Japan at the far eastern

edge of the Eurasian landmass to the borders of Hungary and Poland, and from the Arctic to the Bay of Bengal. This second empire is especially significant since it represents an empire in which an Asian-based dynasty sought to extend its control of, among other territories, European lands. At its greatest, it was an empire that controlled 20 per cent of the earth's surface, within which lived approximately 100 million people.

Second, we can think about the existence of non-European empires during the modern period. Japan, for instance, in the period during the 1930s and 1940s, sought to extend its influence over lands controlled by China and the Soviet Union and engaged in two wars with the former (1894–5 and 1937–45) and one war with the latter (1938–41). It succeeded in annexing Korea between 1910 and 1945, controlled parts of Eastern Siberia between 1918 and 1922 and also exercised formal political control over the Chinese region of Manchuria in the period between 1931 and 1945 (Young 1998). We can also view the US as a state that has possessed imperial aspirations. In some respects, the US has often positioned itself as a state that has not engaged in imperial practices. Indeed, for some, as Smith (2004: xviii) notes, 'the use of the language of "empire" to describe US globalism may seem strange or strained'. There is, nonetheless, according to Smith (ibid.: xix) a real sense in which the twentieth century has witnessed the creation of an 'American Empire' in which 'global power [has been] disproportionately wielded by a ruling class that remains tied to the national interests of the United States'. The work of the American historical geographer, Donald Meinig (1993: 22), bears some testimony to such assertions. He has argued that the Louisiana Purchase of 1803 should be viewed as an 'imperial acquisition – imperial in the sense of the aggressive encroachment of one people upon the territory of another, resulting in the subjugation of that people to alien rule'. The case for the existence of an American Empire is particularly compelling when one remembers that the US held formal control over territories outwith its official boundaries for much of the twentieth century. Following the Spanish–American war of 1898, for instance, the US was given control over Cuba, the

Philippines, Puerto Rico and Guam (Trask 1996). It still retains control over a number of colonies, including Puerto Rico, Guam, the United States Virgin Islands, the Northern Mariana Islands and American Samoa. An additional form of American empire exists in the context of the various military bases maintained by the US throughout the world; at present, these bases are located in over fifty countries worldwide (see Figure 8.2). While these military bases cannot be viewed as direct evidence of a formal American empire that encompasses these various states, their existence signifies the potential for an unequal military and political relationship between the US and a series of 'client states' throughout the world.

In this section, we have discussed the territorial extent of empires. While much work has been conducted on the European empires created during the modern period, we also need to appreciate the way in which other states have also sought to promote the formation of empires. The key question that arises, in this context, relates to the various processes and mechanisms used to create and maintain empires and we focus on this issue in the following section.

Creating and maintaining empires

In this section, we discuss the various techniques that have been employed by metropolitan states – in other words, those states that sought to control extended empires – to facilitate the formation and reproduction of empires. A variety of different methods have been used to achieve this aim and these range from the use of direct military force to the use of more informal and yet no less effective economic and cultural forms of domination. Given the relative lack of space to discuss such issues in one chapter, we attend to the more geographical means through which states attempted to promote their imperial project. We begin by elaborating on the geographical knowledges and imaginations that helped to underpin the process of imperialism. We then proceed to outline the ecological context for the creation and maintenance of empires before finishing this section with a discussion of the technologies of exploitation that made the creation of global systems of political domination feasible.

Geographical knowledges of empire

Much has been made in recent years of the significance of the circulation of geographical knowledges for the creation and maintenance of empires. The relationship between geographical knowledges and empires works in two directions. In the first place, instances of exploration and the drive to create empires have come about, at least in some measure, as a result of the need to expand geographical knowledges. Driver (1992: 23), for instance, makes reference to Joseph Conrad's (1926) three-fold division of the phases of imperialism into 'Geography Fabulous' (an epoch of fantastical assertions about the character of distant places), 'Geography Militant' (a phase of physical exploration) and 'Geography Triumphant' (a period in which scientific knowledge of the whole world is complete and thorough). The most significance epoch in the context of the current argument is that of 'Geography Militant' since this is the period of imperial expansion, which was embodied by a 'rigorous quest for certainty about the geography of the earth' (Driver 1992: 23) and whose emblematic figure was Captain James Cook, the eighteenth-century explorer who was the first European to map Newfoundland and who also led three expeditions to the Pacific Ocean. The activities of individuals such as Captain Cook were a testament to the imperial need to expand geographical knowledges of the world.

At the same time, the creation of empires was also based on the geographical knowledges that existed of different places. Captain Cook's efforts during the eighteenth century to map Newfoundland and to explore the Pacific Ocean generated new geographical knowledges of these places – whether in the forms of qualitative descriptions of the people living in them, quantitative data about the weather, flora and fauna that characterised them, or maps produced of unfamiliar landscapes – and these geographical knowledges, in turn, helped to fuel further imperial practices. Joseph Banks, for instance, was a botanist on Captain Cook's first voyage to the Pacific and became an ardent advocate of British colonisation of Australia (Fara 2004). Of course, these geographical knowledges produced by explorers were written from a particular perspective and sought to portray these newly 'discovered'

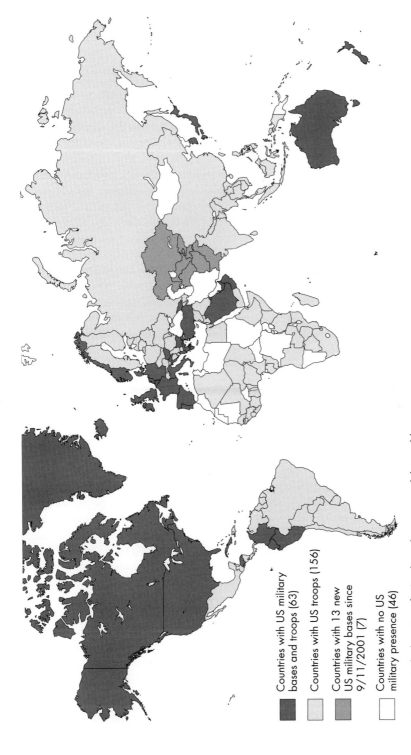

Countries with US military bases and troops (63)

Countries with US troops (156)

Countries with 13 new US military bases since 9/11/2001 (7)

Countries with no US military presence (46)

Figure 8.2 The location of US military bases around the world

lands and people in a particular light. Edward Said (1978) has written extensively about the process of 'othering' that characterised the generation of new (geographical) knowledges by Europeans about different parts of the world (see Box 8.1).

As noted earlier, the kinds of geographical knowledges produced by imperialists could take a number of forms. At one extreme we can think of the formal geographical knowledges produced by state organisations within various imperialist countries. Heffernan (1994), for instance, has discussed the significance of the *Service des Missions*, the research council created by the French state in 1842 as a means of encouraging 'voyages directed towards physical and geographical research or studies of languages, history and all that is of general interest to our [French] civilization' (Antoine 1977: 38, quoted in Heffernan 1994: 24). The geographical knowledges garnered as a result of these expeditions formed the basis of numerous (usually dry) academic treatises concerning other lands and people. But in addition to these formal publications, other detailed reports of the expeditions provide a wealth of more interesting information about the nature of the expeditions and the scientific knowledges produced as a result. The dossiers, according to Heffernan (ibid.: 25), offer particularly insightful information about the 'petty rivalries, feuds and political struggles which characterized French scholarship during this period'. So, for instance, the dossiers demonstrate the way in which the work of particular French geographers was valorised by the *Service des Missions*, while others were marginalised: expeditions by the key geographers Paul Vidal de la Blache and Émile Gautier were funded, while the equally prominent yet radical geographer Élisée Reclus did not feature in the *Service's* dossiers (ibid.). Such inconsistencies highlight the fact that the geographical knowledge produced by so-called 'national schools' of Geography is very rarely unified or homogeneous (Bassin 2000).

At the other extreme to these formal accounts of unfamiliar lands and places lie the more popular accounts of empire produced for mass consumption. Historians have traditionally tended to underplay the

---

BOX 8.1   EDWARD SAID AND THE NOTION OF 'OTHERING'

In his now famous book, *Orientalism* (1978), Edward Said has attempted to outline the interrelationship that exists between empires and ideas of cultural and social difference. Key to this whole process of domination is the idea of 'othering'. In describing the alleged different qualities of peoples encountered in the periphery – which, in the majority of cases, involved emphasising their perceived weaknesses – Europeans could morally justify their efforts to civilise and exploit them. This process helped to essentialise racial and ethnic categories in both non-European and European lands. What this means is that particular racial or ethnic categories were perceived as homogenous groupings of people, with internal differences being underplayed. There are many instances of this process. Black people, for example, could be described in a gentleman's magazine of the late eighteenth century as follows: 'The Negro is possessed of passions not only strong but ungovernable; a mind dauntless, warlike and unmerciful; a temper extremely irascible; a disposition indolent, selfish and deceitful; fond of joyous sociality, riotous mirth and extravagant show . . . a terrible husband, a harsh father and a precarious friend.' The significance of such a description, according to Said, is that the alleged negative characteristics of the black person were seen to represent the direct opposite of the white person's alleged strengths. For Said, it is this cultural act of domination that justified many of the worst excesses of European state imperialism of the modern period. (qouted in Jackson 1989: 135)

Key readings: Said (1978, 1993).

popular support for imperial projects or at least have maintained that this was solely a feature of a particular period of time. And yet, as Mackenzie (1986) and others have shown, there was a popular support for, and a real appetite for geographical knowledge of, empires in many European states during much of the nineteenth and twentieth centuries. In Britain, for instance, nationalism 'took a distinctively imperial form in the defence of real and imagined colonial interests' (ibid.: 3) and this popular support for the nation and empire was nurtured through a variety of different aspects of popular culture. An imperial cult was formed in Britain, which was communicated to the masses through: the music of nationalist composers such as Elgar; the compositions of nationalist writers such as Rudyard Kipling; the songs and jokes of the music hall and working-men's clubs; the everyday practices of youth movements such as the Boy Scouts and the Girl Guides. Adventure novels formed an additional arena for the communication of imperial geographical knowledge. Phillips (1997: 1) has examined the significance of these novels and has argued that 'since the eighteenth century, the eclectic literature of adventure has been consumed by mass reading audiences in Europe; it has fed geographical imaginations'. The key adventure novel of the modern period, according to Phillips, is Daniel Defoe's *Robinson Crusoe,* published in 1719. The familiar story recounts the tale of a castaway, who spends twenty-eight years stranded on an island off the coast of Venezuela, while coming across, among other things, Native Americans and mutineers. Phillips (1997: 22) argues that the novel holds a particular significance for imperial geographical imaginations, particularly in Britain, between the eighteenth and twentieth centuries. It 'mapped' – physically and metaphorically – both British identity and the British colonies of the period, thus 'naturalising and normalising constructions of identity and geography' (ibid.). Furthermore, the novel acted as an inspiration for a series of other novels – 'Robinsonades' such as *The Coral Island* by R. M. Ballantyne, published in 1858 and *Canadian Crusoes: A Tale of the Rice Lake Plains,* by Catherine Parr Traill, published in 1852 – which mimicked the themes covered in the original. Taken together, we see how popular novels

such as these helped to produce popular geographical knowledges of the world and of the appropriate role to be played by imperial states within it.

Somewhere in the middle between these two extremes of academic writings and popular literature lies travel writing. The modern period witnessed a great interest in travel writing, which could take on a variety of different forms, ranging from unpublished diaries to ethnographies and novels. Following Said, Blunt (1994) has argued that the travel writing of the modern period reveals imperial understandings of colonised lands as well as illuminating the character of the imperial project being promoted by the metropolitan state. These tensions between the way in which colonisers, and the colonised lands, are understood are revealed most clearly in the character of the travel writer themselves:

> Travel writing is distinctive because autobiographical narrative exists alongside, and seems to gain authority from, observational detail. The journey undertaken and represented can be seen as a psychological journey, relating to themes of the journey of life and self-discovery.
>
> (Blunt 1994: 21).

Blunt's argument here revolves around the interrelationship between the physical journey undertaken by the traveller and the descriptions given of exotic places and the psychological journey undertaken by the traveller as a representative of a colonial power. These ideas are important because they begin to indicate that the representations of empire contained within travel writings – and indeed other accounts of empire such as academic treatises or adventure stories – are never singular and homogeneous. As Gregory (1995) argues, the accounts of distant places provided by travellers from metropolitan states could vary considerably, with factors such as class, culture and gender influencing the representations of empire produced (see Box 8.2).

We have discussed in this section the various way in which different kinds of geographical knowledge were implicated in the imperial project during the modern period. Two issues need to be addressed before

BOX 8.2 CONFLICTING REPRESENTATIONS OF EGYPT IN 1849–50

Gregory (1995) discusses the differing imaginative geographies of Florence Nightingale and Gustave Flaubert as they travelled through the Nile Valley to Cairo in the middle of the nineteenth century. Florence Nightingale arrived in Egypt in November 1849, aged 29, accompanied by the experienced travellers Charles and Selina Bracebridge and her maid, 'Trout'. Florence wrote long letters to her immediate family during her extensive travels in Egypt and these are the main source of her geographical imagination of the country. Two key literary devices were used by Florence to describe the Egyptian landscape. The first device focused on the unnaturalness of the landscape that she encountered, which proceeded to produce a 'feeling of unease, even of dislike' within Florence (ibid.: 35). The second device revolved around the imagination of a supernatural landscape. Egypt's landscape was shorn of its contemporary population by Florence, only to be repopulated by an imaginary population drawn from Egypt's distant past. Overall, Florence's geographical imagination was underpinned by ambivalence towards its people and the landscape that they inhabited. Gustave Flaubert, travelling in Egypt during the same period as Florence, committed a different kind of geographical imagination to paper in the form of two notebooks and some thirty letters to his family and friends. Flaubert's main motivations for visiting Egypt were to find a degree of exoticism, which would help to challenge the blandness of life in Europe. And yet, on arriving in Egypt, Flaubert found an exoticism that was tainted by its contact with European modernity. When viewing ruins in Egypt, for instance, Flaubert asked 'Are they going to become like the churches in Brittany, the waterfalls of the Pyrenees' (quoted in ibid.: 43)? In other words, Flaubert here is bemoaning the increasing commodification of the Egyptian landscape under European influences. Taken together, Gregory (ibid.: 50) maintains that Florence Nightingale and Gustave Flaubert's geographical imaginations of Egypt shared some similarities. They 'shared the same viewpoint – the commanding heights of European modernity – but their prospects were vastly different'. Whereas Florence Nightingale focused on the legacies of Egyptian antiquity contained within the landscape, Flaubert concentrated on Egypt's present and the way in which this was being corrupted by European influences. The geographical imaginations that appeared in the letters of these two individuals, therefore, were influenced heavily by their own identities and predilections.

Key reading: Gregory (1995).

we conclude this section. First, the work of Gregory (1995) begins to show how the geographical imaginations produced by European explorers and travellers were never unified and homogeneous and, thus, echoes the desire of post-colonialists to demonstrate the way in which a European imperial project could be complicated and fractured. We need to appreciate, for instance, that the geographical knowledges produced by European imperial states contained important elements that were derived from the colonised lands and people themselves. As Driver (2001: 8) maintains, 'the published narrative [of a travel writer] was the end-product of a sequence of stages in the writing of travel, a process through which the explored in the field was translated into the published author'. The lands and people, which contributed to the experiences of the explorer in the field, therefore, are inscribed in some ways into the geographical imaginations contained within such narratives. Being aware of these presences can help contribute to a post-colonial goal of illuminating the 'heterogeneous, contingent and conflictual character of imperial project' (ibid.).

Second, while we have considered different kinds of geographical knowledge in isolation in this section, we need to appreciate the fact that each of these combined with each other in complex ways to produce a totality of imperial geographical knowledges. In the same way that Crang (1996) has sought to demonstrate the myriad connections that exist between popular and academic knowledges with regard to food, we need to realise the important interrelationships that existed between the different kinds of geographical knowledge produced as part of the imperial project. In general terms, we can think of the crossovers between different literary traditions, as travel narratives, academic treatises and adventure stories would begin to 'speak to each other'. Driver (2001: 10) has also attempted to blur the boundaries between these different forms of geographical imagination. He has noted, for instance, that the same explorers would write different accounts of the same journey, with popular stories and exploration narratives being targeted at women and children, and men, respectively. More fundamentally, certain narratives defied categorisation into these neat categories: the renowned Victorian explorer, H. M. Stanley, for instance, wrote narratives that blurred the boundaries between exploration narratives and adventure fiction (ibid.).

The production of geographical knowledges was, therefore, a key product and facilitator of the processes of imperialism that characterised the modern period. We turn in the following section to examine another important geographical context for imperialism; namely the connection between empire and ecology.

Imperial ecologies

We saw above how the process of imperialism was driven forward by the development of particular kinds of geographical knowledges, imaginations and practices. One important way in which these knowledges, imaginations and practices unfolded was with regard to the ecologies of colonised lands and their relationship with the metropolitan core of the empire. A recurring theme has been the attempt made by metropolitan states to develop or impose a sense of *ecological order* on the empires that they governed. A sense of ecological order could be promoted in discursive and more material contexts. In discursive terms, imperialists have sought to comprehend the environments of various parts of the world and, particularly, the flora and fauna that were located within their empires (see Box 8.3). One example of this effort to understand world ecologies is the

---

BOX 8.3 CREATING A 'TROPICAL' PARAGUAY

For much of the modern period Paraguay escaped the attention of European explorers and scientists. While other South American countries were charted and described in great detail by Europeans, Paraguay still remained under-explored, partly as a result of the isolationist stance taken between 1814 and 1840 by its President José Francia. There began a new interest in surveying the geography of Paraguay following his death and particularly after the cessation of hostilities after the War of the Triple Alliance (1865–70). But while Europeans and others may not have been able physically to chart the nature of Paraguay during this period, they were still engaged in efforts to imagine the character of the country. Pendle (1954: 4) has argued that Paraguay has over the long term existed as an imagined idyll or Utopia for Europeans. Writing in 1910, Marcel Hardy, for instance, described Paraguay as follows: 'It is very likely that there is no tropical land in the far interior of a continent which in its undeveloped state offers so many advantages, a rich soil, a healthy climate, and immense riches for both stock farming and agriculture' (quoted in Naylor 2000: 53). This quote is significant since it draws attention to the idyllic qualities of Paraguay as a country – even 'in its undeveloped state'. At the same time, Hardy outlines the potential for the climate and ecology of Paraguay to be developed in a potentially productive manner, promising as it does 'immense riches for both stock farming and agriculture'. Indeed, the country became known as 'that very garden of South America' (quoted in Naylor 2000: 54) and, being a garden, could productively be cultivated by Europeans. In a later study, Hardy, for instance, chose to focus his attention solely on the productive agricultural potential of Paraguay, particularly with respect to oil, cotton, jute, tobacco and tropical fruit. As well as being imagined in particular ways, the ecologies of Paraguay, according to Naylor (ibid.) were recorded by authors such as Hardy in ways that portrayed them as productive within imperial agricultural networks.

Key reading: Naylor (2000).

emergence of the term 'tropic' as a way of describing the regions located near to the world's equator. While we might assume that this term has always existed as a way of referring to such regions, the historian David Arnold (1998: 2) has shown how the term has a particular history and that it has been used primarily as a way of 'defining something environmentally and culturally distinct from Europe'. The tropics, according to Arnold (1998), were viewed by Europeans as Edenic, threatening and, more importantly, exploitable. At heart, therefore, the term 'tropic' – particularly in its more popular connotations – reflects another instance of a long-running attempt by Europeans to 'other' people and places outside of Europe. Geographers, following in Arnold's wake, have also attempted to problematise the term examining its 'enrolment in a variety of scientific, aesthetic and political projects' (Driver and Yeoh 2000: 2; see also Driver and Martins 2005).

At the same time, the imperial projects prosecuted by European states had also been predicated upon a material understanding and transformation of ecologies throughout the world. The environmental historian Alfred Crosby (1986) has drawn attention to the concept of ecological imperialism: namely the way in which the creation and maintenance of empires during the modern period was predicated on a biogeographical logic. The effort made by European states to create global empires was shaped by the various environments that they encountered and, more importantly, the extent to which these new lands could be used to replicate the ecologies found in the metropolitan state. Crosby's argument can be crystallised as follows: 'One

would expect an Englishman, Spaniard or German to be attracted chiefly to places where wheat and cattle would do well, and that has indeed proved to be the case' (ibid.: 6). 'Portmanteau biota' or, in other words, scaled-down versions of the flora and fauna that were to be found within particular European states, were shipped out to the 'neo-Europe' being created by colonisers. The overall aim of doing so, of course, was to be able to produce foodstuffs and other related goods for burgeoning markets back in the metropolitan state.

We can consider grass and grassland as a significant example of a particular kind of ecology that was reproduced throughout European empires during the modern period. Brooking and Pawson (2011), for instance, have examined the significance of grass and grassland for the British Empire, particularly in the context of its relationship with its colony in New Zealand. Significantly, this was a relationship that was played out in discursive and material contexts. Brooking and Pawson draw attention to the way in which the unfamiliar landscapes of colonised countries were represented using familiar motifs. Paintings of New Zealand during the mid-nineteenth century, for instance, emphasised the existence of empty and verdant grasslands, thus stressing their potential agricultural productivity. As Brooking and Pawson (2011: 419) rightly note, 'such acts of representation translated the foreign into the familiar, naturalising the landscape and implying both the inevitability of transformation and the rightfulness of ownership'. But in addition to being exhibited in discourse, it is noticeable that this ecological imperialism also appeared in more material contexts. For much of the nineteenth century, the lack of appropriate technology meant that the only real use that could be made of agricultural land in New Zealand was for the export of wool and wheat but with the introduction of refrigerated shipping in the late nineteenth century, the New Zealand state sought to convert more Maori lands into pasture, which could be used to support an export trade in meat and dairy products. A good example of the transformation of New Zealand's ecology can be found in the Banks Peninsula on South Island. Up until the mid-nineteenth century, the native trees located on the peninsula had acted as a source of timber for colonisers but these were felled in the period between 1860 and 1890. The remains of the forest were burnt and the land sowed with grass seeds in order to create pasture for livestock (ibid.: 421).

The process continued apace throughout New Zealand until the early decades of the twentieth century. By the 1920s, nearly 20 million acres of New Zealand's land area 'had been converted to introduced grassland' (ibid.). The significance of Brooking and Pawson's work is their claim about the broader contexts within which this ecological transformation took place. First of all, it was an ecological transformation that led to the establishment of the primacy of pasture within New Zealand as indigenous forests were cleared in the service of a pastoral economy: pasture represented 'a kind of totalitarianism that marginalised other forms of landscape' (ibid.: 425). Second, Brooking and Pawson maintain that until the early twentieth century, the role played by the imperial British state was 'minimal'. Private seed merchants took the lead in promoting particular kinds of seed that could help to transform New Zealand's ecology. While some merchants were aware of the significance of grass seed 'in the development of imperial agro-commodity chains' (ibid.: 426), it was not until the 1920s that the New Zealand state took a formal interest in shaping the ecological networks that helped to bind it into the British Empire. The Department of Scientific and Industrial Research, replete with a Grasslands Division, was created in 1926. The significance of the organisational should not be underestimated since it was a

formal recognition of ecology as a science in the service of empire, in which New Zealand's role in the imperial division of labour was to be that of grassland farming specialist.

(Brooking and Pawson 2011: 9)

If Brooking and Pawson's work begins to draw our attention to the ecological transformations that were the producers and products of European empires, then we also need to appreciate the key role played by botanical gardens within this whole process. Botanical gardens were the museums and clearing houses of this transnational flow of flora: museums since they centralised, classified and exhibited ecological artefacts; clearing houses since they helped to collect, store and distribute ecological artefacts throughout the empire's territories. Botanical gardens were a product of the Enlightenment, namely the age of reason that was concerned with promoting scientific knowledge, rationality and moral progress and improvement. France and Great Britain took the lead in creating botanic gardens from the late eighteenth century onwards. At the beginning of this period, for instance, only four institutional botanic gardens existed in the British Empire (two in the West Indies, one in India and one on a South Atlantic island). By the end of the nineteenth century, this figure had increased to over a hundred (McCracken 1997: 2).

The two key botanic institutions in France and Great Britain in the eighteenth and nineteenth centuries were Les Jardin des Plantes and Kew Gardens, respectively. Both institutions sent plant hunters abroad so that they could collect interesting or potentially useful flora and bring them back to the parent organisation to be catalogued and studied. Similar developments took place in Schönbrunn in Vienna at the heart of the Austro-Hungarian Empire. As McCracken (ibid.: 3) notes, 'ships were commissioned by the Habsburg emperor to bring home living plants from the West and East Indies', which were then housed in enormous conservatories. Some of the reasoning behind the creation of such gardens can be discerned in the activities of Sir Joseph Banks, a key advisor to King George III on Kew Gardens. As well as commissioning Francis Masson to act as Kew's first plant hunter in 1772, he also encapsulated an attitude towards botanical gardens as 'primarily economic depots to encourage and service plantation agriculture' (ibid.: 2). In addition to promoting the role of Kew as a key ecological and economic node within the British empire's agricultural network, Banks was also an individual who promoted 'a sense of botanical nationalism' within Britain (ibid.: 3). Collecting, cataloguing, exhibiting and re-distributing flora within botanical gardens such as Kew, therefore, was never merely an objective scientific enterprise. It was intimately connected to broader ideological and economic processes of empire- and nation-building.

Technologies of modern empires

One important theme that we have alluded to at certain points during the above discussion is the significance of the development of particular technologies for the creation and maintenance of empires. One characteristic of modern empires is their geographical scale (Ferro 1997) and, as such, it has been argued that these empires could not have been formed in the first place – nor indeed could have maintained their coherence – without considerable technological developments. Headrick (1981: 4), for instance, has argued that 'the real triumph of European civilization [read colonisation] . . . in short . . . has been a triumph of technology, not ideology'. Headrick maintains that new technologies were inter-linked with the imperial project in two important ways. First of all, the development of new technologies enabled large-scale imperialism to take place in an effective manner. As Headrick (ibid.: 200) notes, 'Europeans who set out to conquer new lands in 1880 had far more power over nature and over the people they encountered than their predecessors twenty years earlier had; they could accomplish their tasks with far greater safety and comfort'. Second, the development of new technologies also acted as a spur to imperialism, as industrialists and capitalists within European states sought out the raw materials that were needed in order to create new goods, further technological developments and wealth.

One important technological advance, which facilitated the imperial project, was the development of

more effective forms of communication between the imperial state and its colonised lands. Ogborn (2002, 2007), for instance, has examined the development of technologies of writing and the impact that they had on the emergence of imperial forms during the seventeenth century. Ogborn's study focuses on the networks of communication that existed within the English East India Company and, specifically, on the materiality of the technologies used to communicate between the Company and its mercantile ventures dotted around the Indian Ocean. He draws attention, for instance, to the voyage undertaken by Alexander Sharpeigh as a representative of the Company in 1608. Sharpeigh took with him a variety of different letters and other forms of correspondence to the Red Sea, India and Indonesia, including his own commission from the Company and the King of England, some letters from the King to 'Princes in the Indies', 'sailing directions, an invoice for his cargo and a list of weights and scales' (Ogborn 2002: 155). These documents can be viewed as technologies, which facilitated the control by the English East India of lands dotted around the Indian Ocean. The letters, when they arrived in places such as India, took on the status of symbolic representations of the power and influence of the imperial force. And yet, as Ogborn shows, by thinking about the materiality of the correspondence transported by individuals such as Sharpeigh – in terms, for instance, of the slippage between the meaning ascribed to the text by the writer and the way in which this is translated by the messenger and the reader, or in terms of how the letters themselves could begin to decay and decompose on a long journey – one can also destabilise such an argument. When one considers the 'production, carriage and use of texts as material objects', one can also further the post-colonial goal of 'foreground[ing] the active and collective making of global geographies [of empire] as a contested enterprise involving multiple agents in a variety of sites' (Ogborn 2002: 155).

In addition to the development of new communications technologies, we also need to consider the way in which specifically geographical technologies were implicated in the production and maintenance of empires. The arts of surveying and mapping were especially crucial in this respect. Geography – and specifically the act of mapping – was key to the creation of a system of territorial and political order within empires (Harley 1992). Through the process of mapping, land, resources and people could be configured in ever-more systematic and scientific ways. And yet, we also need to realise the limitations of this ambition. While it is true that major technological developments during the modern period made it easier to survey and map territories – with crucial developments including the invention of the first modern theodolite during the late sixteenth century and the sextant during the first half of the eighteenth century – we should realise also that other factors impinged on the ability of explorers to make geographical sense of the lands that they encountered. Clayton (2000b) has examined the way in which George Vancouver sought to survey and map the Northwest Coast of the Americas during the late eighteenth century. Vancouver was sent in 1792 as a representative of the British state to map this territory. As part of his voyage, he circumnavigated Vancouver Island. According to nineteenth-century commentators, Vancouver is said to have 'truly

sorted out the geography of the Northwest Coast' and this view has been confirmed more recently by other scholars (ibid.: 190). And yet, Clayton's insight has been to show how facile these claims actually are. Admittedly, the survey of some parts of Vancouver Island, such as Puget Sound, had been effective. The flat land had enabled Vancouver and his party to erect triangulation bases in order to map the terrain in a precise manner. Other parts of the island, however, were not mapped as effectively, partly because of low cloud on those particular days (ibid.: 197). Moreover, as Clayton (ibid.) notes, 'the absence of point symbols for land topography on this chart, and the fact that the tiny crosses dotting the coastline appear only on the sea side of the shoreline . . . suggests that Vancouver saw his cartography as a record of maritime surveying'. In effect, Vancouver's map had little to say about the interior of Vancouver Island. The empire's eyes could only extend so far from the coast, whatever the technologies that it possessed (see also Box 8.4). But of course, despite the limitations that were apparent in Vancouver's maps and surveys, they still represented

BOX 8.4 THE GREAT ARC AND THE MAPPING OF INDIA

One of the most significance acts of colonial mapping occurred during the first half of the nineteenth century in India. India had long been considered the 'jewel in the Crown' of the British Empire but little scientific information was available about its geography. The Survey of India, a branch of the British colonial state in India, sought to remedy that deficiency. Starting out in the south of India at the beginning of the nineteenth century and gradually working their way northwards, the staff of the Survey of India used the principles of trigonometry and the latest scientific equipment, including theodolites, in order to chart the whole of the Indian subcontinent, in terms of longitude, latitude and elevation. The scale of the enterprise was enormous since it extended for 1,600 miles. One consequence of the whole project was the determination of the elevation of the highest mountain in the world and its (re)naming as Everest, in recognition of the efforts of George Everest, one of the key figures in the Survey. At one level, therefore, the tale is one of scientific ingenuity as the British state in India sought to bring a more calculating and rational perspective on the geography of the peninsula. While this narrative is certainly appealing, it fails to take heed of the incredible difficulties faced by the Survey of India in completing its task. As Keay makes clear, whole parties of surveyors died as a result of malaria and dysentery and – more uncommonly – tigers and scorpions. More technical difficulties were also encountered. Once again, fog and air pollution – for instance in the region around Delhi – could make it difficult for the surveyors accurately to make out survey points. The terrain and the vegetation could also make it difficult for the surveyors. Carrying a theodolite, which weighed half a ton, through dense forests and across swampy land was not easy. Nor was winching the theodolite up to the top of a ninety-foot platform in order to take sightings of other trigonometry points. Such tales demonstrate, at one level, the heroism involved in the whole enterprise. At another level, they illustrate the difficulties involved when colonial states sought to make their colonised lands more 'legible' (Scott 1998). While the development of new technologies certainly helped this endeavour, their application in unfamiliar and frequently hostile territories was fraught with difficulty and meant that the imperial project of understanding colonised space was always tentative and uncertain.

Key reading: Keay (2001).

an appropriation of the Northwest Coast by colonial powers. Fisher and Johnston (1993), for instance, note that the maps created by Vancouver and his party led to an economic and cultural exploitation of the Northwest Coast, especially by the British imperial power.

One final technology that needs to be considered in this section revolves around the mechanisms of governance that were adopted by imperial states as a way of controlling their far-flung territories. We are concerned here, therefore, with the technologies of rule that were used to govern colonies. The term governmentality (Foucault 1991), introduced in Chapter 2, is a useful way of thinking about these technologies of rule. In general terms, governmentality refers to the technologies used to render society governable, but the concept comprises several different elements as identified by Legg (2007) (see Table 8.1).

Legg (2007) uses this multi-layered understanding of governmentality as a way of describing the forms of colonial rule used in Delhi, the imperial capital of

India between 1911 and 1947. We do not have the space or time to deliberate upon all of the aspects of Legg's fascinating study but choose instead to concentrate on the more spatial aspects of his study or, in other words, his focus upon colonial visibility.

Delhi is significant, of course, because of the effort made by the British to re-order the spatial configuration of the city. New Delhi, the ordered and rational new city, was 'one of Britain's most spectacular showcases of imperial modernity' (ibid.: 1). Indeed, the contrast between the new and old city are stark; whereas the new city is planned and spacious, with wide open streets and green squares, the old city comprises a warren of inter-locked streets and densely populated neighbourhoods (see Figure 8.3). The difference between the two cities can be seen to represent the alleged contrast between the rational and ordered colonial bureaucracies put in place by the British state and the more haphazard and ad hoc Indian society. Furthermore, the two cities were separated by a so-called *cordon sanitaire*: namely a barrier that was

Table 8.1 Various aspects of governmentality

Colonial episteme. The particular ways of thinking and framing the truth that are promoted by colonialists and which are, in some cases, contested.
Colonial identities. The way in which particular identities are ascribed to the governed and those who govern.
Colonial visibility. The visual and spatial representations of colonial space produced by the colonial power.
Colonial techne. The specific techniques used to govern colonial people and territories.
Colonial ethos. The particular values and morals that help to inform colonial government with regard to notions of duty, benefit and so on.

Source: Legg (2007: 12)

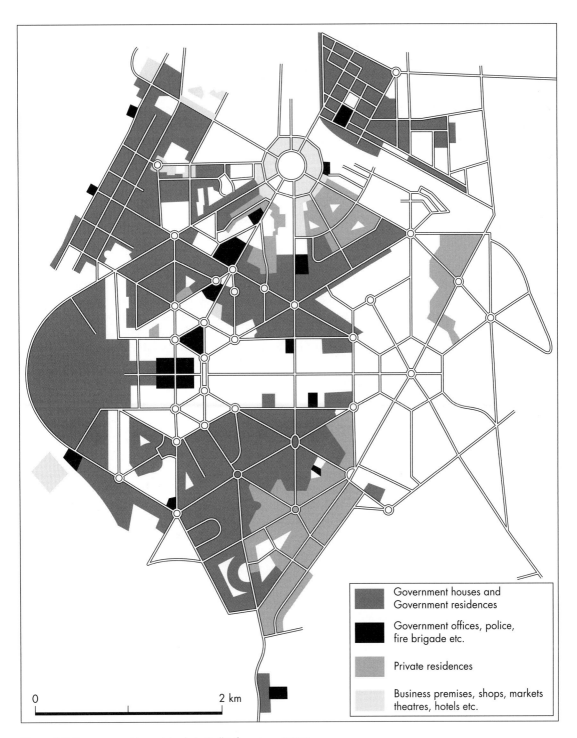

Figure 8.3 The spaces of imperial rule in Delhi (from Legg 2007)

supposed to reduce the contact that took place between the two cities. Once again, this barrier can be viewed as an echo of the boundaries that existed in India at the time between the representatives of British imperial rule and the subjects under their control. At face value, therefore, the colonial visibilities that existed in Delhi at the beginning of the twentieth century reflected the existence of: a particular colonial episteme in which the rationality of British colonial rule was contrasted with the chaos of Indian society; contrasting colonial identities; spatial techniques of colonial rule; a particular moral argument about the superiority of British colonisers and the requirement for them to distance themselves from their Indian subjects.

And yet, the value of Legg's argument is to show the way in which the differences between the two cities – and by extension the differences between the two cultures that existed within imperial India – were actually connected in complicated ways and, indeed, fractured from within. As Legg notes, the *cordon sanitaire* between the two cities is actually connected by a number of 'multiple, well-worn tracks'. It is significant, too, that particular streets in the old town had been widened and some of the buildings within it had been demolished. These internal contradictions and external connections in the townscape of Delhi speak of much broader contradictions and connections in the city. For instance, whereas there was an effort to ensure that government workers were housed in the new city, a report compiled in the 1920s showed that only 159 railway workers actually lived in the new city but over 2,200 lived in Old Delhi (ibid.: 65). In another context, we also need to consider how government spaces were actually subverted. Legg describes a report of 1937, which showed that the electrical supplies of government offices and buildings were used informally and unofficially during festivals as a source of cheap electricity. In addition, government buildings were used on a regular basis to house cows and buffaloes!

Legg's work is important, therefore, because it shows, once again, how the various technologies of rule employed by imperial powers were fractured, uncertain and tentative. This has emerged as a recurring theme within our discussion, whether with regard to the technologies of rule or the kinds of geographical knowledge that were available to imperial powers. The existence of such flaws and weaknesses within an imperial project is a theme that is taken up by post-colonial scholars and it is to this theme that we turn in the final substantive section of this chapter.

Post-colonialism and the break-up of empires

Of course, this formal political control of non-European lands was not to last. Lands in the Americas gained their independence from the Spanish, Portuguese and British empires by the early nineteenth century, and a second round of decolonisation took place in Australasia, Asia and Africa during the mid-twentieth century. Today very few territories remain under the formal political control of metropolitan states. And yet, despite the dissolution of formal empires, there is a sense that the world is still characterised by a series of informal empires that are still as pervasive and still as oppressive as the formal empires of the past. First President of independent Ghana and exponent of pan-Africanism, Kwame Nkrumah, argued in *Neo-colonialism: The Last Stage of Imperialism* (1965), that the end to formal colonialism had not meant an end to colonialism. As he noted, ex-colonial powers and emerging superpowers continued to colonise Africa, for example, through international monetary bodies, multinational corporations and a variety of educational and cultural organisations. These informal instances of imperialism are not as obvious as their historic counterparts but they are just as effective (see Box 8.5).

As Nkrumah perceptively noted back in 1965, and as the case study in Box 8.5 testifies, the creation and maintenance of informal empires can take place as a result of a range of factors: the exertion of political influence over client states; the creation of unequal economic or trading relationships, which help to maintain one state's influence over another; the existence of cultural norms that help to perpetuate popular understandings of the relative status of different states.

BOX 8.5 THE UNITED STATES' INFORMAL EMPIRE IN CENTRAL AMERICA

As noted earlier in this chapter, the United States has acted historically as the hub of a formal empire that extended into parts of Central and South America but, equally, it has sought to extend its influence over a number of other states in more informal and subtle ways. The most infamous examples of these more covert efforts by the United States to influence the internal politics of other independent states have been in Guatemala, Nicaragua and Cuba. The US government has been directly implicated in these informal empires in some instances. The famous Iran–Contra affair of the mid-1980s, for example, was centred on the US state's attempts to sell arms to Iran – which was at the time subject to an international arms embargo – as a way of funding the Contra rebels seeking to undermine the democratically elected Sandanista government. A more indirect influence of the US state on various states in Central America can be found in the workings of the United Fruit Company (UFC). UFC was a multinational corporation that traded in tropical fruit in a variety of Third World countries in the period between 1899 and 1970. Its influence in Central America, in particular, was extensive. It has been accused of exhibiting the worst of colonial practices, including bribing government officials in developing states in order to ensure preferential treatment for UFC, in terms of tax rebates and the relaxation of labour and environmental laws. In addition to working for its own ends, it is clear that there were close links between the UFC and the US state. The UFC lobbied the Truman and Eisenhower administrations during the 1940s and 1950s concerning the perceived communist threat posed by the Guatemalan government of Colonel Jacobo Arbenz Guzmán. An invading force was trained by the CIA and, in 1954, managed to overthrow the democratically elected government of Guatemala. Little evidence has emerged concerning the alleged links between the Guzmán regime and communism. The UFC's interest in toppling the Guzmán regime may have been more to do with the fact that the latter was proposing a series of land reforms in the country, which would have led to UFC losing 40 per cent of its land holdings. Such a case study illustrates the complicated nature of informal empires; in terms of the blurring of the boundaries between states and private corporations and, indeed, between the more subtle aspects of imperialism and its more coercive forms.

Key reading: Striffler (2002).

Contemporary geographers seek to draw attention and to criticise these unequal relationships of domination that exist between states. We have already alluded in previous sections to the significance of post-colonial scholarship as one way of critiquing empires and imperialism. The meanings of post-colonialism are numerous and sometimes contested. The main issue revolves around the way in which the 'post' is conceptualised within 'post-colonialism'. Post-colonial scholars are, on the whole, less concerned with the period or time after colonialism; although this has been a focus of enquiry. Rather, they seek to highlight the way in which colonial ways of thinking – whether those that existed during the period of formal empires or those that help to underpin informal empires – can be questioned and undermined (see Blunt and McEwan 2002). Post-colonial scholars, at heart, seek to emancipate those that have been marginalised and disenfranchised because of colonial practices. Moreover, there is an effort to demonstrate that the binaries that characterise conventional understandings of imperialism – between black and white, oppressor and oppressed, coloniser and colonised – are, in fact, more complicated and fractured than that. As noted earlier, Legg (2007) has attempted to shown how the divisions between coloniser and colonised and between white and black in early twentieth-century Delhi were compromised in numerous ways. Similarly, in a study of British missions in western Canada, Christophers (1998) has shown that certain missionaries (such as the

Anglican, John Booth Good, the subject of his study) refused to regard natives as their racial or colonial others, but instead saw all people in the region – Indian, Asian or European – as humans, and potentially as Christians. For Good, the most significant distinctions were to be made between Christians and non-Christians, as opposed to the alleged racial differences between Indians, Asians and Europeans.

Post-colonialism has the potential to be able to show the persistence of colonialism in more contemporary ways of structuring world politics. And yet, it has also been criticised for two main reasons. First, there has been a tendency for post-colonial scholars to focus solely on complicating historic manifestations of colonialism. Witness, for instance, the tendency for post-colonial geographers to study the historical geographies of colonial encounters in Africa (Blunt 1994; Driver 2001), South America (Scott 2009; Lovell 2005) and the Indian subcontinent (Legg 2007; Duncan 2007). Second, there has been a tendency for post-colonial scholars to examine the discursive constructions that have helped to underpin colonialism. In this respect, Marxist scholars have criticised the lack of attention given by post-colonial scholars to the material inequalities that shape colonial relations (Abrahamsen 2003).

Post-colonial scholars have begun to respond to such criticisms. Some have used post-colonial ideas as a way of interrogating our understandings of contemporary economic relationships. Pollard *et al.* (2009), for instance, have drawn attention to the way in which contemporary understandings of economics and economic geography derive from particular western/ northern – or, in other words, privileged – viewpoints. Their aim is to promote a post-colonial economic geography, which is 'more conscious of its own perspectives and more open to embracing different perspectives through which to view economic practices' (ibid.: 139). Other post-colonial scholars have sought to examine how colonial legacies still influence the kinds of economic relationships and trade relations. McEwan and Bek (2009), for instance, have shown how recent efforts to develop more ethical forms of trading – specifically in the context of the South African wine industry – have been hampered by the legacies of

colonialism and apartheid within the country. Difficulties were found in reconciling the need to remain competitive within global free markets in wine and the need to transform and improve the working and living conditions of South Africans. In this work, we see efforts to bring post-colonial insights to bear on more contemporary forms of inequality that are expressed in the material conditions of human lives.

Other post-colonial scholars have used the recent 'war on terror' as a way of re-examining notions of empire and colonialism in more contemporary contexts. Phillips (2009), for instance, has explored the way in which the 'war on terror' has energised new forms of political protest in the UK, which are based upon multiple networks of resistance. In addition to showing in broad terms how post-colonial ideas can be applied to more contemporary instances of (resistance) to colonialism, Phillips' work is important since he seeks to examine the way in which resistance to the 'war on terror' has not been played out in simple binaries between black and white or between Muslim and non-Muslim. While Muslim activists have been heavily involved in promoting resistance to the 'war on terror', there have also been efforts to develop 'bridges between east and west' within the protests against the war. Once again, such work illustrates the usefulness of questioning binaries, by showing that: not all Westerners have been in favour of the 'war on terror'; not all Muslims have used the 'war on terror' as an excuse for isolating themselves within Western countries; not all Westerners view Muslims as political and/or social threats.

Of course, the 'war on terror' has also encouraged others – not usually or explicitly aligned with a post-colonial academic project – to write anew about notions of 'neocolonialism' or 'new imperialism' (e.g. Smith 2004; Ferguson 2004; Gregory 2004; Harvey 2003; Hannah 2010). Yet, it is worth stressing one point here; namely that the use of such language to describe the recent 'war on terror' and the related military incursions in Afghanistan and Iraq testify to academic concerns about the re-emergence of more formal kinds of empire in the contemporary world. If the US is able to place military bases in a large number of independent states, then should we refer to the existence of a US informal sphere of influence or a more formal US empire? If companies such as Halliburton, the US oil services' company, are able to gain favourable concessions for oil exploration and extraction in places such as Iraq and Kuwait, then should we view this as a consequence of the US' informal influence in these states or as a direct product of the existence of a US formal empire? Such questions illustrate the blurred conceptual boundaries between imperialism and globalisation and it is a theme that we take up in the next chapter.

Conclusions

Our aim in this chapter has been to examine the geographies of imperialism. We noted, to begin with, the territorial reach of empires before concentrating on some of the key processes that helped to create and maintain empires: the development of geographical knowledges of empire; the use of particular ecologies within imperial networks of power; the emergence of new technologies. While many of these developments helped to create and maintain empires, we also showed how a number of different challenges helped to undermine them; whether with regard to fog making difficult for George Vancouver to chart Vancouver Island or British workers in Delhi choosing to live in the old town rather than in the areas that had been prescribed to them in New Delhi. Such themes illustrate the way in which empires are always fractured and riven with contradictions. This is, of course, a key theme within post-colonial studies of empire. We concluded our chapter by examining the break-up of formal empires, the emergence of post-colonial ways of thinking and the possible re-emergence of more formal kinds of imperialism in the context of the 'war on terror'. We develop some of these latter themes in the following chapter on geopolitics.

Further reading

The study of imperialism and colonialism in Geography has been widespread and there are numerous books and

academic articles that provide an excellent introduction to the concepts of imperialism and colonialism, as well as an indication of how Geographers might approach these concepts. Up until relatively recently, the majority of work has been done under the sub-disciplinary badge of Historical Geography, although these studies are also patently concerned with understanding ideas and processes that are linked to Political Geography. Felix Driver's *Geography Militant* (Wiley-Blackwell, 2000) provides a rich account of the link between exploration and geographical knowledge in the context of the British Empire, while Blunt and McEwan's (2003) edited volume, *Postcolonial Geographies* (Continuum, 2003), contains a comprehensive series of chapters exploring various aspects of the imperial venture: knowledges and networks; urban order and spectacle; home and nation.

Much historical and political work has focused specifically on the attempts made by colonial regimes to govern spaces at different scales. Dan Clayton (2000) 'The creation of imperial space in the Pacific Northwest', *Journal of Historical Geography*, volume 26, pages 327–50, examines the attempt made to make imperial sense of North America while Stephen Legg (2006) 'Governmentality, congestion and calculation in colonial Delhi', *Social and Cultural Geography*, volume 7, pages 709–729, examines the significance ascribed to governing urban spaces within empires. Another interesting geographical debate centres on the tendency for imperialists to designate land as 'empty' spaces, ready to be conquered and colonised. This theme is discussed in an interesting case study of South Africa in Guelke and Guelke (2004) 'Imperial eyes on South Africa: reassessing travel narratives', *Journal of Historical Geography*, volume 30, pages 11–31.

Recent years have witnessed a mass of work that has examined more recent manifestations of imperialism, most notably in the context of the war on terror. This work is more explicitly political in its focus and is, in many ways, reflective of a renewed interest in geopolitical themes within Human Geography. Some of the key works in this area include David Harvey's (2003, Oxford) *The New Imperialism*, Derek Gregory's (2004, Blackwell) *The Colonial Present* and Neil Smith's (2005, Routledge) *The Endgame of Globalization*. The themes discussed in these books are taken up at greater length in the following chapter.

Web resources

A useful website, which provides access to a number of key theoretical texts on imperialism, texts on specific instances of imperial practice, as well as some interesting primary sources on imperialism, can be found at http://www.fordham.edu/halsall/mod/modsbook34.asp. Similar themes are covered by http://www.history.ac.uk/ihr/Focus/Empire/web.html. Specific websites focus on the imperialist practices of particular states. http://www.britishempire.co.uk/, as the name suggests, discusses different aspects of the British Empire. Two websites, which provide an account of the United States' imperialist tendencies, particularly during the late nineteenth and early twentieth centuries, can be found at http://www.smplanet.com/teaching/imperialism/ and http://www.besthistorysites.net/index.php/american-history/1900/early-imperialism. There are numerous websites that seek to use imperialism as a lens to understand more contemporary forms of domination and resistance. One of the more interesting and accessible is http://anti-imperialism.com/.

Political geographies of globalisation

Introduction

Globalisation is one of the most important geographical trends of the late twentieth and early twenty-first centuries. The definition, conceptualisation, history, workings, impacts and merits of globalisation may all be debated (see Box 9.1), but there can be little dispute that the society in which we live is more 'global' than it was one hundred, fifty, or even twenty years ago. Global trade has expanded such that more of the commodities we buy are sourced from increasingly distant locations, and more of the market is dominated by a smaller number of large transnational corporations whose operations span the globe. The movement of people around the world has intensified, facilitated by advances in transport but also by increased wealth and by political reforms, such that we are more likely to travel as tourists to different continents, or to consider working or retiring in a different country, but also that the populations of the towns and cities where we live are likely to have become more culturally diverse. Communications technologies, from television to the internet, mean that we find ourselves consuming the same cultural experiences as individuals in places on the other side of the world with very different cultural traditions, but also that we are better informed about what is going on in distant regions of the globe.

These trajectories have presented a challenge to the settled assumptions of political geography, in particular the centrality of nation-states. As discussed in Chapter 2, the power of the nation-state has conventionally been associated with its capacity to control its own territory and to seal its own borders. However, with globalisation borders have become increasingly permeable and the heightened mobility of people, commodities, capital and other things increasingly able to elude the control and regulation of the state. This is evident in examples such as irregular and illegal migration across national borders (for instance, between Mexico and the United States, or from Africa into Spain, Portugal, Italy and Malta across the Mediterranean Sea), the transnational reach of organised crime, drug trafficking and terrorism, or the capacity of a transnational corporation to switch on and off investment in a particular country. Yet, globalisation is not only manifested materially, it is also articulated in discourses that represent the global as a single entity and which promote the idea of a global economy, a global culture, a global society or a global village. Larner (1998), for instance, shows how a 'globalisation discourse' was employed to justify radical economic reforms in New Zealand by interpreting the economy in terms of its perceived readiness to compete in a 'taken-for-granted' global economy.

Globalisation, therefore, does not just happen. It is facilitated, assisted and steered by political policies and practices. In this way, the particular expression of globalisation that has become dominant in the early twenty-first century is associated with neoliberalism as an ideology that seeks to expand markets, limit state power and remove obstacles to capital accumulation (see Chapter 3). As such, some writers, including geographer Neil Smith, have positioned globalisation as a geopolitical strategy, as an attempt by the United States to impose its liberal, capitalist ideals on the world and to construct a new 'American empire' through the reach of transnational corporations rather than by direct colonisation (Smith 2005). Certainly, American diplomatic and military power has been used to promote neoliberalism and to advance globalisation, but Smith's argument can be critiqued on at least two grounds. First, it takes a very narrow view of globalisation as something that is neoliberal in character and emerged in the late twentieth century. From a broader perspective, processes that we identify with globalisation, such as the extension of trade networks, the mobility of individuals and the hybridisation of cultures, can traced back over many centuries (Moore and Lewis 2009), and have been driven at different times and in different contexts by diverse ideologies including imperialism, religion and socialism. Second, theses such as Smith's tend to over-state the top-down, directed nature of globalisation, and under-state its bottom-up, unplanned and incoherent dimensions.

Indeed, it is the incoherence of globalisation that is particularly interesting to political geographers because it allows for multiple sites and spaces of contestation. In contrast to assertions by popular writers such as Thomas Friedman that globalisation has erased geography to make the world 'flat' (Friedman 2005), or Paul Virilio's suggestion that geographical space has been replaced by time as the dimension across which society operates (Virilio 1999), empirical research has repeatedly demonstrated that processes and impacts of globalisation, and responses to them, are still shaped and differentiated by geography. What has changed is the way in which relations between places are enacted, and how we perceive them. Towns and cities that were once described as peripheral in national territories now sit at the centre of new cross-border regions; whilst businesses in provincial cities can connect directly with partners in other countries by-passing national institutions and structures. This has led geographers to variously re-imagine geographical relations in terms of networks, and to re-think concepts of place, scale and territory (see, for example, Amin 2002; Elden 2005; Massey 2005; Murray 2006).

In this chapter we will examine some of these emerging political geographies of globalisation. First, we will explore the politics of place in the context of globalisation, drawing on writers including Amin and Massey to propose that globalisation is grounded in and contested through particular places. Second, we will discuss and critique the concept of 'global governance' as both an aid to and a response to globalisation, as well as the related articulation and performance of 'global citizenship'. Finally, we investigate alternatives to neoliberal globalisation and the mobilisation of resistance to neoliberal globalisation by social movements that have their own transnational spatialities.

BOX 9.1 GLOBALISATION

Globalisation can be defined, following Steger (2003), as 'a multidimensional set of social processes that create, multiply, stretch and intensify worldwide social interdependencies and exchanges while at the same time fostering in people a growing awareness of deepening connections between the local and the distant' (p. 13). This is not the only definition, and other definitions vary in their emphasis, but Steger's approach is particularly helpful for thinking about the political geographies of globalisation because of the way in which he breaks down globalisation into four key components. First, he proposes, globalisation involves the *creation* of new networks and relations, and the *multiplication* of existing networks and relations, that overcome traditional geographical and political boundaries. An international terrorist network such as al-Qaeda is an example of this, but so is a transnational corporation. Second, globalisation involves the *expansion* and *stretching* of social relations, activities and interdependencies over increasing distances. Commodity chains, for example, have been stretched to source, process and sell food or manufactured goods around the world, but political institutions have also been stretched over space in moves towards global governance, as discussed later in this chapter. Third, globalisation involves the *intensification* and *acceleration* of social exchanges and activities, evidenced, for example, by the instantaneous communication possible via satellite or through the internet. Fourth, Steger notes that

> we must not forget that globalization also refers to people becoming increasingly conscious of growing manifestations of social interdependence and the enormous acceleration of social interactions. Their awareness of the receding importance of geographical boundaries and distances fosters a keen sense of becoming part of a global whole.
>
> (Steger 2003: 12)

As Steger observes, this emergent global consciousness has an effect on how people act in the world, as reflected in the notion of 'global citizenship' discussed in this chapter.

Social scientists do not only disagree about the definition of globalisation, but also on how to interpret and explain it, falling into three broad camps (Murray 2006). *Hyperglobalists* regard globalisation as a natural and unstoppable progression towards a borderless world with a single economic market, which they welcome as the triumph of capitalism and the market over the nation-state. In contrast, *sceptics* or *traditionalists* (Cochrane and Pain 2000), question whether globalisation has advanced as far as the hyperglobalists claim. They identify the emergence of new integrated regional blocs, but critique globalisation as a discourse employed to further the economic and political interests of powerful states. As such, sceptics argue that nation-states still have a role as potential sites of resistance to globalisation. The third group, *transformationalists*, take an intermediate position, recognising that processes of integration and interdependence are occurring with transformational effects, but that the outcomes of these processes are not pre-determined.

Acknowledging the unfinished condition of globalisation is important for understanding its politics. As Murray (2006) observes,

> a useful exercise in seeking a definition [of globalisation] is to consider what a completely globalized world would look like: single society, a homogenized global culture, a single global economy, and no nation states? That is clearly not what we have. We are therefore living in a *globalizing* world.
>
> (p. 16)

Steger's definition of globalisation describes a direction of travel not a final state of complete global integration and interconnection, which he refers to as 'globality' (p. 7). Furthermore, as the processes identified by Steger are not intrinsically tied to any particular political project, he suggests that it is possible to imagine numerous different forms of globality: a capitalist globality, a socialist globality, an Islamist globality and so on. As such, globalisation is necessarily a contested process.

Key readings: Steger (2003); Murray (2006).

Globalisation and the politics of place

There is a popular myth that globalisation acts like a steamroller – an overbearing, homogenising force that is imposed from above to squeeze out any local distinctiveness from a place, leaving behind a bland, flat terrain. In this formulation, the global is positioned as the opposite of the local, with the global associated with corporate power and cultural standardisation, whilst the local is characterised by tradition, authenticity and diversity. Globalisation is therefore represented as something that happens *to* the local. Massey (2005), however, notes that there are two problems with this perspective. First, it understands 'the global, implicitly, as always emanating from somewhere else. It is therefore unlocated; nowhere' (p. 101). Second, it presents local actors as being powerless to prevent the advance of globalisation, and local place as therefore 'inevitably the *victim* of globalization' (ibid.).

Massey challenges these assumptions by arguing that globalisation is not an all-embracing single movement that rolls out from some indefinite 'global' centre, but rather that it is 'a making of space(s), an active reconfiguration and meeting-up through practices and relations of a multitude of trajectories, and it is there that lies the politics' (Massey 2005: 83). In other words, globalisation is reproduced in places, through the interactions of global and local actors, and through the capture, disruption and rearrangement of the social and economic relations that constitute place in a relational perspective (discussed in Chapter 6) (see also Massey 2004; M. P. Smith 2001; Woods 2007). This means that local actors can influence the outcomes of globalisation processes (although the capacity to do so may be limited by structural factors, such as wealth and financial capital, geographical location, access to political and communication networks, and social capital); but it also means that globalisation is contested in places. Local political conflicts over the opening of new fast food outlets, coffee bars or supermarkets by transnational corporations, the closure of branch plants or take-over of local companies, new tourism developments or the sale of houses as second homes for international amenity migrants, the introduction of new genetically modified crops, the designation of new protected areas according to international conventions, and challenges to local cultural traditions, all have the potential to shape the local impacts of globalisation processes and, in their small way, resonate back to the overall progression of globalisation.

An example of these local politics of globalisation can be observed in the mountain resort of Queenstown in New Zealand, as documented by Woods (2011). During the 1990s, Queenstown developed from a largely domestic winter sport centre to an internationally renowned resort for year-round adventure tourism, attracting visitors from around the world. This transformation was facilitated by global trends, including faster and cheaper air travel, the liberalisation of controls on financial and property investments, the search for new opportunities by international travel firms, and the promotional role of the global media; but it was also driven by local developers and entrepreneurs who opened new tourist attractions and built new hotels, restaurants and shops. The resulting economic boom fuelled a rapid increase in the local population, from just under 10,000 in 1991 to nearly 23,000 in 2006, with the accompanying demand for housing met by new construction aided by local government policies. Furthermore, a significant proportion of the new housing was being sold to international amenity migrants, including affluent international business leaders and entertainers.

However, by 2000, a coalition of local environmentalists and wealthy in-migrants had mobilised to oppose new developments, expressing concern about the impact on the local landscape and rural character of the district, which had been the attraction for amenity migrants and which was represented as being globally unique. At the same time, continued development was defended by other local actors, including the mayor, who argued that it enabled local people to benefit from globalisation and prevented the district become the preserve of a global elite. As the mayor commented, 'We don't want to become the Aspen of the South Pacific. We . . . shouldn't become a community of millionaires and multi-millionaires' (quoted by Woods 2011: 378).

The case of Queenstown hence demonstrates that the politics of globalisation cannot be reduced to local versus global conflicts, but are entwined with the politics of place in complex ways. The particular circumstances of Queenstown may be exceptional, but similar dynamics of the interplay of local politics with global networks and processes are evident in many localities. Stahre (2004), for example, describes the negotiation of globalisation processes by local political actors in Stockholm, with elected politicians and urban social movements acting to resist certain pressures and to modify the outcomes of globalisation in the city.

Politics of propinquity and politics of connectivity

Further conceptual tools for analysing the relational politics of place in globalisation are provided by Amin (2004), who proposes the notions of a *politics of propinquity* and a *politics of connectivity*. The politics of propinquity refers to 'the intense everyday negotiation of diversity in most cities and regions, associated with the exposure to cultural, social, experiential and aspirational difference among those who share a given regional space' (p. 38). As Amin deliberately positions the politics of propinquity so it 'rules in everything that vies for attention in a given location' (p. 39), and includes 'struggles over roads and noise, public spaces, siting decisions, neighbourhoods and neighbours, housing developments, street life and so on' (ibid.), the politics of propinquity need not necessarily involve global actors, and may be broadly equated with conflicts over different discourses of place discussed in Chapter 6.

However, it is also evident that globalisation has intensified the politics of propinquity by bringing into close contact in specific places different cultures, value systems, social groups and economic interests that were previously spatially separated. Such tensions are perhaps most explicit when concerned with the effects of global migration in reshaping geographies of cultural proximity. Dunn (2005), for example, examines protests against the construction of mosques in Sydney as expressions of a xenophobic politics that contests the idea of multiculturalism and the co-existence of Muslim and white Anglo-Celtic communities in the city's neighbourhoods. In a different context, Hubbard (2005) discusses opposition to a centre for processing asylum seekers in rural England, revealing a racial politics that draws on crude stereotypes and misperceptions to present the asylum centre and its residents as 'out of place' in the English countryside.

Amin's second concept of a politics of connectivity more directly invokes globalisation in recognising that 'a politics of place, whether we like it or not, has to work with the varied geographies of relational connectivity and transitivity that make up public life and the local political realm in general in a city of,

region' (Amin 2004: 40). In other words, local politics cannot be isolated within a particular territory. For Amin, this has implications for the practice of regional politics and governance, suggesting for example that:

> In a relational politics of place, decisions concerning what is good or bad for the local economy would not be decoupled from scalar or territorial assumptions, which in the new regionalism routinely hold that 'local' autonomy is empowering while 'external' control is disabling, that local agglomeration increases local returns, while global commodity chains seep profits away, that homegrown institutions are locally oriented, whilst distant institutions are predatory or indifferent. Instead, judgement over economic worth would be based on public scrutiny of alternative models of economic prosperity and well-being (e.g. neoliberal versus social democratic) and competing visions of the economic good life, which are approved or not by residents on the basis of how well a vision fits with their interests that may well be locked into spatial connectivities beyond the region.
>
> (Amin 2004: 41)

This assertion reflects the normative component of relational politics for both Amin and Massey, serving not only as an analytical tool but also as a way of advancing a more progressive politics of place in response to globalisation, that rejects xenophobic nationalism and reactionary localism and embraces diversity and lines of global solidarity and responsibility – an agenda that we will return to later in this chapter (see also Featherstone *et al.* 2012; Massey 2004, 2005).

More analytically, the concept of politics of connectivity can be developed and deployed as a framework for examining how local political movements recognise the global context of their struggle and seek to follow and incorporate transnational connections into their campaigns. An example of this form of the politics of connectivity is a long-running campaign against the construction of a gas pipeline between the Corrib off-shore gas field and the Erris peninsula in north-west Ireland. Whilst at one level a very situated land use conflict involving environmental and health concerns, the dispute has assumed international significance as pipeline opponents in the 'Shell-to-Sea' campaign have pursued and enrolled connections across both time and space (Gilmartin 2009). As Gilmartin describes:

> The campaigners regularly draw comparisons between their situation and that of the Ogoni in Nigeria: so much so that a nickname for the protestors in Mayo is the Bogoni. A documentary on the Corrib Gas pipeline, called *Those Who Dance*, makes this connection explicit, intercutting footage from Ireland and Nigeria. Ogoni campaigners, such as Owens Saro-Wiwa, have been regular visitors to Ireland and to Mayo. In this way, Shell to Sea has become part of an international coalition of groups opposed to the activities of resource exploitation companies, particularly Shell. However, Shell to Sea has also targeted Statoil, and activists have travelled to Norway, met with politicians and trade union activists, and attempted to influence public opinion in that country. The story of the Corrib Gas pipeline has also attracted interest and attention from international environmental groups, such as Global Community Monitor and the Goldman Environmental Prize.
>
> (Gilmartin 2009: 277)

The mobilisation of a politics of connectivity in struggles such as the Corrib pipeline effectively collapses the global and the local as spheres of political action (Plate 9.1). This compression is also observed by Magnusson and Shaw in studies of forestry conflicts in British Columbia, Canada, and in particular two sites of struggle against industrial deforestation at Clayoquot Sound (Magnusson and Shaw 2003) and in the Great Bear Rainforest (Shaw *et al*. 2004). In both cases, coalitions of environmentalists and indigenous communities mobilised in opposition to transnational forestry companies, developed transnational solidarity networks and consumer boycotts, and staged protests in regional, national and international centres of political and economic power. Yet, Magnusson (2003) cautions against reading this as the globalisation of local campaigning, noting that

it is not really that the politics of Clayoquot (or other places) involves a movement from the local to the global, or even the other way around . . . . Rather, the politics of places such as Clayoquot puts traditional distinctions between local and global, small and large, domestic and international – and much else – into serious question.

(p. 1)

Global governance and global citizenship

Globalisation inherently involves a redistribution of power. To make this assertion is not necessarily the same as claiming that power is becoming increasingly concentrated as a result of globalisation, nor that power is being moved up a global scalar hierarchy – as discussed in the previous section, local actors still retain agency to affect the outcomes of globalisation processes and as such power within globalisation remains diffused. However, the multiplication, stretching and intensification of social and economic relations across the globe does mean that individuals, companies and communities are increasingly exposed to the consequences of decisions made in increasingly distant places, and that the ability of institutions such as national governments or city councils to exert authority within specified territories has been compromised.

Some commentators such as Held (1991) and Waters (2001) have accordingly argued that globalisation threatens the sovereignty of nation-states and is moving towards forms of global governance. Held (1991) identifies a series of steps in this trajectory, starting with transnational cultural and economic flows becoming increasingly difficult for nation-states to regulate, leading to the ceding of authority by nation-states to international and intergovernmental organisations, creating new supra-national territorial units that constitute an emerging geography of global governance, and which in turn provides the seeds for a world government with coercive and legislative power. Yet other researchers have questioned and critiqued the assumptions in this globalist model, pointing instead to a more complex interplay of power relations between actors at regional, national and

Plate 9.1   Making global connections in publicity for the Shell-to-Sea campaign, Ireland

Photograph: Michael Woods

supra-national scales within the global systems. In this section we consider the evidence for the emergence of 'global governance' and discuss its possible limitations.

The global economy and global governance

The rise of global governance is often associated with the challenges to the nation-state posed by economic globalisation. Historically, regulation of the national economy was a core function of the nation-state, with governments seeking to manage supply and demand, increase national wealth and improve the relative economic power of the country by controlling money supply, imports and exports; setting interest rates, taxes and price controls; regulating financial transactions, employment conditions and wages; and through direct interventions such as national ownership of key industries. The globalisation of economic relations has constrained the capacity of nation-states to act in this way. Intensified trade flows and stretched commodity chains mean that supply and demand cannot be effectively controlled at a national scale; whilst transnational corporations (the largest of which have a greater value than many nation-states) can elude coherent regulation by national governments and pick and choose between the regulatory environments of different countries in selecting favourable sites for investment (see Box 9.2).

BOX 9.2 POLITICAL GEOGRAPHIES OF TRANSNATIONAL CORPORATIONS

A notable feature of contemporary globalisation is the concentration of market power with a relatively small number of large transnational corporations (TNCs) with operations that span the globe and, in many cases, several economic sectors. The largest TNCs have a value greater than the economies of many countries, and TNCs present challenges to governance by all nation-states because of their capacity to move resources and activities between countries. For example, in 2012 several high-profile TNCs including Starbucks and Amazon were revealed to be paying little or no corporation tax in the UK because of the geography of their corporate structures. At the same time, TNCs are major employers and major investors, and as such are often courted by national governments and enjoy privileged access to policy makers. For instance, if retailer Wal-Mart were a country it would be the world's twentieth largest economy by GDP, and the *Los Angeles Times* has noted that 'its business is so vital to developing countries that some send emissaries to the corporate headquarters in Bentonville, Ark., almost as if Wal-Mart were a sovereign state' (quoted by Hugill 2006:3).

TNCs have their own internal networks of global governance. Branch plants and outlets – and the jobs and local suppliers that are tied to them – are dependent on decisions made in corporate headquarters that are disproportionately located in a handful of 'global cities'. The head offices of the 500 largest TNCs are concentrated in just 125 cities, and especially in New York, Tokyo, London and Paris (Dicken 2011). As the same corporations have increasingly relocated and out-sourced manufacturing production, customer services and other routine activities to low-wage economies in Asia, they are reproducing a quasi-colonial geography of power.

Furthermore, TNCs also exert power over their suppliers through 'commodity chains' or 'value chains' (Hughes and Reimer 2004; Murray 2006) by setting not only quality controls on goods and products, but also standards for employment conditions, health and safety, environmental protection and so on. In sectors such as agriculture and food, these private networks of governance have arguably become at least as important as regulations enforced by national governments (see Challies and Murray 2011; Konefal *et al.* 2005; and Neilson and Pritchard 2009 for examples).

Key readings: Challies and Murray (2011); Hughes and Reimer (2004).

The extent of interdependence in the global economy was graphically demonstrated by the financial crisis of 2008. Problems in the banking sectors of Britain and the United States spread through transnational corporations and a complex web of international financial relations triggering failures in the banking and financial systems of other countries. As the supply of credit became squeezed, the crisis escalated into wider economic sectors, pushing the global economy into recession. With national governments limited in their capacity to respond effectively, the leaders of twenty of the largest economies in the world – known as the G-20 – met in Washington DC in November 2008 to agree a coordinated Action Plan, including measures to reinvigorate national economies, to regulate global finance and to assist the poorest countries in overcoming the effects of the crisis. The Action Plan assigned tasks to international organisations including the International Monetary Fund (IMF), the World Bank, the United Nations Development Agency and a new Financial Stability Board, all of which grew in stature and influence as a result (N. Woods 2010).

In this way, institutions of global governance such as the IMF, World Bank, World Trade Organization and others (Box 9.3), have been presented as *responses* to globalisation. Yet, a more critical analysis reveals that they are also *drivers* of globalisation. Take, for example, the World Trade Organization (WTO). The WTO didn't invent global trade, and its predecessor body – The General Agreement on Tariffs and Trade (GATT) – was established to achieve greater stability and equality in international trade and avoid a repeat of the retaliatory trade wars of the 1930s (Narlikar 2005). However, in its principles and operation, the WTO has become an advocate of global trade liberalisation, with agreements reached through WTO summits facilitating the multiplication and intensification of trade flows, and its dispute settlement mechanisms empowered to force states to drop policies and practices judged to discriminate against international trade and competition. Similarly, conditions attached to assistance provided to countries by the IMF and World Bank have promoted neoliberal globalisation by requiring states to abolish controls on imports and currency exchanges,

BOX 9.3 NOTABLE GLOBAL GOVERNANCE INSTITUTIONS

United Nations (UN) – Established in 1945 as a forum to facilitate international cooperation in security, development and conflict resolution and to promote peace and human rights. The UN is based in New York and had 193 member states in 2013. Its main decision-making bodies are the General Assembly and the Security Council comprising five permanent members (China, France, Russia, UK and USA) and ten elected members. The UN also sponsors a number of subsidiary organisations operating in different areas of global governance, including the United Nations Development Program (UNDP), the Food and Agriculture Organization (FAO), the UN Educational, Scientific and Cultural Organization (UNESCO), the International Labour Office (ILO), the World Health Organization (WHO), the International Monetary Fund (IMF), the UN Childrens' Fund (UNICEF) and the International Court of Justice. Website: www.un.org

G-20 – A grouping of twenty major economies that collectively account for more than 80 per cent of world GDP and 80 per cent of world trade. Its origins are in the 'Group of Six' (G-6), comprising France, Germany, Italy, Japan, the United Kingdom and the United States, which first met in 1975 to discuss global economic policy. This became the 'Group of Seven' (G-7) with the addition of Canada in 1976, and the 'Group of Eight' (G-8) with the addition of Russia in 1997. The G-20 was formed in 1999 as an expanded forum to discuss governance of the international financial system, including rising economies such as Brazil, China, India and South Africa. At first the G-8 continued to be the most important grouping, and still meets annually, but since the economic crisis in 2008, the G-20 has formally become the primary forum for global economic discussion. The G-20 has no secretariat and functions through annual summits of heads of government and meetings of finance ministers and central bank governors. Its members are Argentina, Australia. Brazil, Canada, China, France, Germany, India, Indonesia, Italy, Japan, Mexico, Russia, Saudi Arabia, South Africa, South Korea, Turkey, United Kingdom, United States and the European Union. Website: www.g20.org

World Trade Organization (WTO) – Established in 1995 to replace the General Agreement on Tariffs and Trade (GATT; est. 1948), the WTO is responsible for regulating international trade. The WTO enforces agreements made at periodic ministerial conferences, including in Singapore (1996), Geneva (1998), Seattle (1999), Doha (2001), Cancún (2003), Hong Kong (2005) and Geneva (2009). These agreements are binding on WTO members and a Dispute Settlement Understanding permits the WTO to adjudicate on breaches of the agreement and trade disputes between member states. The WTO had 157 members in 2013. Website: www.wto.org

International Monetary Fund (IMF) – Established in 1945 following the Bretton Woods Conference to promote global economic stability and monetary cooperation. One of its main functions has been to provide loans to member states to help them meet balance of payments demands, but the IMF has been criticised for imposing radical neoliberal policies on countries as a condition of this assistance. The IMF had 188 members in 2013, with subscriptions and votes weighted by quotas that reflect relative economic power. The United States has the largest quota and the IMF is based in Washington DC, but the Managing Director is conventionally appointed from a European country. Website: www.imf.org

World Bank – The 'World Bank' is an umbrella term for two institutions, the International Bank for Reconstruction and Development (IBRD) and the International Development Agency (IDA). It was originally created in 1944 to fund the postwar reconstruction of Europe, but since the 1960s its emphasis has been on providing loans to developing countries for capital programmes, with the goal of reducing poverty. However, like the IMF it has been criticised for imposing neoliberal reforms as a condition of assistance. The IBRD had 188 members in 2013 and the IDA, 172 members. The World Bank is based in Washington DC and conventionally headed by an American. Website: www.worldbank.org

European Union – The European Union (EU) is the most advanced of several regional economic blocs to develop as part of globalisation. Initially established as the European Economic Community (EEC) with six members in 1957, it has expanded in size through the accession of new members (including the UK and Ireland in 1973, and eight former Communist states from Central and Eastern Europe in 2004), with Croatia becoming the twenty-eighth member state in July 2013. It has also deepened its activities from a free trade area through political and economic integration to include a single market (with standardisation of related business, social and environmental regulations), cooperation on justice and domestic security, and a common foreign policy. These developments were reflected in the adoption of the title 'European Union' in 1993. Not all member states are involved in all activities, however. A passport union introduced in 1995 by the Schengen Agreement does not include Britain and Ireland (but does include non-member states Norway and Iceland), and the single currency, the Euro, introduced in 2002, is used in only seventeen member states. The EU employs a combination of supra-national and intergovernmental decision-making. It has a permanent administration in the European Commission, along with other permanent institutions including the European Parliament. However, the primary decision-making body is the Council of Ministers comprising government ministers from the member states. Most decisions in the Council are made by Qualified Majority Voting, weighted by population size, but on some issues individual states can exercise a veto. Website: www.europa.eu

encourage foreign investment and generally adopt neoliberal economic policies (Peet 2003).

Limits to global governance?

Organisations such the IMF and WTO, as well as the United Nations, are highly visible symbols of attempts to move towards global governance, but we are still a long way from a world government. Assertions about the significance of global governance should be qualified on at least four grounds. First, relatively few institutions of international governance operate on a truly global scale and the most notable feature to date of political globalisation has been the creation of regional economic blocs such as the European Union, the North American Free Trade Agreement (NAFTA) and Mercosur in South America (Figure 9.1). If globalisation is understood as a direction of travel, then these regional groupings are key drivers of globalisation, at least with respect to international trade; yet, they are also justified politically as responses to globalisation. The European Union, for instance, has been the primary agent for the multiplication, intensification and stretching of social and economic relations in Europe in the postwar period, through the creation of a single market, the facilitation of free movement by citizens and the standardisation of regulations concerning many aspects of economic, employment and environmental activity, as well as the introduction of a single currency for seventeen of its member states. At the same time, supporters of European integration argue that it is a necessary response to globalisation, required to counter the power of the United States, China and Japan in the global economy (McCormick 2008).

Second, nation-states remain integral to the operation of global governance. International organisations tend to have either an *inter-governmental* or a *supra-national* model of decision-making: in an inter-governmental model decisions are made by member nation-states through multi-lateral negotiations, whilst in a supra-national model authority is delegated to an executive body that can make decisions on behalf of members. Most international organisations, such as the G-20 or the WTO, follow an inter-governmental model, with key decisions made at periodic summits. The European Union operates through a mixture of inter-governmental and supra-national decision-making, with some decisions delegated to the European Commission and the European Parliament, but major decisions made at inter-governmental summits or by national government ministers in the Council of Ministers. As such, nation-states decide the policies of international organisations, although the influence of individual states may not be equal: votes in the IMF, for example, are weighted relative to the subscription paid by the state (Peet 2003), whilst the five permanent members of the United Nations Security Council (China, France, Russia, UK and USA) have a right of veto over resolutions.

Third, nation-states still retain considerable authority over many areas of policy, and are still the political institution that most influences the everyday life of citizens. Indeed, it has been argued that the 2008 global financial crisis re-asserted the importance of the nation-state, as national governments stepped in to rescue troubled companies such as the Royal Bank of Scotland in Britain and General Motors in the USA (Dicken 2011). Furthermore, the liberalisation of national controls in some respects has been balanced by a tightening of regulation in others. For instance, international travel for citizens of many countries has become easier as passport unions have been established (as in continental Europe) and visa requirements lifted; but for citizens of certain countries, notably those in the global south, mobility has been constrained by a strengthening of visa restrictions, especially for visiting Europe and the United States (Neumayer 2006).

Fourth, globalisation is associated not only with a redistribution of power from nation-states to international organisations, but also a redistribution of power to particular sub-national regions and 'global cities'. This occurs because globalisation and processes and practices are located in specific places, as discussed in the previous section. Economic globalisation, for example, involves the multiplication and intensification of global economic networks and relations, but these are disproportionately anchored and articulated in global cities such as London, New York and Tokyo. Financial decision-making has hence become increasingly concentrated in these places, and local actors in these cites – banks, financial institutions, local regulators and municipal authorities – are able to exert

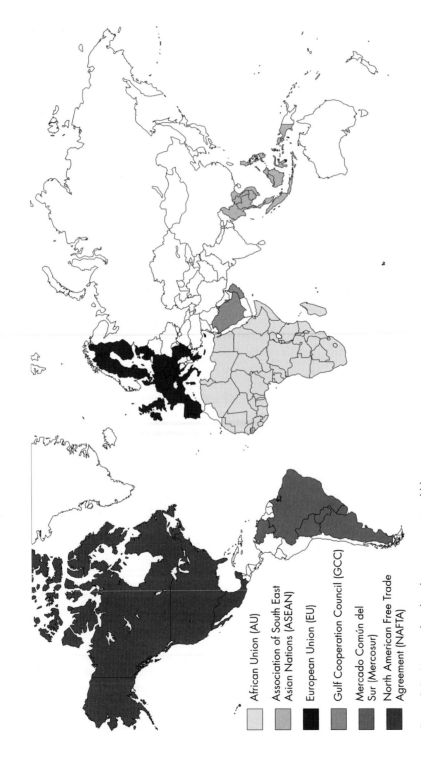

African Union (AU)

Association of South East
Asian Nations (ASEAN)

European Union (EU)

Gulf Cooperation Council (GCC)

Mercado Común del
Sur (Mercosur)

North American Free Trade
Agreement (NAFTA)

Figure 9.1 Major political and economic regional blocs

a disproportionate influence over globalisation outcomes (Taylor 2004). The offices of international political organisations such as the United Nations and the European Union are similarly concentrated in certain cities (notably Brussels, New York, Rome and Washington DC), thus locating power within political globalisation (Taylor 2005). Moreover, global cities are increasingly by-passing nation-states in forming networks and negotiating partnerships and alliances between themselves (Archer 2012; Taylor 2005), whilst aspirations to enhance a city's position within global networks can be a driver for urban politics (Paul 2005).

Practices of global citizenship

The development of global governance institutions and networks has been accompanied by the concurrent evolution of a concept of 'global citizenship'. Global citizenship does not refer to a legal status, as national citizenship does (see Chapter 4), but rather encapsulates a way of thinking about the relationship between individuals and global society that emphasises ideas of universal justice and shared obligations. Global citizenship is a product of the growth of global consciousness promoted by travel, migration and the global media (Steger 2003), from which has emerged a cosmopolitanism that holds that

> the well-being of faraway strangers should be no less of a concern than that of our immediate neighbours, for we are all, as free and equal individuals, participants in a universal political culture that entitles us to the same rights and protections regardless of our specific circumstances or identities.
> (Kurasawa 2004: 236–7)

Global citizenship converts this ethos into political practice in two ways.

First, global citizenship is associated with the globalisation of values – the replacement of value systems embedded in different religions and local cultural traditions with a universal discourse of common rights and ethical standards. The idea of universal human rights is the prime example of this, but the trend also extends to questions of animal welfare and

environmental management. The United Nations has a been a key agent in the globalisation of values, especially with respect to human rights, environmental awareness and the rights of women and children, but the inter-governmental structure of the UN can limit the scope and implementation of resolutions. As such, transnational non-governmental organisations (NGOs) such as Amnesty International, Oxfam, Greenpeace and the World Wide Fund for Nature, play a critical role in promoting universal values, lobbying governments to agree to international treaties and conventions, and enforcing global standards through campaigning and protests (Arts 2004; Desforges 2004). As Arts (2004), describes, NGOs achieve this influence by jumping and collapsing scales, bringing local issues into international negotiations and using global values to challenge local practices, thus enacting the politics of connectivity discussed earlier in this chapter.

Indeed, the globalisation of values is frequently contested, with perceived threats to local cultural traditions and identities (as well as vested political and economic interests) met by fierce opposition. China's resistance to what it perceives to be the imposition of Western liberal values on human rights and democracy is a source of geopolitical tension; whilst conflicts have been sparked by attempts to enforce global environmental values in banning or restricting established local traditions such as seal-hunting in the Arctic (Dauvergne and Neville 2011).

Second, global citizenship is articulated through the actions of individuals as an expression of responsibility towards the world and its peoples. At a relatively passive level, this might include donating to charities, or to disaster relief or aid appeals, or wearing a wristband to support a campaign such as Make Poverty History (Desforges 2004). A little more actively, the practice of global citizenship might involve making ethical consumption choices, such as buying Fairtrade goods, changing environmental behaviour as a response to concerns about global warming or becoming involved with a transition town movement (Mason and Whitehead 2012) or local actions to combat racism or welcome asylum seekers (Darling 2010) – in other words, performing global citizenship within place. Other activists make a more direct connection with local struggles in other parts of the world by joining support networks, with Davies (2012), for example, describing the role of Tibet Support Groups in extending Tibetan resistance to Chinese control into the international arena; whilst global citizenship has also been associated with international volunteering to work in aid, education or conservation projects, especially in developing countries (Baillie Smith and Laurie 2011; Lorimer 2010). As practised through this range of incarnations, the politics of global citizenship are ambiguous. On the one hand, global citizenship can involve participation in campaigns that challenge tenets of neoliberal globalisation such as free trade; yet, on the other hand, by suggesting that issues of global injustice can be tackled through individual volunteering and consumption choices, global citizenship can be critiqued for diverting attention from more radical political transformations, and as Baillie Smith and Laurie (2011) argue, contributing to the neoliberal redistribution of political responsibility from the state to 'active citizens' (see Chapter 4).

Spatialities of alternative globalisations

The ideology of neoliberalism is the dominant political force driving contemporary globalisation, but the neoliberal vision of a single global marketplace is not the only possible outcome of globalisation. As Steger (2003) observes,

> we could easily imagine different social manifestations of globality: one might be based primarily on values of individualism and competition, as well as on an economic system of private property; while another might embody more communal and cooperative social arrangements, including less capitalistic economic relations.
>
> (p. 8)

Accordingly, we should conceive of globalisation as being contested not so much in binary terms – being for or against – but in terms of the power dynamics embedded in its processes, the social and political consequences of its effects, and the type of global society that might eventually result.

Although there are nationalist movements that seek to reverse or block globalisation by reintroducing protectionist trade policies and strengthening immigration controls (see Box 9.4), many of the social movements that get popularly described as the 'anti-globalisation movement' might be more accurately described as an anti-capitalist, alternative globalisation, counter-globalisation or alter-globalisation movement. As captured in the slogan of the World Social Forum – 'Another World is Possible' – this movement embraces cosmopolitanism and the principles of global solidarity or citizenship, but rejects the injustice and subordination of local cultures and economies that it identifies with neoliberalism:

> Its worldview is deeply engrained in the right to cultural difference and the idea that strength lies in diversity, principles that are strategically useful as rhetorical antidotes to the generic culture spawned by global neoliberalism – a culture that would flatten out variations among peoples in the name of cultivating non-descript consumers for the planet's shopping malls and docile labour for its workplaces.
>
> (Kurasawa 2004: 240)

Moreover, the alternative globalisation movement is itself a global movement that transcends space, jumps scales, and makes use of global communications technologies, media and transport networks to organise its activities, propagate its message and mount protests. In this final section of the chapter, we examine the geographies of the alternative globalisation movement, its use of space and its embeddedness in place.

Spaces of the alternative globalisation movement

The alternative globalisation movement is in practice a loose constellation of many different groups, rooted in different local situations, with diverse interests and concerns spanning across environmental issues, land

BOX 9.4   ANTI-GLOBALISATION AND RIGHT-WING POLITICS

Opposition to globalisation and fear of its impacts have been motivating factors in support for populist right-wing movements (such as the Lijst Pim Fortuyn in the Netherlands, Vlaams Bloc in Belgium, One Nation in Australia and the Tea Party movement in the United States), as well as more extreme right-wing parties (such as the Front National in France or the British National Party). Such movements commonly draw on fear of cosmopolitanism and the dilution of national cultures and identity through migration, calling for tighter controls on immigration and in extreme cases the repatriation of non-nationals. In some cases this is explicitly framed as a rejection of the globalisation of values – the Lijst Pim Fortuyn and more recently the Freedom Party in the Netherlands have gained support by raising fears of the 'threat' to Dutch liberal values from Islam. However, support for right-wing parties can also be motivated by economic concerns, including competition for jobs from immigrants, but also perceived threats from free trade, transnational corporations and regulations imposed by supra-national bodies. Accordingly, right-wing movements commonly call for protectionist trade policies, controls on foreign ownership of companies, and withdrawal from supra-national organisations such as the European Union.

As such, support for right-wing parties tends to rise in periods of economic uncertainty, as evidenced by the emergence of the neo-Nazi Golden Dawn party in Greece with its anti-EU platform. The connection is also apparent in the geography of voting for right-wing parties, with support often drawn from working-class communities and districts with increasing immigrant populations (for example, the election of British National Party councillors in parts of East London). However, fear as much as experience can be a motivation for support for right-wing movements. Schuermans and De Maesschalck (2010), for example, describe how the far-right Vlaams Belang party in Belgium has drawn support from rural communities by stoking fears that immigration might spread crime from cities into the predominantly white countryside. Antonsich and Jones (2010), though, propose from analysis of voting in a referendum to ban the construction of minarets in Switzerland that patterns of prejudice and anti-cosmopolitan opinion cannot be reduced to simple binaries of religion, wealth or urbanity, but need to be investigated in the context of complex and social and geographical conditions.

Key readings: Antonsich and Jones (2010); Schuermans and De Maesschalck (2010).

rights, labour relations, indigenous cultures, poverty, corporate practices and others. As Routledge and Cumbers (2009) note, 'whilst the constituent parts of this "movement" can agree on what they are against – a rapacious form of development driven by an unregulated market ethos that has come to be known as "neoliberalism" – there are many fissures and fault-lines that divide it' (p. 2). Yet, in spite of these tensions, the alternative globalisation movement has managed over the period since the turn of the century to project an appearance of cohesion and to consolidate a presence on the international political stage. This achievement has been geographical as well as political, tapping into

the politics of connectivity to make connections between different localised struggles (Featherstone 2008), adopting organisational forms that have enabled mobilisation across non-contiguous spaces (Cumbers *et al.* 2008; Funke 2012; Routledge and Cumbers 2009), and occupying key sites of geopolitical significance, such as the margins of international summits (Lessard-Lachance and Norcliffe 2013; Wong and Wainwright 2009; Zajko and Béland 2008).

The political and geographical coalescence of the alternative globalisation movement has been facilitated by the use of what Routledge (2003) labels 'convergence spaces'. Convergence spaces comprise diverse social movements articulated around a shared vision that generates 'sufficient common ground to generate a politics of solidarity' (p. 345), and as such act to facilitate multi-scalar and translocal political action by participant movements. For example, People's Global Action (PGA) has been one of the key agents in the alternative globalisation movement, created as a network of ten social movements in 1998 and later reorganising into regional (continental) networks as its membership expanded. Participants in PGA converge around a common vision that includes the 'rejection of capitalism, imperialism and feudalism; and all trade agreements, institutions and governments that promote destructive globalization' (PGA website, quoted by Routledge 2003: 338), as well as resistance to domination and discrimination, and commitments to a confrontational attitude, direct action, civil disobedience and a decentralised organisation. As a convergence space, the PGA network facilitates communication, information-sharing, the expression of solidarity through demonstrations and letter-writing campaigns, the coordination of conferences, meetings and collective protests, and the mobilisation of resources of skills, people and money (Routledge and Cumbers 2009).

One of the hallmark techniques employed by PGA is the 'inter-continental caravan', which deliberately distorts spatial relations to take people marginalised by neoliberal globalisation to centres of economic power. The first caravan, in 1999, brought 400 Indian farmers to Europe to protest against trade liberalisation and the imposition of genetically modified crops, undertaking a series of demonstrations at key financial centres (including the City of London, the WTO offices in Geneva and the G-8 Summit in Cologne) and sites involved in GM crop development (including corporate offices in Amsterdam and Leverkusen, and rural GM test sites) (Featherstone 2003).

However, convergence spaces also feature uneven processes of facilitation and interaction and comprise contested social relations (Routledge 2003). Participants may temporarily coalesce around shared interests, but because they have differing core concerns (or 'militant particularisms'), tensions can and do arise in networks such as PGA, as do unequal power relations:

The place-specific hopes and dreams of marginalized folk are not necessarily realized in the collective visions of a network. For, while the PGA hallmarks reject all forms of domination, some continue within the convergence. For example, unequal gender relations persist within several of the peasant movements that participate in PGA. The leadership of such movements continues to be dominated by men, and patriarchal attitudes and actions – particularisms that masquerade as universal – persist within the functioning and organization of the movements.

(Routledge 2003: 344)

Moreover, different participant movements have differential access to financial and other resources, are of differing size, and operate in differing political and economic contexts, and consequently some are better able to shape the direction and strategy of networks than others. As such, the dynamics of mobilisation of the alternative globalisation movement cannot be separated from the places in which it is grounded.

Places of alternative globalisation

The alternative globalisation movement relies on organisation strategies that enable it to transcend space, but it is also grounded in particular places. The places from which alternative globalisation groups emerge shape the character, aspirations and trajectories of the movement, reflecting the cultural, political and economic contexts in which grievances are first articulated and the resources available to local activists (see also Chapter 7). As well as influencing the capacity of local campaigns to jump scales and connect their concerns to wider struggles, the context of place is translated into cultural representations of the local struggle, described by Routledge (2000) as the 'geopoetics' of resistance, which can '"travel" within and between civil and political societies, and potentially relay messages to a variety of audiences (including movement supporters and opponents), across a variety of scales (from the local and the global)' (Routledge and Cumbers 2009: 83–84). For example, the Zapatista peasant insurgency in Chiapas, Mexico, against land reforms imposed following the creation of NAFTA in 1994 is deeply embedded in the cultural and political heritage of the region, but the struggle has been mythologised and transported into geopoetic representations that have travelled around the world, both initiating Zapatista support groups and inspiring the wider alternative globalisation movement (Plate 9.2).

Similarly, the hosting of the first World Social Forum in Porto Alegre in Brazil in 2001 was not accidental, but reflected the catalytic role of Brazilian social movements, and especially the Movimento dos Trabalhadores Rurais Sem Terra (MST, or Landless Workers Movement), in convening the meeting as well as the cultural and political heritage of the city:

Porto Alegre was chosen for the first meeting, as a symbol of a more democratic alternative politics, through its much celebrated participatory budgeting process, which has been imitated by radical local authorities and activists elsewhere. It is also a city with a strong radical tradition as a centre of left resistance during the long years of Brazil's military rule, based upon vibrant neighbourhood associations that had strong links to the Workers Party . . . . As such it became an 'articulated moment' in the enactment of the WSF, where opposition to neoliberalism as well as alternative visions could be voiced.

(Routledge and Cumbers 2009: 178)

Plate 9.2 Globalising the narrative of the Zapatista struggle, on a Norwegian film poster

Photograph: Michael Woods

The WSF has formed a critical convergence space for the alternative globalisation movement, bringing together activists from all continents around a common programme of opposition to neoliberalism, but the location of the first WSF permitted Latin American participants to exert unequal influence over the vision and principles of the movement, drawing on their own context-specific heritage of interaction with Catholic liberation theology, anti-militarism, indigenous rights and trades unionism. As Routledge and Cumbers (2009) observe, 'even the "South" perspective remains predominantly Latin American and Brazilian in character, with African and Asian influence only emerging when the social forum process takes place in those continents' (pp. 181–182).

A further example of geographical context shaping alternative globalisation mobilisation can be found in the protests of radical farmers in Larzac, southern France, led by José Bové. Reacting to a trade dispute between the European Union and the United States, the farmers embraced the politics of connectivity by protesting about their specific concerns about tariffs on exports of Roquefort cheese by dismantling a branch of McDonald's as a symbol of globalisation, and calling alternative globalisation activists from around the world as witnesses at their subsequent trial. The trial also attracted a gathering of alternative globalisation supporters, which was followed-up with a social forum meeting (Larzac 2003). However, as Williams (2008) documents, the emergence of the Larzac as 'a focal point in the alter-globalization movement' (p. 52) was conditioned by its history and geography. Bové and several of his co-activists were not conventional farmers but seasoned campaigners who had come to the Larzac to protest against a proposed military base in the 1970s and had settled and bought smallholdings. The presence of this group together with the marginal geographical location of the Larzac plateau cultivated a politically infused agrarian society, considered as

> a place where people live their activism in daily life: in the communes that existed until recently; in the quest for simplicity and autonomy from consumer society; in the attempt to cut ties of dependency on

> Roquefort; in the de-intensification of farming and the rise of organics; in farmers' markets and direct selling. Everyday life, for many, is an expression of political commitment and awareness.
>
> (Williams 2008: 149)

Through the physical manifestation of convergence spaces in places such as Porto Alegre and the Larzac, sites of regional social forums in Mali, Pakistan and Venezuela, and the Zapatista-organised 'encounter' in La Realidad, Mexico, the alternative globalisation movement has constructed a counter-geography of globalisation that highlights places of subordination and contestation (Figure 9.2). Yet, the alternative globalisation movement has also been articulated through protests that have subverted the sites and spaces of international summits. The 1999 WTO meeting was held in Seattle at the invitation of Microsoft chief Bill Gates, but its location in a city with a vibrant counter-culture and history of radical protest, in a region with a high level of environmental consciousness, and easily accessible by air travel, allowed its agenda to be challenged and subverted by over 50,000 protesters (Thomas 2000). Subsequent international summits have tended to be held in more remote geographical locations, where surrounding space can be ordered and controlled to separate and police protesters (Lessard-Lachance and Norcliffe 2013; Zajko and Béland 2008).

At the same time, the alternative globalisation movement has also progressed by reclaiming spaces for a different purpose, not as sites of protest but as intentional communities designed to demonstrate alternative forms of social and economic organisation. This has been a core strategy of the MST in Brazil, which since the 1980s has occupied over 9,000 sites of 'unproductive' land and established agrarian based colonies on communal principles (Garmany 2008). Other examples include communes and low-impact communities in Western countries (Meijering et al. 2007). In rejecting neoliberal globalism, these places are also arguably products of globalisation, and also frequently present a challenge to nation-state authority by dissenting from prevailing political-economic discourse and sometimes operating outside planning and property laws.

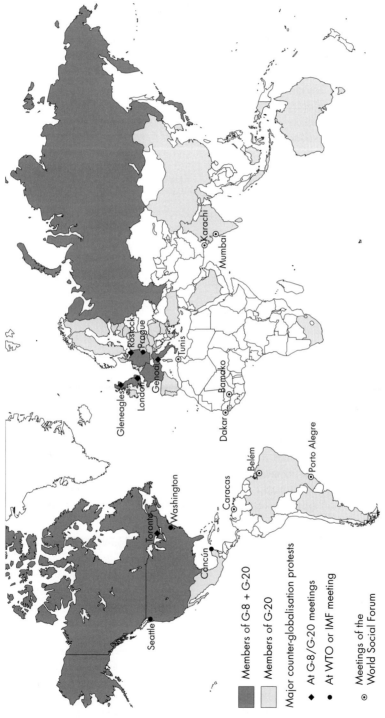

Figure 9.2 The geography of the counter-globalization movement

Members of G-8 + G-20

Members of G-20

Major counter-globalisation protests

♦ At G-8/G-20 meetings

● At WTO or IMF meeting

⊙ Meetings of the
   World Social Forum

Seattle

Toronto

Washington

Cancún

Caracas

Belém

Porto Alegre

Gleneagles

London

Genoa

Rostock

Prague

Tunis

Dakar

Bamako

Karachi

Mumbai

Summary

This chapter has explored the complex political geographies of globalisation. By adopting a relational perspective, we show that globalisation is not a monolithic, coherent force of domination, but rather a complicated array of diverse and often contradictory processes, trajectories and discourses. Globalisation is therefore not imposed from above, but is reproduced through place and as such local actors can have a degree of agency to influence the outcomes of globalisation processes – albeit conditioned and constrained by structural factors. Consequently, globalisation is routinely contested through local conflicts. Some of these reflect what Ash Amin describes as the 'politics of propinquity', the tensions arising from the mixing of different cultural groups or economic interests in a locality, or the competing claims made on space from 'global' and 'local' discourses. Furthermore, local struggles against globalisation processes have the capacity to 'jump scales' in order to take on transnational actors and enrol transnational support, embracing the 'politics of connectivity'.

The identification of lines of connection and shared interest between local social movements has facilitated the mobilisation of an alternative globalisation movement that has challenged the primacy of the neoliberal model of globalisation, and asserted that 'another world is possible'. Networks such as People's Global Action and the World Social Forum have formed 'convergence spaces' in which diverse social movements have formed temporary alliances around a common vision and coordinated action against neoliberal globalisation, including protests at major international summits. At the same time, the alternative globalisation movement is embedded in particular places, with the concerns, principles and resources of participating social movements informed by the geographical context in which they have been formed.

The interests of neoliberal globalisation, meanwhile, have been advanced through international agreements and conventions, and the formation of international organisations such as the World Trade Organization and the European Union. The WTO, in particular, has played a key role in promoting the liberalisation of global trade, constraining the autonomy of nation-states.

Yet, claims that we are moving towards a system of global governance should be challenged and critiqued. Not only do nation-states continue to exert considerable influence, including through inter-governmental decision-making processes, but as transnational institutions and power centres are all located in actual places, the apparent redistribution of power to the global scale may be more accurately described as a concentration of power in key 'global cities'.

Further reading

Manfred Steger's book, *Globalization: A Very Short Introduction* (2003), is a quick and accessible overview of key processes and interpretations of globalization, and Warwick Murray's *Geographies of Globalization* (2006) provides a fuller introduction to a geographical analysis of globalisation. The relational perspective on globalisation is developed by Doreen Massey in *For Space* (2005), and by Ash Amin's paper, 'Regions unbound: towards a new politics of place', *Geografiska Annaler B*, 86, 33–44 (2004)[move], which introduces the concepts of 'politics of propinquity' and 'politics of connectivity'.

The role of transnational institutions in promoting neoliberal globalism is critiqued by Richard Peet (2003) in *Unholy Trinity: The IMF, World Bank and WTO*. The concept of global citizenship and its performance in international NGOs and volunteering is discussed by Baille Smith and Laurie (2011) 'International volunteering and development: global citizenship and neoliberal professionalization today', *Transactions of the Institute of British Geographers*, 36: 545–559; Desforges (2004) 'The formation of global citizenship: international non-governmental organizations in Britain', *Political Geography*, 23: 549–569; and Lorimer (2010) 'International conservation volunteering and the geographies of global environmental citizenship', *Political Geography*, 29: 311–322.

The geographies of the alternative globalisation movement are examined in a series of publications by Paul Routledge, Andy Cumbers and colleagues, including Routledge and Cumbers (2009) *Global Justice Networks: Geographies of Transnational Solidarity*; Cumbers, Routledge and Nativel

(2008) 'The entangled geographies of global justice networks', *Progress in Human Geography*, 32: 182–201; and Routledge (2003) 'Convergence space: process geographies of grassroots globalization networks', *Transactions of the Institute of British Geographers*, 28: 333–349.

Web resources

Websites for major global governance institutions are provided in Box 9.3. The People's Global Action website can be found at www.nadir.org/nadir/initiativ/agp/en and the World Social Forum website at www.forumsocialmundial.org.br . Reports from the 1999 Intercontinental Caravan are archived at www.all4all.org/2004/08/1057.shtml. An extensive online archive for the Clayoquot Sound conflict is maintained by the University of Victoria at http://web.uvic.ca/clayoquot/home.html whilst the Rossport pipeline campaign's website can be found at www.shelltosea.com.

Political geography and the environment

Introduction: what are environmental politics anyway?

Can I (Mark Whitehead) let you into a little secret? This is, perhaps, not the best secret for someone writing a chapter on the form and nature of environmental politics to admit, but I must confess to finding it very difficult to effectively define and/or discern where environmental politics begins and other manifestations of political life end. What I can say with some certainty is that this definitional difficulty I experience is not a product of a lack of effort. I teach lecture courses, write newspaper articles, convene seminars, and regularly blog and micro-blog on environmental politics. But the more I think about, talk on and practise what I believe to be environmental politics, this elusive phenomenon (or more accurately collection of phenomena) appears to slip my grasp.

My problem does not derive from the fact there are no clear definitions of what environmental politics is all about. For many, environmental politics is synonymous with environmentalism: that collection of interconnected political movements that first started to emerge in the eighteenth and nineteenth centuries. While there are various shades and strengths of environmentalism (ranging from *strong* to *weak*, and *dark green* to *light green*), what appears to unite this movement is a desire to ensure that environmental considerations are a part of political debate, and a factor within related forms of decision-making (see Pepper 1984, 1996; Dobson 1995). On these terms environmental politics is often reduced to the political defence and protection of nature. Understanding environmental politics on these terms is helpful to the extent that

it enables it to be situated alongside, and differentiated from, other political movements such as liberalism (which is animated by the defence of personal freedoms), socialism (which is inspired by the needs of the working classes) and feminism (which takes up the struggle for equal rights for women). The defence of Nature, in essence, becomes the touchstone upon which environmental politics can be identified and assessed. But this association between environmental politics and Nature is also unhelpful, and problematic, on a number of fronts (Castree 2012). At one level, this perspective tends to assume that there is a pristine world out there, which is somehow separate from human activity, and can thus be defended (N. Smith 1984). But this idea of *first* (unspoilt) nature is increasingly being challenged by scientific studies that suggest there are now actually very few places or environmental processes that have not, in some way or another, been affected by human action (see Crutzen 2002; McKibben 1990; Ruddiman 2001, 2005). At another level, an environmental politics that focuses on the rights and abuses of nature tends to disconnect it from a range of important environmental issues that affect people's everyday life. Gottleib thus observes that:

> If environmentalism is seen as rooted primarily or exclusively in the struggle to reserve or manage extra-urban Nature, it becomes difficult to connect the changes in material life after World War II—the rise of petrochemicals, the dawning of the nuclear age, the tendencies to overproduction and mass consumption—with the rise of new social movements focused on quality of life issues.
>
> (Gottleib 1990: 7)

The desire to connect environmental politics to the rise of the petrochemical economy, nuclear technology, overproduction and mass consumption, urban air pollution, hazardous building materials, and human exposure to toxic waste, inter alia, has given rise to what is often referred to as a the *New Environmentalism*. The New Environmentalism started to emerge during the 1960s and 1970s, and sought to connect environmental politics to a range of interconnected social issues. Eco-socialists, for example, exposed the environmental problems associated with the workplace; eco-feminists considered the processes that connected gender exploitation and environmental chauvinism (Plumbwood 1993); and the Environmental Justice Movement exposed the racial injustices associated with environmental pollution (Bullard 1990). The New Environmentalism has always, to some extent at least, been caught between a desire to extend the parameters of what environmental politics should include, and the problems of constructing a political movement that is indistinguishable from any other (Whitehead 2005). Many have been critical of newer manifestations of environmentalism. They claim that the New Environmentalism has seen early calls for the protection of nature, and more recent attempts to develop a radical realignment of both social and environmental policies, being replaced by a conformist environmental politics, which is ultimately compliant with the broader needs of a capitalist, industrial economy (see Bernstein 2000).

This chapter charts a course through the complex terrains of environmental politics. While it does not provide a definitive definition of what environmental politics is, or does, it offers an insight into the transformative impact that environmental considerations have on some of our longest held political assumptions. The first section of this chapter considers the connections between the availability of environmental resources, political conflict and the organisation of political communities. The second section introduces work on Political Ecology and state theory that has sought to move beyond environmentally determinist accounts of politics in order to understand how systems of political and economic power shape human relations with the environment. The third, and final, section of

this chapter considers the way in which emerging environmental concerns have challenged the spatial, temporal and human parameters associated with citizenship and membership of political community. While exploring multiple perspectives on the mixing of politics and the environment, this chapter draws particular attention to the implications of environmental concerns for the study and practice of political geography.

'Arenas of scarce resource': environmental politics in the shadow of Malthus

Politics is often referred to as the 'arena of scarce of resources'. This understanding reflects the fact that the practice of politics is intimately connected to the ways in which limited sets of resources (both material and logistical) are distributed within a population. But this connection between politics and resources can be understood in more literal terms. Many have argued that the availability and distribution of resources have played a central role in the formation and reformation of nation-states; the military conflicts that are waged between and within states; and in the very constitution of political society itself (see Whitehead *et al.* 2007). The work of those who explore the connections between political conflict and resource availability is often referred to as Malthusian (or neo-Malthusian). This eponymous school of thought takes its name from the eighteenth- and nineteenth-century writings of the English clergyman Thomas Malthus. In his famous *Essay on the Principles of Population*, Malthus argued that as national populations increased in size it would become increasingly difficult to provide sufficient resources (whether that be food or fuel) to meet their demands. As resource scarcity grew, Malthus argued that societies would be plunged into a state that combined hunger, misery and warfare.

While human technological development appeared to have warded off the most dire of Malthus's predications, so called neo-Malthusians have suggested that many political struggles and armed conflicts occurring in the contemporary world are connected to questions of resource scarcity (Homer-Dixon 1999;

BOX 10.1 MEMPHIS AND THE BIRTH OF A NEW ENVIRONMENTAL POLITICS

In many ways, the story of New Environmental Politics begins in the American City of Memphis, Tennessee on 18 March 1968. It was on this day the American civil rights activist Martin Luther King visited the city in order to deliver an address to a rally. Dr King was in Memphis in order to support strike action being undertaken by black sanitation workers in the city (Figure 10.1). The sanitation workers' strike action was initiated in order to draw attention to the unsafe working environments that they were being subjected to (two sanitation workers had lost their lives while at work in Memphis in the February of 1969). On 28 March Martin Luther King led a march through the streets on Memphis in support of the black sanitation workers. Much to Dr King's dismay this march turned violent, with one person being killed and sixty injured. Although Memphis in 1968 may seem like an unlikely place to begin our discussion of environmental politics, in many ways it represents an historical significant juncture between politics and the environment. Many have argued that the sanitation workers' strikes and protests mark the beginning of a new Environmental Justice Movement. This movement suggested that the struggle for civil rights was about more than desegregation and legal equality; it was also a struggle for safe living and working environments and access to environmental amenities (Bullard 1990). But the events of Memphis reflect more than merely an environmentally inspired redefinition of what political struggle should be about, it also embodied a political recalibration of how socio-environmental relations were understood. Suddenly, the environment was no longer an external object on the fringes of politics: it was a central component of the political field. Martin Luther King's visit to Memphis marks an important point in time, following which the qualities of the environment and access to environmental amenity were no longer attributed to nature and chance, but to the workings of socio-economic power.

Key readings: Bullard (1990); Gottleib (1993).

Kaplan 1994; Klare 2001). Resource conflicts can take many different forms, including struggles over access to water, oil, agricultural land and minerals. In his famous study of resource conflicts in West Africa – entitled *The Coming Anarchy* – Robert Kaplan went as far as to claim that military and civilian struggles over environmental resource were resulting in the unravelling of formal political societies and the redrawing of state boundaries. Many have been critical of neo-Malthusian perspectives on resource conflicts, claiming that it overstates the role of resources in stimulating political conflicts of different kinds, and depicts a far too simplistic account of the relationship between resource scarcity and warfare.

In his analysis of the relationship between armed conflict and natural resources, Le Billon (2001) reveals the geographical factors that shape this relationship. According to Le Billon, it is not that absolute shortages of resources generate conflict (the so-called *scarce resource war hypothesis*), but rather that particular forms of armed struggle are produced by the nature and geographical form of the resources in question. One factor that does appear to commonly drive resource conflict is when a national economy is overly dependent on a single resource (such as oil) for its economic growth (see Watts *et al.* 2004). This resource dependency tends to see political power concentrated in the hands of those who are able to control, exploit and tax the resource in question. The concentration of power in this way often leads to great social disadvantages emerging between the resource controlling elite and the rest of society, and thus increases the likelihood of civil war (Le Billon 2001). The geographical form of the resource is also, however, a key factor in shaping the nature of the conflicts that may develop around them. So-called *point sowa resources* tend to be heavily

Plate 10.1   Sanitation workers protesting in Memphis, 1968

concentrated in particular areas (like mining or oil extraction sites) and are thus more easy to command and control by government forces. At the other end of the resource spectrum are *diffuse resources*. These resources, as the name suggests, are spread over large areas and include forestry and agricultural land (ibid.). The nature of point resources means that they tend to be connected to struggles over the control of the full apparatus of government and associated *coup d' etat* (a kind of all or nothing conflict scenario). *Diffuse resources* on the other hand are much more likely to be associated with the establishment of enclaves of rebellion and warlordism; or, if the resource is located on the peripheries of a state with breakaway, secessionist movements (ibid.: 572–573). What is most significant about Le Billon's work is that it illustrates the complex ways in which natural resources become a part of political struggle and armed conflict. In essence the impact of resources on conflict is heavily conditioned by prevailing political, economic and geographical circumstances.

The work of William Ophuls suggests that natural resources are not only connected to politics through armed conflicts, but are actually much more fundamental to the nature of political society and some of our most treasured political values (see Ophuls 1977,

1997, 2011). In his most recent book, *Plato's Revenge*, Ophuls (2011) argues that the political and economic freedoms that have become the hallmarks of liberal, market economies have been facilitated by the relative abundance of natural, non-renewable resources during the nineteenth and twentieth centuries. According to Ophuls, this age of resource abundance has enabled liberalism to prosper to the extent that one person's, corporation's, or nation's use of a natural resource has not significantly impinged upon the abilities of others to also avail themselves of the natural resources upon which their development may depend (see Quilley 2013). The most significant implication of the work of Ophuls is his claim that as non-renewable resources become scarcer in the twenty-first century two threats will emerge to liberalism. On the one hand, more authoritarian governmental systems may be needed to regulate resource access and to ensure its equitable distribution. On the other hand, we may witnesses the increasing incidence of resource conflicts – as state and communities become more protective of their resources – and the associated breakdown of the systems of international trade and exchange upon which liberal economic systems are based (Ophuls 2011).

Political ecologies and environmental government

Many accounts of the role of scarce environmental resources in shaping political conflict and practice have been criticised for exhibiting forms of *environmental determinism*. Environmental determinism exists when key features of a society, culture or political community are attributed to prevailing environmental conditions (including resource availability, climate, soil quality inter alia) (see Sluyter 2003). Critiques of environmental determinism argue that such accounts of the world tend to naturalise the often unjust socio-economic and political conditions people around the world find themselves in, and disconnect these conditions from the broader political and economic systems that contribute towards their production (ibid.).

In response to the problems associated with environmental determinism, Political Ecology has

offered an alternative, and more critical, account of the relations between the physical environment and politics. Political Ecology finds it antecedents in the 1980s with the pioneering work of Piers Blaikie. In his 1985 book *Political Economy of Soil Erosion in Developing Countries*, Blaikie challenged Malthusian notions that soil erosion was caused by over-population, and environmental determinist perspectives that argued it was the consequences of climatic fluctuations. Blaikie argued that soil erosion was actually the consequence of prevailing political and economic structures, which often saw peasant farmers having their land seized, being poorly paid and becoming economically marginalised (see Blaikie 1985; Blaikie and Brookfield 1987). It was these conditions, according to Blaikie, which forced farmers to work commercially unviable land, and which ultimately led to soil erosion and the breakdown of the peasant agricultural economy. Blaikie's early work has provided the basis for a rapidly expanding portfolio of work on Political Ecology. What unites this now varied school of thought is a desire to develop an account of environmental change and transformation, which recognises the role of physical environmental systems (such as climates, soils, fish stocks), but which also incorporates an analysis of political and economic power (see Bryant and Bailey 1997; Robbins 2004).

Bryant and Bailey (1997) have argued that Political Ecology is primarily concerned with how and why the benefits and disadvantages associated with various forms of environmental transformation (whether it be an agricultural harvest or the production of air pollution) are distributed, often unevenly, among a population. Such questions have inevitably led political ecologists to consider the relationships that exist between nation-states and the environments they attempt to govern. Since the emergence of early manifestations of the nation-state, governments have asserted sovereign control over their natural resources (Scott 1998). However, as states have evolved they have assumed an increasingly important role in mediating collective social relations with the environment (see Whitehead *et al.* 2007; Whitehead 2008). While it is commonly accepted that states are important political actors in environmental affairs, there is much debate about the nature of states' role in such matters. At one end of the spectrum, Anarchist thinkers argue that the rise of large-scale nation-states has resulted in the increasing alienation of people from the natural world, as governments assume increasing levels of responsibility for managing environmental affairs and common ecological resources (see Bookchin 2004). Others, however, perceive a more positive role played by states in managing the environment (see Johnston 1996: 131–2). Related work claims that the relative political autonomy of state institutions from the varied interests of society mean that they are able to make decisions that serve the collective interest. On these terms, it is argued that while there exist strong economic incentives for individuals and corporations to selfishly exploit the environment for their own ends – and with little thought for the needs of others or future generations – states can act as environmental referees, arbitrating between the varied demands that are placed on the environment. As environmental referees, states often have a role in ensuring that the environmental activities of one group do not adversely affect other social groups (this is most obviously apparent in the regulation of air and water pollution, which often affects groups who are downstream and downwind from the polluting activity itself).

Marxist geographers have, however, consistently questioned the ability of states to act as neutral arbiters within human-environmental affairs (Castree 2007a, 2007b). Related work has explored the ways in which states attempt to resolve some of the worst contradictions that the capitalist exploitation of nature generates, but ultimately serve to underwrite the expanded transformation of the environment in pursuit of commercial profit (see Prudham 2007a, 2007b, 2008; McCarthy and Prudham 2004). From an eco-Marxist perspective, it is not just that governments fail to effectively protect ecosystems and people from the adverse consequences of economic development, but that through the provision of subsidies, the building of infrastructure (such as roads and bridges), and the marshalling of environmental protests, governments actively support the exploitation of the environment and the uneven distribution of costs and benefits associated with this process. In this context, Steven

Bernstein (2001) has gone as far as to argue that contemporary environmental agreements forged between nation-states at international levels (and in particular concerning the delivery of sustainable forms of development) have been informed by a *liberal environmental* orthodoxy. At the heart of this liberal environmental agenda is the assumption that by opening up more of the world's environmental resources to capitalist economic exploitation and trade, new technological capacities will emerge that will enable the better use and protection of environmental resources in the future. In light of these contrasting theories of the role of states in environmental affairs, it remains important to study the operations of different governments on their own terms, and to consider the different ecological records of national governments around the world.

Citizenship, new environmental subjects and ecological belonging

Although the real environmental politik of nation-states may have uncertain consequences for socio-environmental relations, states and the political communities they delimit, provide other important perspectives on the mixing of politics and environmental concerns. Throughout much of modern history, nation-states have defined the subject that sits at the centre of political analysis. The features of the modern political subject who inhabits the nation-states are, of course, familiar to us all. In the democratic world citizens of nation-states are bearers of certain rights: including the right to vote, free speech, the right to own property, the ability to avail themselves of legal counsel. Citizenship also comes with responsibilities: not to cause harm to others, to obey the law and, perhaps, to partake in some form of military service. Whilst, as our previous discussion of the civil rights movement illustrates, political history has been marked by a constant struggle to ensure that the rights and responsibility of each citizen are equal, citizenship itself has been defined by three key parameters: 1) it relates to the defined political territory of a nation-state, beyond which the rights and responsibilities of a citizens may not be recognised; 2) it applies to the political here and now; and 3) it is centred on the human subject.

Our reason for outlining the basic parameters of modern political subjectivity and citizenship is to provide a context within which to explore the impact that environmental concerns are having on emerging understandings of citizenship (see Dobson 2003; Bullen and Whitehead 2005). In his book *Politics of Nature* the prominent French philosopher Bruno Latour reflects on the challenges that environmental issues are now presenting nation-states with. Latour observes that,

> Political philosophy did not anticipate that it would end up administering the sky, the climate, the sea, viruses, or wild animals. It had thought that it could limit itself to subjects and their right to property; science would take care of the rest. Everything changes with the end of modernism, since the collective may have as its ambition bringing together the pluriverse.
>
> (Latour 2004: 204)

Latour's point is that as human influences on the environment have grown so too has the ecological scope of politics. Latour's reference to the climate and sea is significant in this context. Many contemporary environmental issues (such as air and water pollution, and climate change) have consequences that reach beyond the narrow confines of any one political community (see Andresen *et al*. 2000; Eckersley 2007). In this context, the polluting activities of one place can have serious consequences for distant places that may well not engage in practices of environmental pollution themselves. On these terms, many argue that the global nature of contemporary environmental problems requires a spatially expanded political community of rights and responsibility (see Bullen and Whitehead 2005; Christoff 2006; Dobson 2003; Eckersley 2007; Evans 2012).

There have, of course, been various initiatives, which have sought to create international political regimes for environmental government: these range from the United Nations' protocols on climate change to the work of the International Whaling Commission.

But Dobson points out that on spatial terms, environmental concerns mean more than the globalisation of political communities:

> The space of ecological citizenship is therefore not something *given* by the boundaries of nation states or of supranational organizations such as the European Union [. . .]. It is rather, *produced*, by the metabolistic and material relationship of individual people with their environment.
>
> (Dobson 2003: 106)

What Dobson is essentially claiming here is that effective eco-political communities are ones that not only incorporate large-scale environmental systems (such as the global climate, or Antarctica), but also enable people to feel connected to the environmental systems upon which they depend. A popular manifestation of the ecologically inspired redrawing of political boundaries is expressed in the bio-regionalist movement (see Scott Cato 2012). Bio-regionalists argue that political (and economic) boundaries should be determined not on the basis of culture or economics, but on the basis of ecological systems. Popular manifestations of bio-regions range from river-catchments to mountain ranges. Bio-regions are believed to enhance environmental governance on two levels: 1) they make it easy to administer environmental systems as integrated wholes, and thus avoid the arbitrary division of eco-systems in to political subdivisions; and 2) they tend to empower the environmental knowledge of those living within such regions and who, perhaps, have the best understandings of the how these ecological systems work. The principles of decentralisation that are associated bio-regionalism are echoed in the work of the prominent environmental philosopher Murray Bookchin. Bookchin (2007) argues that political communities, and associated decision-making, should be moved from the national level to the level of city, or municipality. Bookchin labels his proposals as *libertarian municipalism*, and suggests that by rescaling political life at an urban level it will be easier for people to make face-to-face decisions about how to use and protect the environmental systems upon which they depended (ibid.).

BOX 10.2 ECOLOGY AND NATIONAL SOVEREIGNTY

The political scientist Robyn Eckersley recently sparked an intriguing debate within the field of environmental politics concerning the relationship between ecological harm and national sovereignty. This debate provides some important insights into just one of the ways in which environmental concerns can challenge the established boundaries, and associated systems of rights and responsibilities, that we associate with nation-states. In an article published in the journal *Ethics and International Affairs*, Eckersley explored the potential ways in which environmental threats could provide the basis upon which outside interventions into a nation's sovereign territory could be justified. On these terms Eckersley's paper is not so much concerned with the creation of international systems of collective environmental government, but how one state's mismanagement of its own environmental systems/resources could place a responsibility upon other states to intervene.

At the heart of Eckersley's argument is the question of whether forms of humanitarian intervention, which, following the Holocaust and the Nuremberg Trials, are often justified on the basis of evident crimes against humanity, could be extend to include 'crimes against nature'. Eckersley considers certain hypothetical, environmental situations when military intervention within the sovereign territory of another state may be justified: perhaps following a nuclear accident, which is not safely addressed; or because of the potential extinction of rare species; or the destruction of rainforest ecosystem on which the whole planet depends. Within each of these scenarios, it is important to note that the environmental harm being perpetrated within one state has consequences for other, so-called *victim states* (Eckersley 2007: 300).

In the context of these scenarios, Eckersley considers the various legal and ethical bases upon which military intervention into ecological affairs could be justified. At one level, Eckersley identifies the possibility of developing an eco-humanitarian doctrine, within which military intervention is justified on the basis that deliberate forms of human harm are being perpetrated through the destruction of the ecosystem upon which target groups depend (such as Saddam Hussein's draining of March Arab's homelands in Iraq) (ibid.: 301) (Plate 10.2). At another, more radical level, Eckersley considers whether 'crimes against nature', or ecocide could provide a basis for the use of military force within another state's territory. Intervention on the basis of 'crimes against nature' presents a more complex ethical challenge as it requires us to rethink who and what belongs to political communities of responsibility, and who and what's rights can be defended through systems of international justice.

Key readings: Eckersley (2007); Dalby (2007).

In essence the re-thinking of political boundaries, which has been prompted by emerging environmental issues, has, first and foremost, been about enabling the environment to feature as a more prominent aspect of political consideration in the communities in which we live and organise our lives. At a global scale, the creation of supra-national units of governance has enabled the global climate, ozone layer, and oceans to become a part of political deliberation. At the other end of the spectrum, attempts to redraw political boundaries around cities and bio-regions have enabled local environmental knowledge and ecological systems to enter political debate and discussion. These processes provide a link back to a second important aspect of the Bruno Latour's aforementioned quote. While the political administration of ecosystems may require the reconfiguration of the geographical boundaries associated with political communities, Latour's reference to wild animals raises another set of boundary issues. As we have previously mentioned, citizenship

Plate 10.2 Marsh Arab settlement in Iraq

has historically been constructed as the exclusive domain of the human subject. While the emergence of the animal rights movement has questioned the boundaries that are constructed between the rights of humans and other sentient creatures, there remains a clear division between the legal and ethical rights and responsibilities of humans and other inhabitants of the global ecosystem (Whatmore 2002: 12–57). The exclusion of non-human animals and other parts of natural environment from communities of citizenship, in part, reflects an enduring anthropocentrism within politics. Increasingly, however, this human-centric definition of who/what can be a citizen is being challenged by those who emphasise the interconnected, cross-species and trans-material nature of political communities (see Latour 2004). On these terms, it is argued that the existence of modern societies is dependent upon complex networks of human and non-human actors who all require political protection and the defence of their rights to exist and function.

A final challenge to prevailing notions of citizenship generated by environmental concerns relates to the temporal definition of a political community (Bullen and Whitehead 2005). Generally the rights and responsibilities associated with belonging to a political community are reserved for the current generation of living citizens (although debates around human abortion raise important biological and ethical questions about precisely when membership of the 'current generation' begins). Discussion of the use of environmental resources and the preservation of ecosystems does, however, inevitably raise questions of fairness and justice that go beyond the timespan of the current generation. It is on these terms that environmental politics is often involved in questions of intergenerational justice and exchange (see Barry 1997). If current oil reserves are predicated to last to the end of the twenty-first century (this is a hypothetical scenario) debates about how to best conserve these supplies, and how to manage the gradual transition to a different energy economy show a concern for unborn generations. At the same time, many claim that the worst consequences of climate change lie in a near future that may not affect those living now. Current attempts to address climate change are thus just as much about tackling the environmental concerns of future generations as they are about preserving our own needs. Since the United Nation's adoption of the principle of sustainable development – with its emphasis on 'meeting the needs of the present generation without undermining the ability of future generations to meet their own needs' (World Commission on Environment and Development 1987: 8) – concerns about the environmental future have led to discussion about intergeneration justice and a broadening of the political community to incorporate the future. However, as Brian Barry has pointed out, extending systems of justice into the future is a difficult ethical task (not least because it inevitably involves the current generation projecting its own sense of what is just and normal into an unknown future, which may see things very differently). What is clear, however, it that by unsettling the spatial scope, species focus and temporal parameters of citizenship, the mixing of politics and the environment represents one of the most significant challenges to the prevailing assumptions that undergird our contemporary political systems.

BOX 10.3   TRANSITION TOWNS AND THE POLITICAL GEOGRAPHY OF THE ENVIRONMENT

The Transition Town (or Culture) Movement provides some important practical insights into the connections that exist between political geography and environmental concerns. The Transition Town Movement finds its origins in the permaculture tradition (permaculture is a framework for developing human settlements and agricultural systems that are modelled upon ecological processes). At its heart the movement is a response to the twin threats of peak oil production (the point after which the production of oil becomes increasingly difficult and thus more expensive to consume) and climate change. In this context, the Transition Town Movement is a form of pre-emptive response to the threat of resource scarcity and global environmental change. The defining characteristic of all Transition Town initiatives is their *Energy Descent Strategy*. Energy Descent Strategies outline how a community intends to reduce their energy use and associated carbon footprint. Most Energy Descent Strategies tend to emphasise the importance of re-localisation as a starting point for reducing collective energy consumption. Re-localisation prioritises, where possible, the local sourcing of food, goods, and skills and thus seeks to reduce the need for people and things to be transported over long distance (and with great outlays of energy) in order supply an area with what it needs to survive. In order to encourage the process of localisation, some Transition initiatives have introduced local currencies (which can only be traded in the Transition community area) in order support the local exchange of goods and services. Many initiatives have also developed local skills sharing and training programmes (associated with food production, clothes making and repair, and house building/maintenance).

Transition Town initiatives have important implications for how we understand the connections that exist between political geography and the environment. At one level they reveal how, following planned energy descent (rather than imposed energy shortages), it may be possible to develop local political systems that are not authoritarian, or based on the violent appropriation of scarce resources. In part because of the practical need to ensure that the collective insights and knowledge of a local community are being utilised as fully as possible, Transition Town initiatives focus on the implementation of very open forms of democracy. At another level, however, the emphasis that Transition communities place on re-localisation has led some to believe that they could become geographically isolationist in their outlooks and intents (Mason and Whitehead 2012). Related to these concerns, others have criticised the Transition Town Movement for failing to develop a critique of the prevailing political economic power structures associated with capitalism (Trapese Collective 2008). At the heart of this critique is a concern that relocalisation movements are not being political enough in their activities (Figure 10.1).

Figure 10.1 Publicity material for Aberystwyth Transition Town Initiative

Key readings: Mason and Whitehead (2012); Trapese Collective (2008).

Summary

This chapter has offered an introductory overview of the nature of environmental politics and how it connects to a series of geographical concerns. At the heart of this chapter has been a desire to show the breadth of issues that are associated with the politics of environment, which collectively means it is about more than simply the defence of nature. This chapter began by outlining the ways in which neo-Malthusians have suggested that there is strong connection between the availability of natural resources and political conflict. Related to this we also considered the ways in which resource abundance may have enabled the emergence of liberal political societies, and how future resource shortages could signal the rise of more authoritarian brands of community. The following section introduced the work of political ecologists and the ways in which they have challenged environmentally deterministic accounts of ecological change. In contrast to neo-Malthusians, political ecologists assert that it is not so much environmental issues that drive politics, but that prevailing political system that determine how environments are treated and transformed. It is this insight that has essentially informed the emergence of the New Environmentalism that has emerged over the last forty years, and which seeks to address environmental issues alongside related questions of social and economic injustice.

The final section of this chapter outlined the ways in which environmental issues are challenging established forms of political citizenship. In this section we considered how the emergence of environmental concerns onto the political agenda has been associated with the rise of spaces of politics that operate both above and below the level of the nation-state. Analysis also explored the ways in which a concern for the environment is increasingly stretching communities of citizenship beyond humans in order to include the non-human world of animals and ecosystems, and beyond the here and now in order to incorporate future generations. While discussing all of these themes, this chapter has illustrated how the mixing of politics and environment has not only transformed the parameters of politics and how we understand ecological change, but has also had key geographical implications. Thus, through our discussion of transition cultures and bioregions, and supra-national governmental agreements and threats to national sovereignty, we have seen that the politics of the environment is always a geographically mediated process and a process that has important consequences for our collective political geographies.

Further reading

For a more detailed discussion of the approach and field of Political Ecology, see Robbins (2004) *Political Ecology: A Critical Introduction*. The relationship between the environment and state power, in terms of control over natural resources and of the responsibility of the international community to act to prevent environmental disasters, is discussed in Ophuls (2011) *Plato's Revenge: Politics in the Age of Ecology*, and Eckersley (2007) 'Ecological interventions: prospects and limits' in *Ethics and International Affairs*, 21, 293–316, respectively. The development of environmental notions of citizenship are explored further in Bullen and Whitehead (2005) 'Negotiating the networks of space, time and substance: a geographical perspective on the sustainable citizen', in *Citizenship Studies*, 9, 499–516. For a broader introduction to contemporary environmental issues and the politics of nature, see Castree (2012) *Making Sense of Nature*, and Whitehead (2006) *Spaces of Sustainability: Geographical Perspectives on the Sustainable Society*.

Geopolitics and critical geopolitics

Introduction

The field of geopolitics – wherein the highly complex 'geo' is added as a prefix to the already complex, not to mention contentious, word 'politics' – can be described as a modern invention insofar as it emerges at a particular time and place. That is, the coining of the term can be traced back to the work of Swedish political scientist and conservative politician Rudolf Kjellén, at the turn of twentieth century (Heffernan 2000; Tunander 2001). Writing at a time when diverse European nation-states were engaged in economic, cultural and military imperialism across the globe, and Western social science largely sought to buttress such efforts via reference to environmental determinism and social Darwinism, Kjellén (1916) argued that states needed to protect and expand their territories at home and abroad so as to advance what he regarded as their racialised populations, or *volk*. Heavily influenced by his teacher, the geographer Friedrich Ratzel, who wrote in 1897 on the state as an organism subject to growth and decay (as opposed to a static, legally bounded entity), Kjellén emphasised state policy as a form of sustenance. A government's *raison d'etre*, he argued, was, or rather should be, the social and economic health of its populace. (For an account of Ratzel, often termed the 'father' of political geography, see Chapter 1).

Geopolitics, then, emerges from a period and setting associated with modernity, including: the rise of industrialism and urbanism, and associated societal ills such as alienation and marginalisation; the far-flung impacts of the globalisation of markets and labour pools in the form of uneven development; the bureaucratisation of warfare and the wholesale marking of civilians as targets; and, finally, the compartmentalisation of academia into specialist subjects and expert cohorts, the latter regarded as capable of describing, but also prescribing for, some aspect of the social or natural world.

There is no doubt, however, that while some of those working within geopolitics have regarded themselves as contributing to a tradition of 'statecraft,' others have taken an avowedly *modernist* approach to the field itself. That is, there have been concerted efforts, particularly over the course of the latter half of the twentieth century, to critically reflect upon this notion of a disciplinary origin and canonical texts, and, moreover, to rework the conceptual and methodological terrain pertaining to geopolitics (Ó'Tuathail 1996). As these scholars variously engage with the character of both the 'geo' and the 'political', texts are reprised, rearranged, rewritten and revisioned into new constellations of the idea and practice of the geopolitical.

In this chapter, we begin by outlining the broad intellectual and political context within which geopolitics as a state-centred field of inquiry *and* statecraft emerged. Underpinned by both realism (which assumes an inevitable conflict between individuals and peoples, and which emphasises *realpolitik* as both a fact of life and an admired form of statecraft), and biological ideas pertaining to both race and the causative role of the physical landscape, 'classical' geopolitics became constituted from three strands: geostrategy, environmental determinism and social Darwinism. Whilst outlining some of the key writers on these thematics, however, we also want to point to a tension lying at the heart of geopolitics. That is, a growing dissonance between the hard-headed flexibility of *realpolitik* and the moral righteousness of much of this biology-inspired work. Whilst the latter was to prove hegemonic in the run up to World War II, as manifest in a German *geopolitik*, there was a sustained effort after the war to disassociate geostrategy from what came to be regarded as its fascist trappings.

The re-emergence of a neo-realist geopolitics in the 1970s and 80s was met by a re-animation of geopolitical theory (see Chapter 1). Loosely assembled under the rubric of a 'critical geopolitics', new lines of inquiry sought to deconstruct assumptions and proclamations as to the innately conflictual character of society. Some pointed to their lack of scientific credibility; others identified decidedly non-realist movements and efforts that gave lie to these assumptions. And others drew attention to the fact that despite its pretensions to objectivity, realist and neo-realist geopolitics emerged from particular constellations of academic theory, military strategy and state policy. Small wonder that geopolitics for so long had served to further some interests over and against others.

There are numerous, compendium accounts of the objectives and aims of a critical geopolitics, as well as reflective reviews of the field, and individual analyses that address (at least): the social construction of texts, from policy documents to films and TV, that 'script' the world in a manner that benefits capitalism and the state; the masculinism of many such texts that purport to achieve both objectivity and neutrality; the deliberate non-depiction of certain events and objects as a means of governing public opinion; the proliferation of realist geopolitical framings of the world through bodies, as more and more people are placed under surveillance and subject to emergency powers; and the engendering of panic and fear as defining conditions of society. Rather than outline the key features of this field, however, most of which have been noted in Chapter 1, in the last third of this chapter we focus on emergent lines of inquiry – namely, subaltern geopolitics, feminist geopolitics and posthuman geopolitics – that, whilst having numerous antecedents and common concerns, nevertheless have distinctive questions to pose. Our intent here is to provide a sense of what these lines of inquiry are interested in, and

how, in building a particular form of knowledge of the world, they are renegotiating what we mean by the term geopolitics.

Realist politics and the state

The naming of geopolitics at the turn of the twentieth century as the study of how and with what effect states compete (in the name of their citizenry) for territory, coincides with the increasing formalisation of similarly state-orientated doctrines such as the 'balance of power' and *realpolitik*. Together, these provided a means of making sense of political events at the height of European imperialism, but they also had a programmatic purpose, insofar as they became valorised, and sometimes condemned, as instruments of state policy. Geopolitics, the balance of power and *realpolitik* became blueprints for state activities, effectively blurring a distinction between the analytic lens used to study objects – in this case the state – and the objects themselves.

In light of their collective emergence, and the manner in which the assumptions underpinning them converge time and again, these terms are often referred to as specific manifestations of a broad-reaching 'realist' epistemology. With antecedents in the work of Greek historians of warfare, Chinese legal scholars and Renaissance deal-brokering amongst Italian city-states, realism is primarily characterised by the presumption that power is a matter of competing for resources and influence, and that in order for individuals and groups to be successful, ideology and ethics need to take a back seat to a 'hard-headed', pragmatic approach to negotiation and governance. How, then, did this realist framing denote the state as an object to be described, certainly, but also to be shaped? And, how did this denotation help produce a geopolitics that centred on the activities of the state?

A balance of power

The Treaty of Westphalia, signed in 1648 by some of the major European powers of the day, and ending the Thirty Years War, is commonly regarded as a critical moment in establishing the sovereignty of nation-states. As Dalby (2013: 35) notes, there remains a prevailing tendency in foreign policy discussions today to view matters through 'Westphalian lenses'; these continue to observe power as pertaining to the military capacity of the state. Key to the 'Westphalian doctrine' was respect for the territorial integrity, and independence from interference, of European states. In effect, the doctrine proffered the state as the principal actor in international relations and, moreover, marked each state as equal in regard to these legal criteria. In the nineteenth century, this valorisation of the state as a recognised entity amongst other states was considerably enhanced via reference to nationalism, by which was meant a populace united by culture, particularly religion and language. The 'nation-state' became the ideal mapping of a homogenous populace onto a sovereign state and territory, and vice versa (as discussed in Chapter 3).

Events in Westphalia also ushered in, however, a particular understanding of how states should apprehend and respond to these new international relations; this was an understanding, moreover, that emphasised a visual rendering of states as laid out across a map, and whose contours could be overlain by military advances. In the decades following the 'Peace of Westphalia', universalism (a doctrine that advocated the reconciliation of diverse population under Christianity) was gradually displaced by the idea of a necessary balance of power. Here, it was assumed that, whilst states were in a state of competition with each other, it was in the interest of each that no one state should have the capacity to overwhelm all others. As such, each state needed to consider its place amongst this new community, and negotiate allegiances such that potential threats from opposing 'power blocs' were curtailed. Formally recognised in the Treaty of Utrecht (1713), which ended the Spanish War of Succession, the balance of power became an axiom of diplomatic strategy, and a reference for military coalitions throughout the nineteenth century. Cartographic renderings of these shifting 'power blocs' emphasised the vulnerability of contiguous borders, the strategic importance of key towns and the utility of transport networks for troop movements.

Realpolitik

Realpolitik, which refers to the pursuit of state interest via a politics of what needs to be done, rather than an idealised version of what should be done, was to become the primary mechanism for achieving such a balance of power. In many ways foreshadowed by Machiavellianism, which, as manifest in the classic essay *The Prince* by Italian diplomat Niccolò Machiavelli (1532), advocated machination over ethics, *realpolitik* was coined by the German journalist and political exile Ludwig von Rochau in 1853. For von Rachau, the term indicated the embeddedness of state policy in a web of power relations, by which he meant the use of might and influence in an instrumental manner devoid of ideological imperatives or moral concerns. Whereas Machiavellianism was to be condemned in the sixteenth century as a plague upon the body of the Kingdom, and critiqued as unbecoming to an Enlightenment ruler in the eighteenth century, *realpolitik* was pronounced a pragmatic adaptation to things as they were in the nineteenth century by such key European players as the Austrian diplomat Prince von Metternich and the Prussian Chancellor Otto von Bismarck. In the absence of appeals to a greater good, the wielding of power via state policy was to be justified solely as a means of buttressing the legitimacy and reach of the state itself, which, by this period, had become synonymous with the national interest. In contrast to the grand spectacle of warfare, however, *realpolitik* was a matter of hidden intrigue.

Classical geopolitics

Both *realpolitik*, and the balance of power doctrine, then, presume not only the sovereignty of states, and their mutual recognition, but also their intrinsic competition with each other. What is more, both presume to operate in the national interest, by which is meant not some kind of future utopia, but rather the defence and (hopefully) extension of the reach of the state itself. Whilst there is a very real apprehension of the importance of territory here, as indicated above, and a strong cartographic impulse, a critical appraisal of the role played by geography in international relations was to be provided by the concept of geopolitics. And, whereas it was the Swedish jurist Kjellén who was to coin the term, there is no doubt that the key ideas pertaining to geopolitics were emerging from the rapidly developing, systematising world of academia then being institutionalised in the modern universities of Germany, Britain and the US. Here, a particular confluence of theoretical analysis, military training and statesmanship was being formulated. Geopolitics was to become constituted from three overlapping strands – geostrategy, environmental determinism and social Darwinism – each buttressed by reference to academic authority, and put into practice as a matter of state policy.

Geostrategy

The governments of Europe had long been accustomed to the 'Great Game'. A phrase made famous by Rudyard Kipling, it referred to the intense competition between the British and the Russian governments for control over central Asia. A sustained acknowledgement of international strategy, and the crucial role of geography therein was, however, initially proposed by Admiral Alfred T. Mahan of the US Navy. Inspired by the naval victories of Britain during the nineteenth century, Mahan, in a series of books, beginning in 1890 with *The Influence of Sea Power upon History, 1660–1783*, argued that if the fledgling United States of America was to achieve its 'manifest destiny' – a term till then associated with a Westward, overland expansion – then strategic control of seaborne commerce in the Pacific as well as the Atlantic was required. Calculating the possible attack strategies of both the United Kingdom and Japan against the US, Mahan advocated that the US build naval bases in the Caribbean as well as at Hawaii in order to achieve a balance of naval power in both spheres. As Sumida (1999) points out, Mahan introduced a number of geostrategic terms in his work such as 'choke points', by which he meant sites of especial import for building capacity and exercising military might, but which were also vulnerable to attack, such as canals and coaling stations. His arguments as to the formative

role of such physical and social geographies were tempered, however, by the recognition of contingency and individual expertise. Somewhat ironically, Mahan's geostrategising was to have a significant impact not only upon the naval policies of the UK and Japan but also Germany (Evans and Peattie 1997). His arguments were to be taken up, for example, by the cultural geographer Friedrich Ratzel, who looked to German naval control over maritime commerce as a means of guaranteeing the country's burgeoning industrial growth and imperialist agendas in Africa.

For the British government, struggling to hold together a vast Empire by the end of the nineteenth century, the nature and probable site of the threat posed by its competitors was of great interest. One such site was Asia, where Britain anticipated and felt competition from the US, Germany and Russia. Geography, according to Sir Halford Mackinder – one of the founding figures of political geography (see Chapter 1) – was well placed to provide a means of analysing this situation. In his 1904 paper to the Royal Geographical Society on 'The Geographical Pivot of History', Mackinder emphasised that Britain's might as a sea power was potentially on the wane insofar as the land-based powers, and in particular the Russian-ruled 'Heartland', which stretched from the Himalaya to the Arctic, and from the Volga to the Yangtze, had immense resources and hence capacity for economic development. This capacity, he argued, could feasibly be projected as military might across Europe. In his 1919 book *Democratic Ideals and Reality*, Mackinder later summed up his argument as,

> Who rules East Europe commands the Heartland;
> Who rules the Heartland commands the World-Island;
> Who rules the World-Island controls the world
> (Mackinder 1919: 106)

For Mackinder, any use of *realpolitik* in the pursuit of curtailing Russia's potential military might, or even of replacing it, would need to bear in mind the geography of the area, by which he meant its physical landscape (ice to the north and mountains and deserts to the south), as well as the presence of transport networks such as railways, which, he believed, helped to open up the possibility of invasion.

Mackinder's strategising resonated with the contemporary description of geopolitical calculations as 'the Great Game'. Field Marshal Sir Arthur Barrett, the senior British officer on the ground during the Third Anglo-Afghan War (1919), for example, depicted everyday operations in the region as reminiscent of a game of chess (Raza 2012). In the map that formed the centrepiece of Mackinder's 1904 paper on the 'Geographical Pivot of History' (Figure 11.1), state boundaries fade in importance; instead, physical features such as coastlines and land masses become central to his exposition. This rhetoric is updated in the *The Grand Chessboard* by Zbigniew Brzezinski (1998), a former advisor to US President Carter during the late 1970s, which is very much inspired by Mackinder's work. For the book's cover image, state borders once again take a back seat, and coastlines and

land masses are literally overlain this time by gigantic play pieces.

Whilst Mackinder's arguments were to have a somewhat negligible impact upon British policy, they were afforded a recapitulation not only in German geopolitics of the 1930s and 40s, a subject that will be expanded upon below, but also US rhetoric on a balance of power doctrine and interventionism during the same period. Key here was the work of political scientist Nicholas Spykman, Professor of International Relations at Yale University. In *America's Strategy in World Politics* (1942), Spykman argued that a policy of isolationism would leave the US, as an ocean-power, both 'impotent' and open to attack, whilst his *The Geography of the Peace* (1944) emphatically addressed the balance of power in the Heartland, and the implications for the US of possible events here. Differing from Mackinder, however, Spykman saw the Heartland as very much open to invasion from the sea-powers. Rewording Mackinder's phrasing of geostrategy, Spykman was to declare:

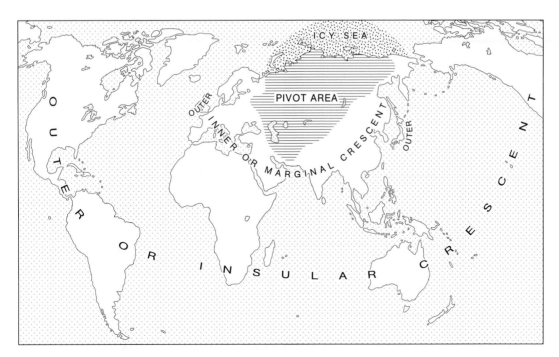

Figure 11.1   The 'heartland' map of Halford Mackinder (1904: 426)

Who controls the rimland rules Eurasia;
Who rules Eurasia controls the destinies of the world.

(Spykman, 1944: 43)

Much more influential than Mackinder in his day was the American geographer Isaiah Bowman, Director of the American Geographical Society, President of John Hopkins University, chief territorial adviser to President Woodrow Wilson at the Paris Peace conference ending World War I, and advisor to the US Department of State during World War II. With a background in Geology, Bowman had, prior to the Paris Conference, been convinced of the causative impact of the physical landscape upon society. Exposed to the heady world of diplomatic negotiation, however, and persuaded as to the lack of new frontiers to conquer, Bowman had come to the conclusion that a country's future well-being depended upon its control over economic resources and markets (see Smith 1986; 2004). And these, he argued, in *The New World*, were accomplished by a policy of interventionism. 'Whether we wish to do so or not,' he wrote in the introduction, 'we are obliged to take hold of the present world situation in one way or another' (Bowman 1921: v).

Environmental determinism

In drawing out their geostrategies, Mahan, Mackinder and Spykman were concerned with balancing their emphasis on the determining role of physical landscapes upon a state's military might via reference to the mitigating character of social geographies of transportation and commerce, and particularly the import of individual leadership, expertise and vitalism of spirit. Eschewing a simple cause–effect relationship, it is, rather, this framing of environment *versus* society, larger forces *versus* individual circumstances, which came to be understood as environmental determinism.

As with realism, there are numerous antecedents one can point to with regard to this field of inquiry, including classical Greek and Roman texts indicating the role of climate in shaping temperament, and the writings of fourteenth century Arab geographers such as Ibn Khaldun, who, in his book *Muqaddimah* (1377) posited

temperature rather than divine will as the causal factor in skin colouration. Again, however, as with realism, this understanding of the environmentally embedded nature of populations was to gain pre-eminence in the nineteenth century as imperial conquest not only continued to make visible a wide diversity of cultural behaviour and physical characteristics, but also raised tempestuous issues around the longer term impacts of colonialism. The key question here was what happened to European bodily types living in non-European habitats? Geographers such as Ellen Churchill Semple, who had studied under Friedrich Ratzel, worried at the possible physical and moral degeneracy of Europeans in such circumstances.

Initially, Semple had become concerned with the relative backwardness (in an otherwise highly developed country) of residents in Appalachia where, she argued, 'Man has done so little to render this district accessible because nature has done so little' (1901: 589). Only those who moved downriver, or who were sent far afield to the penitentiary, were enabled a glimpse of civilisation. This referencing back to the physical landscape as a means of explaining not only behaviour, but also, as she wrote in 1911, bodily morphologies, is perhaps most famously encapsulated in her dictum:

Man is a product of the earth's surface. This means not merely that he is a child of the earth, dust of her dust; but that the earth has mothered him, fed him, set him tasks, directed his thoughts, confronted him with difficulties that have strengthened his body and sharpened his wits, given him his problems of navigation or irrigation, and at the same time whispered hints for their solution. She has entered into his bone and tissue, into his mind and soul . . . . Man has been so noisy about the way he has 'conquered Nature,' and Nature has been so silent in her persistent influence over man, that the geographic factor in the equation of human development has been overlooked.

(Semple 1911: 1–2)

In this and her later work, as Peet (1985) explains, Semple, alongside other geographers, sought to ground her work as scientific via reference to prevailing realist and evolutionary ideas on conflict, thereby helping to legitimate the conquest of some societies by others. As with Kjellén, key here was the adoption of Ratzel's organic theory of the state, which strove to advance in the face of intense competition. 'The need to escape from guilt over the destruction of other peoples' lives,' Peet asserts, 'a guilt that survived even in a racialist view of the world, meant that the motivations for actions had to be located in forces beyond human control – "God," "Nature," or some amalgam of the two' (1985: 311). In order to explain the mechanisms by which Nature shaped society, environmental determinism drew upon social Darwinism.

Social Darwinism

Within environmental determinism, there is the assumption that populations can be considered largely homogeneous by virtue of their racial character and, moreover, that an evolutionary biology can be usefully applied to the reproduction of such groups. This sustained racialisation of humanity, again, can be traced back to numerous antecedents. But, it is in the 1870s that a systematising, academic explanation for the relations between such groupings, and their relation in turn with the environment, is proffered under what came to be called at the turn of the twentieth century (as a critique of such ideas) social Darwinism.

The reference to Charles Darwin ensues not so much from his 1860 classic, *The Origin of Species*, which described the process of natural selection and species' differentiation under conditions of changing environments and limited resources, but rather from his 1882 book *The Descent of Man, and Selection in Relation to Sex*, which specifically addressed the issue of the reproduction of what he called the 'weaker' members of society and their injurious impact upon Mankind as a species. Whilst Darwin himself advocated that the 'strong' exercise sympathy and protect such creatures, a number of other scholars such as the sociologist Herbert Spencer, who coined the phrase 'survival of the fittest' in his *Progress: Its Law and Cause* (1857) decried this noble, yet unrealist, sentiment. Spencer appealed to the demographic principles of Thomas Malthus, and the spectre of overpopulation and scarce food resources, to argue that charity would only exacerbate such problems.

The answer, for many on the Left as well as the Right, to the prospect of degeneracy and even population collapse was eugenics. A racialised reading of Darwin's theory of natural selection, alongside an apocalyptic reading of Malthus' overpopulation model, facilitated and legitimised a welter of domestic and international legislation aimed at halting the weakening of civilised society degeneracy, and purifying populations via a policy of racial hygiene. The physical education of boys, for example, became extolled in the 1880s as a means of strengthening national defence, whilst interest in the disciplining of women's reproductive capacities – manifest par excellence in the pelvic area – became a matter of putting them to effective use in the birth of model citizens (Bohuon and Luciana 2009). Meanwhile, the Eugenics Record Office in the United States opened in 1904, the German Society for Racial Hygiene in 1905 and the British Eugenics Society in 1907, all extolling the gathering of 'medical' data to prove hereditary failures, and presenting a series of legislative solutions (Box 11.1). Whilst the US federal government passed quota laws limiting immigration from Asia and Eastern Europe, US state governments strove to purify their populations. Indiana passed the first sterilisation laws in 1907; by 1931 the total number of such laws in the US reached twenty-eight. The Scandinavian countries all began sterilisation programmes in 1926. In Sweden, these laws were not repealed until 1976, resulting in the sterilisation of an estimated twenty thousand young women. As Foucault argued, the state was urged here to undertake another battle, one that did not address the safety of its borders so much as the health of its populace. 'It is no longer a battle in the sense of a war,' he wrote, 'but a struggle in a biological sense: differentiation of species, selection of the strongest, survival of the best adapted races' (cited in Elden 2002: 132).

Social Darwinism was as committed to the idea of humans as naturally competitive as realism, and it is no surprise to find that nation-states, as the embodiment and will of their *volk*, as Kjellén expressed it, were

BOX 11.1 EUGENICS IN NATIONAL SOCIALIST GERMANY

In post-World War I Germany, massive economic recession in the 1920s triggered the collapse of its welfare system. As a solution to the spectre of mass starvation, the much cheaper option of sterilisation of those considered undesirable residents of Germany's social, as well as physical, space became the choice of the increasingly vocal, and violent, Nazi Party. 'The simultaneous growth of racial anthropology was strongly criticised by the left as a cover for right-wing racism,' observes Wittmann (2004), 'however the massive growth in the right's political power was able to override the opposition. Once Hitler came to power in 1933 the Nazis were able to pass increasingly stringent eugenic laws with little opposition' (2004: 18). The *Law for the Prevention of Genetically Diseased Offspring* was passed in 1933, which required doctors to notify 'Information Centers for Genetic and Racial Hygiene' of women who had a genetic illness, such that they could be sterilised; these included the so-called 'Rhineland bastards', the children of (African heritage) French colonial troops stationed in Germany after World War I. Alcoholics, the physically disabled and the mentally 'feeble' were also to be compulsorily sterilised. Into World War II, the German Nazi Party's policy of racial hygiene was to have devastating consequences for those defined by the state by their Jewishness, Slavishness, homosexuality, mental illness and so on, culminating in the Holocaust, which sought to erase even the memory of those purged. These were victims deemed without the right to exist, who, write Clarke *et al.*, (1996) 'ought not to exist, and who were therefore *obligated* not to exist' (1996: 18). Considered parasitic in nature, such groups were to be cleaved from the body politic.

Key reading: Clarke *et al.* (1996); Wittmann (2004).

urged to expand their territory at the expense of their weaker neighbours in order to increase their carrying capacity. Whilst realism decried a moral imperative, however, social Darwinism, legitimised via reference to the health and betterment of the populace, was rendered a 'just' cause. As Ó'Tuathail (2013: xx) notes, an ensuing 'social Darwinism on a map', was manifest, for example, in the Nazi *Generalplan Ost* (Master Plan East) of 1940, which envisioned the ethnic cleansing of Central and Eastern Europe, and the 'Germanisation' of those remaining, supplemented by colonisation from good, solid Aryan stock, or *Volksdeutsche*.

Records pertaining to the *Generalplan Ost* are fragmentary; these plans were secret, and memos and maps relating to them were mostly destroyed at the end of World War II. Nevertheless, we can gain some idea of their content and consequence from testimony provided at the post-war Nuremberg Trials of SS members and Nazi Party officials. The plans were drawn up by the Reich Main Security Office of the SS, in association with the Planning Office of the *Reichskommissariat* for the Strengthening of Germandom, and utilised a series of academic studies carried out in Central and Eastern Europe by German academics before the war. The Planning Office employed several geographers, including Walter Christaller, better known after the war for his promulgation of an abstract central place theory. According to Barnes and Minca (2013), this central place model 're-created the Nazis' territorial conquests in the geographical likeness of the German homeland'. The *Generalplan Ost* calculated what percentage of the population (those with racial material, or *rassische Substanz*, close to the Aryan) of those countries conquered could be 'Germanised', and what percentage of those deemed racially undesirable would need to be either killed, enslaved or expelled to Western Siberia. This would free up living space, or *lebensraum*, for an estimated 8–10 million Germans. Whilst the plans, clearly, did not come to fruition, in World War II German forces did enslave a vast population of Slavs from Eastern Europe and progressively expelled Poles from their territory. According to Browning (1992), both the Holocaust and the devastation and partitioning of Poland can be considered a 'trial and error' process guided by Hitler's vision of a 'Jew-free Europe' combined with a colonised East.

The rise, fall and rise of geopolitics

In the above, the emergence of geopolitics has been explained via reference to the development, and intensification, in the late nineteenth century of broader ideas pertaining to realism and evolutionary biology. These ideas helped to flesh out the three key strands of geopolitics, namely geostrategy, environmental determinism and social Darwinism. Yet, we can also observe a tension building here. That is, whereas realism, associated primarily with a concern for a balance of power managed via *realpolitik*, and manifest most clearly in the theory and practice of geostrategy, disavowed ideological and moral imperatives, those working primarily with evolutionary concepts did so because they offered a glimpse of (what was argued as) a better future for particular, racialised cohorts. In regard to the latter, then, geopolitics was not only a matter of pursuing a vague notion of the 'national interest', but was thoroughly concerned with ensuring both the purity and increase of a national body, understood as a distinct biological grouping. And, we can see this not only in regard to the foreign policies of National Socialist Germany, but other governments also, from Portugal (Mensagem et al. 2005) to Japan (Fukushima 1997). Whilst there was no clear separation of geopolitics into two camps of thought (the non-ideological and the ideological, the non-determinist and the determinist), there was, nonetheless, a differing emphasis placed upon, on the one hand, hard-headed practice adapted to prevailing and future events and conditions, and, on the other, a commitment to changing the face of Europe and even the world.

There is no doubt that, in the years preceding World War II, an explicitly, ideologically driven geopolitics held sway, and gave animation not only to foreign policy, but domestic policy also. The intellectual centre for such efforts was Germany.

German *geopolitik*

Geopolitik is the name given to a field of inquiry that not only emerged from Germany, but which also sought to centre ideas of progress and civilisation upon the notion of an expanded German homeland. Through the late eighteenth and nineteenth centuries, many of the basic contours of what we now think of as a modern day, Enlightened Geography were outlined by a series of scholars (Alexander von Humboldt, Karl Ritter, Friedrich Ratzel and so on) associated with a region that was a central part of the Holy Roman Empire, later the German Federation, and then the German Empire. Under Kaiser Wilhelm II, the German Empire – a later-comer to the European community of nation-states – was considered under threat from the Anglo-Saxonism of Britain on the one hand, and the Slavs of central Europe on the other, as well as the presence of Jews within the Empire itself. For Wilhelm, a nationalist movement was required to defend Germany's racial inheritance and cultural values; and this movement quickly became an expansionist one in the form of a proposed *Mitteleuropa* on the European continent, and German colonies abroad. World War I was the culmination of this Wilhelmian geostrategy.

Following the defeat of Germany and her allies in 1918, and the partitioning of Europe ushered in by the Peace of Versailles, *geopolitik* was reformulated under the leadership of General Karl Haushofer, Professor of Geography at the University of Munich. As Herb (1997) demonstrates, maps showing Germany's purported vulnerability in the face of a new world order were circulated and debated, whilst the philosopher Martin Heidegger was to observe in 1935, 'We are caught in pincers. Situated in the centre, our *Volk* incurs the severest pressure' (cited in Agnew 1998: 100). *Geopolitik* became, as Haushofer also wrote in 1935, 'the duty to safeguard the right to the soil, to the land in the widest sense, not only the land within the frontiers of the Reich but also the right to the more extensive *Volk* and cultural lands' (cited in Walsh 1949: 48). *Geopolitik* was not only an observational analysis of geostrategy, but a declaration of national destiny. Though Haushofer did not reiterate the anti-Semitism of Chancellor of Germany Adolf Hitler's *Mein Kampf*, there is a pronounced emphasis, borrowed from Ratzel, upon the required living space (*lebensraum*) for a vital German populace, an autarky (or self-sufficiency) based on land and a push for colonialism not as a means of generating capital, but as a cultural and even spiritual campaign.

Whilst Haushofer and his students were primarily concerned with articulating a distinctly German school of geostrategy, other scholars contributed to a growing discourse on the nature of the state engaged in such international relations. One such was Carl Schmitt, President of the Union of National-Socialist Jurists, and Professor of Law at the University of Berlin. In his 1921 book *On Dictatorship* Schmitt was critical of the post-World War I Weimar Republic in Germany, arguing that a faith in open political debate and a liberal politics was at odds with the real power brokering occurring behind the scenes. In Schmitt's articulation of a *realpolitik*, or hard-headed pragmatism, the sovereignty of a state lay not in a system of mutual recognition by equivalent states, as posited by the Westphalian doctrine, but in the dictatorial capacity of its constitution. What Schmitt meant by this was, first, that the origin of modern day states lay in the displacement of the notion of divine authority (a religious sentiment) by the secular notion of the ruler as decision-maker. Second, this decision-making took place in the context of the exercise of a legally defined, everyday political constitution, certainly, but also, importantly, roved beyond this. That is, sovereignty entailed the capacity of a leader to decide to suspend such legal frameworks in times of emergency, and to respond with all types of violence that were, by default, in the right. It is this formulation that underpins Schmitt's famous defence of Hitler's continual suspension of the Weimar Republic's Constitution under the dictum, '*Der Führer schützt das Recht*' ('The leader defends the law').

For Schmitt, one of the tasks (and indeed, a constitutive aspect) of the sovereign state was to preserve a unified civil society. This required a careful monitoring of protest and opposition in case of civil warfare, but also the promulgation of a sense of national purpose via the demarcation of friend and enemy. Indeed, for Schmitt, any domain of activity, from the economic to the religious, could be considered as manifesting a politics insofar as it made such a demarcation. Echoing a realist framing of human beings as fundamentally in conflict, the enemy was, for Schmitt, 'in a specially intense way, existentially something different and alien, so that in the extreme case conflicts with him are possible' (1996: 27).

The return of realism

Following the defeat a Germany and her allies in World War II, geopolitics largely fell into disfavour. The grandiose empire-building associated with *geopolitik*, rampant nationalism and racism under the guise of an expanding, organic state, and the bureaucratisation of purge and holocaust, all combined to portray a picture of a political geography in thrall to fascism. In the US especially, geographers were advised to steer clear of such subject matter, and to examine instead a seemingly more neutral, scientific political analysis that relied on statistical modelling, or to contribute to a chorological analysis of regions (Dodds 2010). And, as many have noted, it was not until the 1970s that the term geopolitics once more came into vogue, though references to a geopolitics as geostrategy were certainly present in the 1950s and 60s not only in the US, but also, for example, in Francoist Spain, Portugal, and Turkey, and throughout South America.

As Kearns (2011) points out, the resurgence of interest in geopolitics in the 1970s took a decidedly retro form, as once again statesmen and their advisors, such as Henry Kissinger (who worked for both the Nixon and Ford administrations) and Zbigniew Brzezinski (who worked for President Carter), looked to the machinations of the Great Game, albeit focused this time on Latin America as well as the Heartland. They were sceptical of international law, despite the instigation of the League of Nations (later the United Nations); were concerned with balancing and containing blocs of power; debated land versus sea power; located choke points such as oil, gas and mineral production sites; and displayed a willingness for military intervention in the pursuit of national interest. It is no small surprise to find that rhetoric of vulnerability and containment appears in both American and Soviet commentaries on conflicts in Korea, Vietnam and Afghanistan during the Cold War. And, that such simplistic accounts of people and place were to have such devastating human and environmental costs.

Whereas in the nineteenth century a balance of power had been concerned with the control of resources and the projection of military might, the Cold War was, of course, animated as much by a desire to promote particular socio-political systems, such as 'democracy', over and against others, as it was a matter of capturing market shares. And, whilst events such as then President Nixon's diplomatic trip to the People's Republic of China in 1972 evidenced a hard-headed *realpolitik* awareness of the need to work with and adapt to extant conditions, for the most part the Cold War period can, perhaps, be more usefully described as a return to realism. Indeed, Dalby (1990) has referred to this period as 'neo-realist' in the sense that a belief in the structural inevitability of competition and conflict was to become axiomatic not only within the professional world of statecraft, but also the mainstream academic world of International Relations (IR). The rewriting of the twenty-first century's 'war on terror' as a 'clash of civilisations' in best-selling books (the key text here being Huntington 2002) has certainly borne out Dalby's contention.

As will be shown below, it is this resurgence of realism that has prompted in turn a determinedly *critical* accounting of geopolitics in the aftermath of the Cold War. That is, numerous scholars in geography as well as IR have been concerned to unpack the assumptions underlying realism as a field of academic inquiry, in addition to bringing attention to how and with what effect it has underlain various forms of statecraft, from the planning of military operations to their legitimation via the media. Nevertheless, it is worth pointing out, as a means of closing this section, that *realpolitik*, and the subjugation of a moral imperative to the national interest, remains a key object of analysis. As some have noted, legal arguments emerging in the US and Europe as to the 'unlawful combatant status' of targets, for example, which remove these subjects from the jurisdiction of the Geneva Convention, as well as the use of interrogation *cum* torture techniques, drone attacks and extraordinary renditions, bear something of a resemblance to Carl Schmitt's writings on all forms of violence as potentially rightful instruments of a sovereign state in times of emergency (Abraham 2007; Kutz 2007). As numerous geographers have observed, whilst the twentieth century was arguably punctuated by military engagements, the twenty-first century, by contrast, seems to be a time of continual, ever-present warfare and emergency.

Reanimating geopolitical theory

Whilst this section deals primarily with the emergence of lines of inquiry that both offer an alternative epistemology to that of realism, and place realism itself under critical scrutiny, it is important to acknowledge that over the course of the twentieth century there were other modes of thought and practice. For Kearns (2009), whose work addresses the writings of Kropótkin, Kingsley, Hobson and Reclus, these are 'progressive geopolitics', insofar as they take on board the notion of a global interdependency but, in contrast with realism, assume a generosity of spirit, manifest in the work of solidarity in face of oppression, and the fundamental claims of human rights. What is more, whilst the academic lines of inquiry outlined here seek to open up debate on what both the 'geo' and the 'political' can and should signify, these are very much inspired by practical efforts outside of what was officially termed statecraft, such as post-colonial struggles for independence, trans-border unionisation efforts, civil and human rights movements, and anti-globalisation coalitions. Indeed, what these examples afford an insight into is the breadth of choice available as to how the geopolitical was to be framed and deployed at the state level: that one hegemonic understanding of the same was to emerge under the guise of realism is testament to the fact that geopolitics did not so much serve a national interest so much as reflect prevailing inequalities of privilege and resource, pander to particular forms of prejudice, and facilitate certain economic interests.

That the field of thought and practice we refer to as geopolitics is very much a complex, often contested social enterprise has been taken to heart by a loose field of inquiry that, whilst drawing on a wide variety of traditions and articulating diverse understandings of the objects of analysis usually associated with geopolitics, nonetheless is unified by a collective concern to make explicit the manner in which such knowledges are produced and disseminated, and with what effect. As noted in the introduction, this can be called a modern*ist* approach insofar as there is a refusal to take realist geopolitics as a naturally given condition of the world; rather, it is considered a product of time and place that has, also, been subject to revision and reappraisal from the vantage point of the present day. The term modernist here is very much associated with a sceptical approach to knowledge production, and an appreciation of how such knowledges can be put to use not as glimpses of the reality of things, but as a means to achieve particular ends such as social control and mastery over the environment.

Yet, the lines of inquiry outlined below are often referred to as *post*modernism. This is because they draw on broader-scale ideas and concepts that take issue with a host of rhetorics associated with modernity, by which is meant a series of social and environmental conditions such as industrialism, urbanism, widespread militarisation and bureaucratisation. There are two main points of contention here. First, these modern rhetorics, which sought to describe but also to legitimise such conditions as the necessary outcome of prevailing trends, tend to revolve around the notion of a linear time-frame along which society (or at least certain elements of it) 'progresses' or 'develops'. The very idea of an 'origin' to fields such as geopolitics, and 'canonical' texts, is an example of such a modern, linear rhetoric; the 'rise' of civilisations, and the 'march' to equality, are similarly phrased. Second, and this is of particular import to geographers, such modern rhetorics are bound up with a particular rendering of space as a two-dimensional field across which locations are sited, and processes – from the building of roads

and the epidemiological spread of disease to suburbanisation – flow and ebb. Sometimes referred to as a 'Cartesian space' or a 'grid system', such spaces are not, critics argue, given features of the world, but a way of framing it such that particular projects that 'pin down' people to place can be planned and enacted. As postmodern theories have pointed out, both of these forms of rhetoric are immensely powerful, insofar as they persuade us as to the necessity, and even the value, of social and environmental conditions today. What is more, they insist that future events are already fated.

Aptly, then, the following lines of inquiry are placed in no particular chronological order. They also are not clearly demarcated from each other, despite the use of subheadings; instead they overlap time and again, establish contingent points of distance from each other, and otherwise become thoroughly entangled.

Subaltern geopolitics

We begin with subaltern geopolitics because it offers a profound critique of geopolitics as it re-emerged in the 1970s at the height of the Cold War, as well as the contemporary emphasis in much of IR on neo-realism, but also because it makes the point that, for much of its history, geopolitics has effectively glossed over four-fifths of humanity. That is, whilst geopolitics has purported to be global in scope, and has certainly under-pinned imperialist projects that profoundly impacted peoples across that globe, there is simply no acknowledgement of how those peoples not only lived with imperialism on a day to day basis, but also how they envisioned, and enacted, their own place in the world.

For Sharp (2011) subaltern geopolitics proceeds as a sustained acknowledgement of post-colonial thought and narratives of people and place. Importantly, whilst post-colonialism recognises an asymmetry in the power relations behind the production of images and texts – an issue encapsulated in Spivak's famous question, 'Can the subaltern speak?' if, in order to be 'heard', there must be the use of imperialist iconographies, phrasings and cadences – there is a disavowal of the subaltern as either intrinsically different (which begs the question of to what?) or resistant (again, the question becomes to what?). Both of these formulations merely serve to place attention back onto the purported centres of activity, namely the imperial Western powers. Subaltern geopolitics, Sharp argues, 'refers to spaces of geopolitical knowledge production which are neither dominant nor resistant, because studying only the dominant accounts and those that absolutely oppose them, can have the effect of reifying this binary geopolitical structure rather than challenging it' (2011: 271). Indeed, for Sidaway *et al.* (2013), it is useful to think in terms of the vulnerability of imperialist logics, and how their subversion, and even their mimicry, led to concerted efforts, often including the use of violence, to reassert their function as a means of ordering people and place.

As a mode of critical inquiry, then, subaltern geopolitics can be said to situate itself not in opposition to a realist geopolitics and traditional modes of statecraft, but as a critique of the same. Critique here refers to an appraisal of how and with what impact such a body of knowledge emerges, as well as an unpacking of its underlying assumptions, where the critique itself is not intended to 'privilege a singular history of knowledge associated with a specific world region . . . or presume conceptions of knowledge that implicitly or explicitly assume their own self-evident universality . . .' (Agnew 2007: 139). Such a description, however, does not do the term full justice. As Sharp goes on to point out in her introduction to a special issue on the topic in *Geoforum*,

> (subaltern) practices, whether strategies of survival or getting on with everyday life in Palestine, newspaper publication in Tanzania, or practices of peace building in the Philippines or Colombia, are all ways of reworking dominant geopolitics not simply through critique, but through offering up lived alternatives.
>
> (Sharp 2011: 271)

There is an attentiveness here to the materialities of everyday life as they constitute the substantive foundations – the bodies, the subjectivities, the practices and discourses – of constantly unfolding tensions and conflicts. These materialities are visceral, emotional, affective and (for some) transhuman. Put these careful engagements with material conditions together with a deep ethical concern to draw attention to the production of vulnerability, and it is not too surprising to find that such analyses are very much concerned with a wide range of present and emerging real world events (see Smith 2011; Woon 2011; and Koopman 2011).

Feminist geopolitics

The emphasis in subaltern geopolitics on the 'lived' dimensions of geopolitics resonates strongly with a feminist geopolitics. Building upon a trans-disciplinary, feminist project that foregrounds, as an entry point for analysis, the bodies of those at the 'sharp end' of various forms of international activity, from immigration to development to warfare, feminist geopolitics covers a range of spaces and concerns (see Kofman 1996). Arguably, what holds these analyses in tension with each other is a collective desire to expose the force relations that operate through and upon those bodies, such that particular subjectivities are enhanced, constrained and put to work, and particular corporealities are violated, exploited and often abandoned. As Dixon and Marston note in their introduction to a themed issue on feminist geopolitics in the journal *Gender, Place and Culture*, 'These articles expose the proliferating bodies of geopolitics, not simply as the bearers of socially demarcated borders and boundaries, but as vulnerable corporealities, seeking to negotiate and transform the geopolitics they both animate and inhabit' (2011: 445).

Just as theoretically and methodologically diverse as post-colonial theory, feminist analysis has worked to bring to light the embodied, everyday, informal practices that make manifest the 'place' of traditionally disempowered people – such as women, children, immigrants, asylum seekers, prisoners and others – within all manner of ostensibly geopolitical land-scapes. In this vein, a feminist approach to the traditional repertoire of geopolitics – nationhood, the state, borders, security, refugees, militarism and warfare – has sought to reconfigure these complex objects and practices, questioning their normativity within statecraft *and* academia (Dowler and Sharp

BOX 11.2 TANZANIA AND PAN-AFRICANISM

Where is geopolitics? For much of this chapter, geopolitics has been situated in the midst of European and American debates on statecraft, rivalry and the projection and legitimation of military and economic might. Sites such as Africa are present by virtue of their representation as geopolitical spaces to be surveyed and controlled, as well as, more recently, spaces of failed states and aid recipients. Sharp (2013) makes the point that the academic field of geopolitics has tended to deconstruct such representations, rather than investigate the emergence of alternate worldly imaginations. In looking to the geopolitical vision of Julius Nyerere, the architect of post-Independence Tanzania, Sharp stands Nyerere's rhetoric of hope and inclusion against the cynical neo-realism of classical geopolitics. Nyerere was to become a much revered advocate of Pan-Africanism, which sought a mutually beneficial solidarity of those of African descent both on the continent and beyond, as well as the Non-Aligned Movement, formed in 1961, which sought to consolidate states lying outside of US and Soviet power blocs as a powerful consortium in their own right. For Nyerere, Sharp argues, a subaltern geopolitics entailed both an acknowledgement of the power of a Cold War *realpolitik*, and its disavowal, made possible by a Tanzanian policy of state ownership of the means of production and a self-sufficiency. Nyerere would embrace traditional communal values that, whilst emerging from a particular, regional heritage, could yet prove inspirational across the African continent and beyond. These were to be realised in 'the bodies and identities of Tanzanian citizens', Sharp writes, but were also to become a 'doing' or 'embodiment' of subaltern geopolitics manifest well beyond the territorial boundaries of this newly recognised nation-state.

Key reading: Sharp (2013).

2001; Hyndman 2007; Secor 2001). What this reconfiguration has led to, at the disciplinary level, is an increased awareness of the diversity of attitudes, emotions and behaviours that make up the 'matter' of the geopolitical. Indeed, numerous scholars have sought to flesh out what Jennifer Hyndman (2003) calls a 'feminist geopolitical imaginary' by: complicating our understanding of key concepts such as corporeality, emotions, and ethics; exploring new objects of analysis such as trauma and violence, terrorism, security and conspiracy; and reaching out to other disciplines, including psychology and literary theory, as well as politics/political science and international relations, and post-colonial studies, to help animate these.

The empirical *depth* of much of this work, often founded upon a committed period of fieldwork, and the careful gathering of lengthy, *in situ* interviews, highlights a complex, feminist ethics of care. That is, at one level, this commitment to field work is very

much embedded within a feminist concern to engage with others, to work through ethical issues of trust, responsibility, empathy and compassion. And, as with subaltern geopolitics, there is another, just as firmly embedded concern here, and that is to unsettle the implied fixity of social categories – *the* marginalised, *the* vulnerable. There is a desire in feminist analyses to allow the conditions of the site in which the researcher is engaged to help specify the subjectivities that are at work, and the ways they shift and settle under different stresses and pressures so that we are able to recognise how space and power are differentially experienced and embodied (see Fluri 2011; Martin 2011; and Mills 2011).

Posthuman geopolitics

Both subaltern and feminist geopolitics have very much taken up the challenge of examining what it is

BOX 11.3 GEOPOLITICS OF LOVE AND DESIRE

How and to what extent can love and desire be managed as factors in geopolitical strategy? Smith (2011) examines the ways that geopolitical conflict in Leh, India (in the contested states of Jammu and Kashmir) reverberates through women's (and frequently men's) bodies as they are caught up in the intimate issues of sex and birth. In Leh District, marriage between a member of the Buddhist majority and the Muslim minority is fraught with tension because religious identity has become so deeply politicised around the tenuous territorial status of the state. 'On the margins of the nation-state', Smith writes,

> J&K is perceived as particularly vulnerable, to be gained or lost in part through population – with each voter marked by their religious identity. This anxiety over the demographic makeup of the state renders the body a key site of contest, and the gendered nature of this territorialisation highlights women's bodies in particular as contested territory.
>
> (2011: 456)

Whilst marriages may well be micro-managed, there yet remains the thorny issue of love and desire in such a context. And, for Smith, working through interviews with and the oral histories of several generations, love and desire become complex objects that are not open to explication, but are more usefully thought of as narrative devices told about young couples, and the various purposes such devices serve, such as rallying co-religionists around a perceived threat, providing a code of conduct for how young women in particular should behave and, perhaps, engendering a frisson around contemporary practices of marriage sanctions and their transgression.

Key reading: Smith (2011).

to live with various modes of foreign (and related domestic) policies. And, in the process, both have problematised the assumptions underlying so-called 'geopolitical' lines of inquiry; these are not abstract renderings of a detached academic inquiry, it is argued, but modes of knowing that serve some interests over and against others. In the last section of this chapter I dwell on a field that, whilst having much in common with the above, nevertheless has its own particular goals. For the most part, these lie in revisioning the conceptual terrain of geopolitics by taking to task the significations associated with the 'geo', and offering alternate understandings of the same.

Whilst, as noted earlier, the relatively simplistic accounts of the physical landscape as determining the bodily morphologies and behaviours of a racialised population (most notably manifest in the work of Semple), were disavowed by generations of geographers, there is no doubt that across the discipline there is a renewed attentiveness to the nature of the 'geo' in geopolitics. This time, however, the geo is not a matter of the cumulative impact of environmental *effects*; rather, all manner of relations, and all manner of objects of analysis, that help constitute the geo have been posited under the loose banner of a *post*humanism. Put briefly, a posthuman geopolitics displaces not the sovereignty afforded the state in classical accounts, but rather the sovereignty of the human. That is, posthumanism queries both the rhetorics and practices (in the life sciences and law, as well as in geopolitics) that take the uniqueness and the agency of human beings for granted. In place of these, we find, for example, reference to the *affect* of the environment upon emotional registers, and emotions as *contagious*; here both affect and emotion do not belong to one side or the other – environment or human body – but rather emerge from the interplay of both. We find reference to the capacity of bodies to undertake work, a capacity enabled and limited by their interaction with other bodies, some human, most not. And, we find reference to the complex assemblage of such human corporealities, and the role of inherited DNA, epigenetic influences, microbial viruses and bacteria, neural receptors, kinaesthetic sensitivities and so on, all of which subvert the idea of a sovereign subject. Importantly for this section, all of these de-centring moments provide a sense not only of the complex inter-relations that continually exceed the enclosed geographies of the body, but also a sense of the deep history, or 'Earth history', within which all life, as well as the materials that enable and aver this combination, is embedded (Dixon *et al*. 2012). As Hird (2010), for example, notes, 'Bacteria are Gaia theory's fundamental actants, and through symbiosis and symbiogenesis, connect life and matter in biophysical and biosocial entanglements' (p. 54). Clark's (2005) concept of the geo is even more far-reaching; here, life, evolves through a 'succession of time-irreversible events which are in turn entwined in the no less-irrevocable movements of the world around it, all the way up to the level of the universe' (p. 165). Power as an analytic concept is

less relevant to such analyses; rather, these speak to an ethics of responsibility, indebtedness and justice.

Posthumanism also finds a world-ly expression in the *umwelt* of organisms, and their aesthetic expressivity. *Umwelt*, a term associated with the theories of biologist Jakob von Uexküll, is sometimes translated into English as a 'self-centred world'. It captures a sense of the importance of various facets of the environment for a particular organism – whether in the form of food, shelter, threat of navigation point – and the manner in which the interactions of organism and these environmental elements are thus laden with signification, a term more traditionally associated with uniquely human language systems. The geo becomes, in effect, a 'semiosphere' (Kull 1998). In pursuing this line of inquiry, some geographers have looked to the signifying role of environmental elements – rivers, for example – more usually described as merely a matter of mass and energy (Hinchliffe *et al.* 2005). Some have also made the argument that a geopolitics can be usefully understood as entangled, yet mutually incomprehensible, *umwelts*, including that of humans (Shaw *et al.* 2013). Such an approach necessarily sees politics as the cut and thrust of competition, but also the exertion of particular capacities to engage with the environment in a manner that does not curtail the capacity of another (Box 11.4).

BOX 11.4 GEOPOLITICAL VIOLENCE AND THE LOSS OF BIODIVERSITY

How do we envision a world profoundly impacted by human beings? For just as nature can be 'de-naturalised', so humans can be considered to work as agents in an Earth history. Yusoff (2012) notes how images, facts and figures pertaining to loss of biodiversity, and especially species extinction, tend to become objects of mourning that 'presence' the dead in particular ways. And this presencing can certainly take the form of a conservation politics. Much of this loss, however, 'is never construed as a loss at all, rather as an achievement of eradication or control' (2012: 580). In part, this inability to apprehend loss is because, despite the excess of life on Earth, human beings are able to sense (see, hear, taste, smell, touch) only a very small part of this. We must acknowledge, Yusoff argues, how the sensible governs the sphere of human experience. But, we must also become much more aware of, 'how this sensibility is extended into the insensible worlds of subjects that are unavailable for our meetings; the great multitude of writhing life that eludes, or fails to collude, collaborate or cohabitate with our technologies of presence' (2012: 580). Care and affinity are two emotion-laded frames via which we apprehend and respond to loss of biodiversity, as well as mourning and violence. These last two are more often than not side-lined in a consideration of posthuman geopolitics. Yet, as Yusoff observes, both offer a way of thinking about the ties that bind across kingdoms and species, challenging us to 'reconfigure our understandings of relations beyond our social to our worldly sensibilities' (2012: 581).

Key reading: Yusoff (2012).

Summary

In this chapter we have given an account of the emergence of what has been termed a 'classical' geopolitics, noting the time and place specific concerns of some of the key thinkers associated with this body of work, and the subsequent critiques advanced of these theories. As we discussed at the start of this book in Chapter 1, much of this history reflects developments in political geography; however, it is useful to note that for some decades the term political was seen to be concerned principally with international relations. *Geo*politics was the leading edge of political geography. We have also noted how and with what effect classical geopolitics became entangled with a practical statecraft, such that an ostensibly academic analysis of the real world itself became a blueprint for how state policy is to be accomplished.

The tendency for classical geopolitics to help create what it purportedly set out to find has very much been taken to task by a *critical* geopolitics. Certainly, such work is critical in the sense that it queries the conditions under which knowledge is produced and disseminated. In outlining some of this work, we have also, however, aimed to show how terms and objects associated with classical geopolitics – from borders to the globe itself – have been, and are being, revisioned, such that new forms of geopolitical inquiry emerge. For some, these developments invoke a concern, insofar as the remit of geopolitics seems to become broader by the day, whilst the diverse expertise that can be brought into play – including now psychoanalytic theory as well as the performing arts, biology and physics – dilute a sense of what *geography* brings to this terrain. Yet, we would suggest that these developments

are welcome insofar as they indicate not so much the relative paucity of a classical geopolitics, but a pattern of decision-making as to what mattered to the field and what did not. This patterning effectively glossed over the lives of much of humanity; if a critical geopolitics is to continue to flourish, then it is this everyday world, and its sheer complexity, that students of political geography need to engage with.

Further reading

For a more detailed overview of the history and development of geopolitics, see Flint (2006) *Introduction to Geopolitics*, whilst good critical commentaries on early geopolitics include Kearns (2009) *Geopolitics and Empire: The Legacy of Halford Mackinder*, Smith (2004) *American Empire: Roosevelt's Geography and the Prelude to Globalization,* and Ó'Tuathail (1992), 'Putting Mackinder in his Place' in *Political Geography* 11, pages 100–118.

The emergence of 'critical geopolitics' in the late twentieth century is reviewed by Dalby (2010), 'Recontextualising violence, power and nature: the next twenty years of critical geopolitics?' in *Political Geography*, 29, 280–288. Jennifer Hyndman puts the case for a feminist geopolitics in 'Towards a feminist geopolitics', *Canadian Geographer* (2001), 45, 210–222, and develops the argument in Hyndman (2007), 'Feminist geopolitics revisited: body counts in Iraq', in *Professional Geographer*, 59, 35–46. A special issue of *Geoforum* in 2011 on 'subaltern geopolitics' (volume 42, issue 3), further expands critical perspectives on geopolitics in contrast to its origins in 'statecraft', with papers examining geopolitics in a range of everyday situations.

Public policy and political geography

Introduction

Political geography has always had degrees of relevance to, and influence over, real-world issues (House 1973). Sometimes this has not been progressive or productive. The military dictator General Augusto Pinochet, for instance, was trained as a political geographer and used this background to remake Chile during the early 1970s. According to David Harvey, 'Pinochet did not approve of "subversive" academic disciplines such as sociology, politics and even philosophy'; geography was his poison for instilling patriotism, regulating culture, and undertaking social engineering (Harvey 1974: 18). As President of the military Junta, Pinochet overthrew a democratic and elected government and undertook brutal reforms on, amongst other things, health and social policy. Harvey talks about the ways in which military control allowed Pinochet to smash the actors and institutions of the progressive Allende regime, which created the space for re-establishing the 'old geography' of a centralised and dictatorial power base.

The relationship between geographers and public policy (defined in Box 12.1) has been critically questioned over the past decade (see Dorling and Shaw 2002; Lee 2002; Martin 2001; Massey 2000; *Scottish Geographical Journal* 1999) and we would argue this is important for political geography. In 1999, Jamie Peck wrote an editorial statement in the journal *Transactions of the Institute of British Geography*. This was partially a response to Brian Berry's (1994) call for more public policy analysis in geography, and also an attempt to provoke a similar debate to that raised by Harvey and others in the 1970s. Peck asked why geographers were not involved in the policy-making process under Britain's Labour Government given our 'unique insights into "real-world" processes and practices' (Peck 1999: 131). Replies followed and the early twenty-first century is a fruitful time to consider if, or how, political geographers can contribute to public policy. If geographers are, as Doreen Massey claims, working 'themselves up into quite a lather' (Massey 2002: 645) then this could be important.

This chapter considers the importance of these debates for political geographers by discussing the different exchanges between human geography and public policy over especially the last thirty years.

BOX 12.1 PUBLIC POLICY AND GEOGRAPHY

According to Ron Martin, public policy involves 'any form of deliberate intervention, regulation, governance, or prescriptive or alleviative action, by state or nonstate bodies, intended to shape social, economic or environmental conditions' (Martin 2001: 206). Ron Johnston offers a similar synopsis, where public policy is the 'study of and involvement in the creation, implementation, monitoring and evaluation' of public initiatives (Johnston 2000: 656).

Key reading: Johnston (2000); Martin (2001).

Although the majority of these debates have taken place in British geography journals and have often not involved leading political geographers, they have implications for the wider discipline of political geography. The chapter starts by looking at links between geography, empire and public policy, and questions whether this was the golden-age for geographers and the policy process. The chapter then considers the 'relevance debate' of the early 1970s, which pushed public policy back onto the geographical agenda. It also discusses the more recent exchanges that are recommending a new 'policy turn' in the discipline. To think about what political geography could offer this debate, the chapter concludes by suggesting that the capitalist state's role in the policy process is a key missing link throughout these exchanges and we discuss how political geography students could consider this in their own work. By the end of the chapter, it becomes evident that public policy questions social science itself (Blowers 1974). Our position on where political geographers can perhaps make a difference is summarised as Box 12.2.

Geography and empire: the golden-age of public policy?

The example of Pinochet can be contrasted with the more balanced interventions made by Sir Halford Mackinder. As we suggested in Chapter 1, the early intellectual foundation of political geography rested on the transition from systematic to regional geography, which at that time made descriptive connections between physical spaces (natural regions), and social and political ('ethnographical') worlds (Mackinder 1902). Political geography at this time was very much an inductive science and it influenced British thinking on the ascendancy of the territorial state, set within a context of rapidly shifting power relations. Mackinder was central to this context and after being Director of the London School of Economics and Political Science – which has historically positioned academics close to Britain's national political machinery – time was spent as Britain's High Commissioner for South Russia. Chapter 1 highlighted that Mackinder's book *Democratic Ideals and Reality*, an interpretation of world power-politics, was presented as a warning to the peacemakers at Versailles (Mackinder 1919).

Was this the golden-age of close relationships between human geographers and policy processes? Mackinder was heavily involved in the Royal Geographical Society, which at that time had close links to the British political system, and also held numerous positions within British commerce and industry. Mackinder certainly had the 'ear of ministers'. Others followed in this line and acted as advisors to British Parliamentary Committees. Sir Dudley Stamp worked closely with the government on mapping agrarian trends and influenced post-war land-use policy in Britain. Stamp was later rewarded

BOX 12.2 POLITICAL GEOGRAPHY AND PUBLIC POLICY: ARGUMENTS AND QUESTIONS

The contemporary political geographer is *not* a policy maker, but they are public: they can only offer academic analysis on public policy, but in doing so can influence those involved in policy-making processes. They 'cannot take decisions' (Blowers 1974: 32). This said we perhaps have a moral obligation to take public policy seriously in our ongoing academic analysis. This involves not only writing critically about public policy and the policy-making process. Where possible, geographers can use their various insights to inform political and policy practice, accepting the tensions between political/policy action and academic/intellectual critique. What do *you* think?

Key reading: Blomley (1994); Blowers (1974); Harvey (1974); Massey (2000); Ward (2006).

for contributions to the 'use of the land' (Dickinson 1976: 7). John House discusses the involvement of geographers in urban and regional planning since the 1930s, both through the involvement of academics in giving policy advice and also through the direct employment of geography graduates within the state's apparatus. Geographers were the driving forces at the Barlow Commission's 1937 to 1940 inquiry into the 'Distribution of the Industrial Population' – a key moment in the evolution of the Keynesian welfare state (see Chapter 3) – and occupied strategic positions within the civil service (House 1973). Brian Robson saw these interventions as effective ways of getting the human geography voice heard (Robson 1972).

After 1945, human geography searched for a stronger intellectual and scientific identity and this pushed political and social necessity into the background. In Britain, for instance, key ties between academia and policy-making were also restructured through the formation of the Institute of British Geographers (breaking away from the Royal Geographical Society). Combined, this created the space for a 'new-style' of geography (House 1973). Instead of serving the needs of Empire and its Colonial Surveys, and the nation and its Regional Surveys, academics developed a university-based intellectual and pedagogic discipline that went hand-in-hand with nurturing human capital in accordance with national socio-economic needs and modernist political priorities (see Harvey 1974; Unwin 1992). Human geography also broadened its intellectual reach by incorporating developments in, and having a critical dialogue with, other disciplines (such as sociology, economics and politics) to capture multiple ways of interpreting 'geographical worlds'. The 'old-style' geography of human-physical-state interactions (House 1973), which policy makers could perhaps understand, was gradually replaced by a diverse and intellectually stimulating set of agendas.

The rise and fall of relevance debates

During the early 1970s, a 'new wind of change' swept across academia and brought with it a more 'radical geography' (Berry 1972) that was not based on traditional concerns with location, classification, regularity and conformity. For David Smith and others airing their thoughts at the Association of American Geographers' conferences, radical geography meant a politicised 'social geography' that practised 'social responsibility' with greater professional involvement in welfare rights, social justice and political activism (Smith 1971). The context for this was the Vietnam War, student riots in Paris, and growing urban poverty and social inequality (see Watts 2001). Michel Dear also suggests that some geographers at this time became alienated by a geography overly focused on quantitative techniques and under-concerned with real-world issues (Dear 1999). To cut a long story short, advances in social philosophy appeared to be bringing with them a renewed sense of academic and political responsibility.

The watchword of these times became 'relevance'— introduced to gauge the degree to which geographers were making a contribution to the analysis and resolution of economic, environmental and social problems (Prince 1971). Such claims were contested and the 1970s witnessed a wealth of debates, with clear disagreements between liberals, humanists, Marxists and others on *how* to tackle geography and public policy. In one intervention, Brian Berry argued that 'an effective policy-relevant geography involves neither the blubbering of the bleeding hearts nor the machinations of the Marxists. It involves working with – and on – the *sources of power and becoming part of society's decision making apparatus*' (Berry 1972: 78, emphasis added). Berry added that academic analysis had to be interdisciplinary and follow a problem-orientated approach, where 'the solution to social problems would be facilitated by careful, clear projection of policy objectives, programme alternatives, and underlying economic and social forces, proceeding towards a solution through experimentation and feedback guided by theory and analysis' (ibid.: 80). This challenged the purely intellectual and scholarly pursuit of human geography, by advocating a more applied discipline.

This created ripples across the Atlantic at the 1974 annual meeting of the Institute of British Geographers. In a presidential address, Terry Coppock argued that, amongst other things, strategic research – utilising computer technology to allow prediction through modelling – was required to demonstrate the geographers' role in formulating alternative public policies (Coppock 1974). For Bridget Leach, this debate provided an opportunity to discuss how policy problems become politically constituted. Leach argued that 'diversionary tactics' were used by policy elites to protect the legitimacy of the political system and by highlighting such tactics academics could inform debate by empowering opposition groups (Leach 1974). Open questions, however, remained as to how this could be achieved, given that – with the exception of David Harvey's interventions – little attention was paid to uncovering the 'decision-making apparatus' and 'sources of power' under capitalism.

For Harvey, understanding the (public policy) world was about using this insight to change things and 'before geographers commit themselves to public policy, they need to pose two questions: what kind of geography and what kind of public policy' (Harvey 1974: 18). The two categories are not separate – they are seen as linked through the moral obligations that geographers have to create a better society. According to Harvey

> relevance in geography was not really about relevance (whoever heard of irrelevant human activity?), but about whom research was relevant to and how it was that research done in the name of science (which was supposed to be ideology-free) was having effects that appeared somewhat biased in favour of the status quo of the ruling class of the corporate state.
>
> (Harvey 1974: 23)

Harvey's Marxist stance encouraged geographers to break out of this loop and challenge the 'corporate state' within capitalism.

Last, and perhaps most interestingly for Peter Hall, geography had much to learn from political science for uncovering the actors, power networks and organisational dynamics at work in the policy process. Hall suggested that a *new political geography* was on the horizon and

> [f]rom this, certain central lines of research seem to follow. The new style of urban political geographer, for that is what he [sic] seems destined to become, will be concerned with the values, the organization, and access to power of groups. He will analyse the relationship of these groups to the decision-making machinery (and the personalities who operate this machinery) at different levels of government. He will study how different agents in the decision process – politicians, bureaucrats, technicians, opinion-formers – interact, how they form alliances and coalitions, how they bargain, promise or threaten each other to obtain objectives. His concern . . . is to analyse what happens, not to postulate what should happen. Yet, by the very fact of exposing the way decisions are taken in practice, I would expect and hope that the political geographer would provide powerful suggestions for future improvement.
>
> (Hall 1974: 51)

Hall's ideas were taken forward by geographers in their work on 'urban managerialism', which was interested in the roles played by different agents (such as local government, central government, builders, estate agents and landlords) in producing cities. Some of this research featured in a special issue of *Transactions of the Institute of British Geographers*, where Robson argued it captured 'a clear reflection of the social concern and the interest in process rather than form which have made geography in the middle 1970s a very different animal from that of a decade ago' (Robson 1976: 1). Simon Duncan's paper on 'social geography' and the city, which questioned how housing systems worked and who benefits from this – so that an 'information base' could be provided to influence 'political will' and 'achieve change in the allocation of houses and housing resources' – was indicative of these process-based concerns (Duncan 1974: 10).

The challenge to managerialism contained some of the explanations for the reduced interest in public policy. Hall's 'new political geography' was continually challenged by those who felt that it could not sufficiently explain the links between the structure of the housing market, the actions of agents and institutions, the spatial arrangement of the city and wider capitalist society (Bassett and Short 1980). But, instead of theorising the links between the capitalist state, class, power and urbanisation – which could be fruitful ways of capturing the policy dynamics within the contemporary city – some Marxists saw this agenda as ultimately reproducing the 'corporate state' by working within the constraints of capitalism, as opposed to transcending capitalism itself (Martin 2001).

The cooling of 'relevance' debates was further aided by the reaction to these trends. So the argument goes (Martin 2001), the 1980s gave birth to humanistic geography and then the recasting of the 'social' within social and cultural geography – both countering structurally determined processes under capitalism (as expressed by certain Marxists) and models of expected behaviour (as predicted by spatial/regional scientists). As these authors grappled with new ways of exploring the action, movement and experience of individuals within geographical settings (see *Area* 1980), attachments to public policy became increasingly more tenuous. To be fair, public policy was not the object of analysis for these scholars, who were more interested in place and the geographies of everyday life. Some continued to work within the tradition of old school 'social geography'. Robson, for instance, developed an urban geography that influenced those formulating British urban policy (see Robson *et al.* 1994). Berry followed a similar career path in the US, through research on housing (see Berry 1994).

Trends in academia were not the only explanation for this movement away from what could be considered 'relevant' geography. The onset of neoliberalism (see Chapter 3), especially in North America and Britain, alienated scholars from undertaking policy relevant research. Policy makers required research that was ideologically relevant for justifying privatisation, deregulation and public sector restructuring, and they turned to economists practising predictive and normative thinking. Economists were the perfect ideological bedfellows for the New Right governments of the 1980s and 1990s.

Public policy in the twentieth century: shallow, deep or just grey?

This relationship between economists and policy makers is one of two entry points for more recent debates (Peck 1999). The other relates to the increasing dominance of postmodernist thinking and according to critics, the perceived irrelevance of these approaches to address real-world issues (Martin 2001). We consider each of these in turn.

The first agenda is clearly evident in the debate initiated by Peck, concerned by the fact that few geographers appeared to be advising Britain's Labour Party in the late 1990s (Peck 1999). Peck's work at this time centred on welfare-to-work – a key political strategy in the first term of Blair's government which, through geographically specific modes of policy transfer, sought to move Britain towards a North American (post-welfare) model (see Chapter 3). The key advisors in the US and UK were right-wing labour market economists, who manipulated the local-embeddedness of policy, and Peck bemoans the roles that these 'intellectuals' play in globalising welfare state restructuring. Given the unique insights that geographers possess, Peck asks: where are the geographers in this policy process; why do economists have 'the ear of the minister'; and why are geographers always placed at the bottom-end of the policy research ladder, examining hard outcomes and being excluded from policy formulation (Peck 1999; also Massey 2001)?

Peck's explanation points to contemporary academic practices, which privilege abstract scientific knowledge over and above more practical and policy-orientated concerns. Because public policy research is not considered scientific or 'top-draw' quality, it is consequently deemed 'bad science'. This process is augmented by targeted research that can gain universities academic excellence and their staff promotion. Consequently, geographers are being encouraged to come up with theoretical innovations and write 'big papers' and through time this has reduced public policy research to a 'grey', boring and somewhat second-rate academic practice (Peck 1999).

Peck challenges these assumptions and argues that public policy research does not have to be this way: it

BOX 12.3 BEYOND GREY GEOGRAPHY: SHALLOW AND DEEP POLICY ANALYSIS

Shallow analysis

This is policy research that is confined to addressing the 'stated aims and objectives' of policies from within an orthodox theoretical position. This often serves the needs of the policy-making system, which it often takes for granted, by licensing quick-fix solutions. Shallow researchers are often closer to the policy-making process that their deep colleagues. Examples of shallow researchers include mainstream economists.

Deep analysis

This is research that sees policy as politicised and contested and questions the 'parameters and exclusions of policy-making'. Deep policy researchers often take a theoretically unorthodox position and question the local embedded and path-dependent nature of public policy. Examples of deep research include those undertaking critical investigations of the policy-making process.

Question: 'which group is conducting the most effective form of policy analysis'?

Key reading: Peck (1999, 2000).

can be theoretically and politically progressive. Policy research, then, 'is a legitimate, non-trivial, and potentially creative aspect of the work of academic geographers, but one that we are currently neglecting and/or undervaluing' (Peck 1999: 131). To take forward this agenda and its potential creativity, Peck makes a critical distinction between 'shallow' and 'deep' public policy analysis (see Box 12.3) and argues that geographers have much to offer a deep approach that engages 'critically and actively with the policy process itself' (Peck 2000: 255).

In a reply, Jane Pollard and colleagues argue that the involvement of geographers in public policy is far greater than that implied by Peck (Pollard *et al.* 2000). They agree that geographers are generally not that involved in national-level social policy debates, but they are involved in other policy areas, which have merits and cannot be written off as trivial. This argument is extended by Mark Banks and Sara MacKian, who urge Peck to take stock of the ways in which geographers are involved in evaluating the themes of 'renaissance', 'partnership' and 'social

capital' within British urban policy (Banks and MacKian 2000). Peck's response emphasises the importance of teasing out relationships *between* different levels of public policy, to overcome what is seen as a danger of failing foul of the rhetoric of localism (Peck 2000). The challenge for geographers, then, 'is to connect together the smaller pictures with the bigger pictures of the policy process, to connect the specific with the general, without undermining the integrity of our particular take on the policy process' (ibid.: 257).

The approach adopted by Ron Martin is somewhat different and situates the decline in public policy research within a broader academic context. Like Peck and others, Martin also feels that human geography is exerting little influence on policy. Rather that questioning the motives behind what we publish or exploring the general crisis in the social sciences in relation to 'relevancy' (see Massey 2001), Martin pursues an internal critique of the geography discipline. Martin argues that 'much of what is done under the banner of human geography is unlikely to be seen by policy makers as being remotely germane to policy issues' because it 'has little practical relevance for policy; in fact, in some cases, one might even say little social relevance at all' (Martin 2001: 191). For Martin, this is why geographers are not being asked their opinions when it comes to consultations on policy-making. We talk the wrong language, are lost 'in a thicket of linguistic cleverness', ask the wrong questions, and generally do not deliver research findings that have 'relevance to real-world issues' (ibid.: 196; also Martin 2002).

Martin takes issue with the contributions being made by postmodernism and the cultural turn to human geography – the so-called 'leading edge'. As we suggested in Chapters 1 and 11, this has been influential in political geography and has sensitised us to the need to consider textual and discursive strategies. Critical geopolitics, for instance, has demonstrated the usefulness of this methodological approach (see Chapter 11). In Martin's opinion, however, concerns with identity and culture have diverted our attention away from the larger social and political problems of today. This is partly explained by the cultural turn's denial of 'extra-discursive reality' – it disengages itself from material power relations, and thereby does not accept that such forces also shape identity and politics (Martin 2001: 196). Consequently, in Martin's opinion, the cultural turn does not engage with the many processes and practices that provide the scenery for social interaction. It does little to challenge the 'structured determinants of sociospatial problems and inequalities' (Martin 2001: 201). We have, of course, challenged some of these assumptions in our discussion on Cultural Political Economy (Chapter 3).

Martin also challenges the research designs that leading geographers use to demonstrate their claims. Too often our theories, which are borrowed from trendy thinkers, are being put into practice with a 'lack of rigour': geographers rely too heavily on selective quotations from a limited number of individuals, located in particular geographical locations (Martin 2001: 197). This in turn leads to 'fuzzy conceptualization' (Markusen 1999) – our claims do not stand up to any serious scrutiny or interrogation. Linked to this, the policy and political implications that flow from

human geography are redundant and this reinforces a lack of political commitment.

Levelling criticism is easy and the 'difficult part is suggesting what needs to be done, how we should move forward' (Martin 2001: 202). We have pulled together some of Martin's suggestions for doing a 'new geography of public policy' in Box 12.4. As you can see, there are a number of key challenges facing geographers tackling this debate and these can be summarised as: developing intellectual cohesion through practical social research; finding imaginative ways of combining qualitative and quantitative data to ensure rigour; and using 'action based' and co-production approaches to influence the direction of policies and their outcomes. Students of political geography will have to make up their own mind, as to whether they agree, or disagree, with Ron Martin's critique. What do you think? Is this debate important? What can political geographers contribute to this agenda?

Practising policy-engaged geography

In recent years, discussion has moved on from debating whether or not geographers should seek to engage with policy makers to considering the methodological and practical issues concerned with doing policy-relevant research. Advocates and practitioners of policy-engaged research have at times been guilty of lacking reflexivity about the research that they do, and of suggesting that the process of informing policy is relatively unproblematic. As Beaumont *et al.* (2005: 119) critically observe

> too often it would seem that geographers are somehow faced with a choice to produce relevant research, assuming that as long as they meet certain conditions (e.g. use specific kinds of language and methods), policy-makers will listen to their recommendations, respond quickly and smoothly and adjust their policies accordingly.

In practice, doing policy-engaged research presents particular challenges both in conducting the research

BOX 12.4 TOWARDS A 'NEW GEOGRAPHY OF PUBLIC POLICY'?

There is no single, all encompassing, universally superior or commonly agreed theoretical framework or methodological approach on which to base our research. Thus, there can be no single approach to policy analysis, no blueprint for how geographers should integrate public policy into their research or how they should evaluate its sociospatial impacts. There are different forms of, and approaches to, policy analysis ranging, for example, from the critical analysis of policy discourses and practices to reveal their underlying ideological and instrumental content, to extensive empirical analyses of policies to evaluate their intended impacts and unintended consequences, to intensive ethnographic type investigations of precisely how particular policies effect specific individuals, groups and localities. Each provides a different 'cut' on policy, and different policy issues will require different methods or combinations of methods. Public policy analysis has to be pluralistic, not monistic. We need more interesting and imaginative ways of combining qualitative and quantitative analysis, and of integrating intuition into our research methodologies and analyses. Above all, for a policy turn to occur in the discipline, our research has to become much more 'action based'. We need to see research not simply as a mechanism for studying and explaining change, but – by following our investigations through to their implications for possible policy intervention and action – as instigator of change, as an activist endeavour . . . . The geography of public policy is not just about evaluating policy impacts. Important though that role is, geographers should also be engaged in fundamental debates over the direction of society, economy and environment, and what policies would be required to achieve different outcomes. But, equally, it is surely as important to research and campaign for achievable reforms as it is to debate ideal transformation which have little prospect of being implemented.

(Martin 2001: 202–203)

Key reading: Martin (2001).

and in negotiating the politics of working with and seeking to influence policy makers.

Guidance on how to meet these challenges has until recently been limited. Pain (2006), in a helpful review of arguments in the policy debate, notes that 'at present there is only a limited body of knowledge, no textbooks, no postgraduate training courses or workshops on how to approach and negotiate policy research' (p. 256). However, more reflexive accounts of working with policy makers have begun to appear, describing a variety of experiences. Bell (2007), for example, critically details the pressures of conducting short consultancy projects for government agencies, producing identikit reports that tick boxes but have little influence on policy; whilst Beaumont *et al.* (2005) recount work on politically sensitive topics that has been undermined by political interference and reluctance to publish results. Yet, more positive

experiences are described by Burgess (2005) and Woods and Gardner (2011), who both point to opportunities to build relationships with government officials and policy makers, and to achieve small, targeted outcomes in the policy process.

Successfully influencing policy requires 'researchers to assume a number of different roles that are not routinely part of an academic's repertoire: as negotiator, diplomat, salesperson, strategist and counsellor' (Woods and Gardner 2011: 207). Political geographers may arguably be better placed than other human geographers to do this, as an understanding of power, the state and politics can be helpful in framing realistic expectations of policy engagement and identifying strategies for achieving influence. The engagement of contemporary geographers with the policy process stands in contrast, however, with the over-zealous political ambitions of early political geographers, discussed in Chapter 1. Whereas individuals such as Mackinder and Haushofer sought influence by pursuing political careers themselves and tailoring their arguments to support political objectives, the reflexive approach of contemporary political geographers helps to maintain the critical integrity of their research and the differentiation between research and politics.

The distinction between research and politics has been blurred, however, in efforts by some critical geographers to utilise their skills and knowledge not to work for the state, but to assist and empower marginalised groups in challenging authority. Such 'participatory geographies' involve policy-engagement from the bottom-up, working with and involving communities and social movements in doing research that can make power relations visible, equip subordinate populations to contest state policies and develop progressive alternatives. Examples include Cahill's (2007) work with young women confronting gentrification and racial stereotyping in New York city, projects building participatory citizenship among marginalised groups in India (Williams *et al.* 2011) and the active involvement of political geographers in counter-globalisation, environmental justice and anti-neoliberal movements (Clough 2012: Mason 2013; Routledge 2012). These engagements resonate with Koopman's (2011) call for an 'alter-geopolitics', working with subaltern populations not the state, and complete the rotation of political geography over the last century to stand diametrically opposed to the imperialist agendas of its founding figures, described in Chapter 1.

Towards 'deeper' engagements with the policy progress

We conclude this chapter with a still missing-link – the policy process itself – which is linked to how we conceptualise the state's changing institutional forms, functions and modes of intervention. According to Christopher Ham and Michael Hill, analysing policy-making depends on some appreciation of the institutionalisation of power and representation in society, which in turn, requires some understanding

of the state under capitalism (Ham and Hill 1993). Rather than finding ready-made answers from within our discipline, political geography might benefit from adopting a 'post-disciplinary' stance on the state and its politics, whereby a dialogue is opened up with, and ideas are drawn from, social and political science.

To engage 'critically and actively with the policy process itself' (Peck 1999) political geography could offer an insight into the 'deeper' political arena. We could focus on how government responds to and represents its wider social environment. Public policy is not just political; it also has profound impacts on society by framing socio-spatial relations. Indeed, the two go hand in hand – the social and the political are mutually reinforcing, constructed and embedded in each other. By understanding the social situations and politics that go hand in hand with forms of state intervention and the multiple terrains through which this occurs, political geographers could begin to understand what makes public policy tick, why changes take place, and we can also begin to highlight access points for those individuals and campaign groups wishing to practise 'activism'.

The public policy debate, then, warrants a reconsideration of the capitalist state. Murray Low (2003) has suggested that the state remains an important missing-link in contemporary political geography and a decade later this remains the case. As we suggested in Chapter 2, the state is everywhere and nowhere: it is the backdrop to almost everything that we experience. In the public policy debate, however, the state rarely makes more than a cameo appearance. One way forward, and acknowledging Martin's (2001) point that there are many 'cuts' into the cake, is to think about a *régulation* approach take on public policy. As we suggested in Chapter 3, the *régulation* approach presents the state as a complex and broad set of institutions and networks that span both political society and civil society in their 'inclusive' sense. From this perspective, state intervention, state functions and public policy concerns relate to the 'micro-physics' of power.

Bob Jessop's 'regulationist state theory' remains interesting for thinking about the links between political geography and public policy. Jessop draws on the work of state theorists and political activists such as Antonio Gramsci, Nicos Poulantzas and Claus Offe to think about the changing institutional forms and functions of the capitalist state. For Jessop, the state needs to be thought of as 'medium and outcome' of policy processes that constitute its many interventions. The state is both a social relation and a producer of strategy and, as such, it does not have any power of its own. State power in relation to the policy process relates to the forces that 'act in and through' its apparatus. According to this view, attempts to analyse the policy process need to uncover the strategic contexts, calculations, and practices of actors involved in strategically selective, or privileged, sites (Jessop 1990a, 2008, 2014). This can be summarised as a framework that demonstrates 'systems analyses' for the undertaking of 'systematic' forms of public policy analysis (Ham and Hill 1993) – drawing attention to the intricate links

between actors and forms of representation, institutions and their interventions and practices, and the range of policy outcomes available. This connects to our argument in Chapter 1 that political geography recognises intrinsically linked entities – power, politics and policy, space, place and territory.

Box 12.5 details the six dimensions of the state that appear in much of Jessop's work on the institutional forms and functions of political economy. Three are associated with institutional relations within the political and policy system. Jessop adds a further three to tease out the ways in which the state interacts with its wider social environment. The capitalist state, then, can be viewed as a strategic and relational concern, forged through the *ongoing* engagements between state personnel, institutions and public policy implementation. This perspective could assist political geographers and their students to delve deeper into

---

BOX 12.5 RE-STATING THE POLICY PROCESS

Institutional relations within the political and policy system

Representational regime

This has a concern with delimiting patterns of representation and the state in its inclusive sense. It uncovers the territorial agents, political parties, state officials, community groups, para-state institutions, regimes and coalitions that are incorporated into the state's everyday policy-making practices.

Internal structures of the state

This is the institutional embodiment of the above and it underscores the distribution of powers through different geographical divisions and departments of the state and its policy systems. This not only allows research to study the apparatus of central government, it also explores the ways in which political strategy helps to create sub-national spaces and scales of policy intervention and delivery. For political geographers, the relationship between the different politically and socially charged scales of governance is important.

Patterns of intervention

This is associated with the different political and ideological rule systems that govern state intervention, such as frameworks of rights and responsibilities, the balance between the public and private, and the perceived roles of the social partners in the policy process. Additional concerns can include the discourses of citizenship, social inclusion/exclusion, universal versus targeted and selective service provision, and equality versus allocation through competition.

Wider social relations and civil society

Social basis of the state

This consolidates the representational regime through civil society, which can be spatially selective, and explores the different ways in which uneven development is mobilised into the political system through targeted state strategies.

State strategies and state projects

This brings some overall coherence to the activities of the state, its forms of intervention and its policy-making priorities. The state is seen here as a political strategy and its various policy and power networks can privilege some coalition possibilities over other and some interest groups over others.

Hegemonic project

This mobilises the state and its multifarious policy-making networks and coalitions, and also tries to externalise/resolve conflicts that can disrupt policy systems, around an ideological programme of action. It thereby considers the ways in which collective actions, forms of knowledge, and discourses become codified and mobilised to advance particular interests. There are links here to notions of governmentality (see Chapter 2).

Source: Jessop (1990a, 2008, 2014); MacLeod (2001); Peck and Jones (1995).

the policy process itself. What kind of political geography for what kind of public policy?

Further reading

For further reading on the public policy debate in geography, see: Ron Martin (2001) 'Geography and public policy: The case of the missing agenda', *Progress in Human Geography*, volume 25, pages 189–210; Jamie Peck (1999) 'Editorial: Grey geography?', *Transactions of the Institute of British Geographers*, New Series, volume 24, pages 131–135; Brian Berry (1994) 'Let's have more policy analysis', *Urban Geography*, volume 15, pages 315–317; Terry Coppock (1974) 'Geography and public policy: challenges, opportunities and implications', *Transactions of the Institute of British Geographer* volume 63, pages 1–16; David Harvey (1974) 'What kind of geography for what kind of public policy?' *Transactions of the Institute of British Geographers*, 63, pages 18–24; Michel Dear (1999) 'The relevance of postmodernism', *Scottish Geographical Journal*, volume 115, pages 143–150; Danny Dorling and Mary Shaw (2002) 'Geographies of the agenda: Public policy, the discipline and its (re)'turns', *Progress in Human Geography*, volume 26, pages 629–646; John House (1973) 'Geographers, decision takers and policy matters' in Michael Chisholm and Barry Rodgers eds *Studies in Human Geography* (Heinemann/SSRC); Rachel Pain (2006) 'Social geography: seven deadly myths in policy research', *Progress in Human Geography*, volume 30, pages 250–259; and Michael Woods and Graham Gardner (2011) 'Applied policy research and critical human geography: some reflections on swimming in murky waters', *Dialogues in Human Geography*, volume 1, pages 198–214.

For arguments on activism and political relevance, see: Nik Blomley (1994) 'Activism and the academy', *Environment and Planning D: Society and Space*, 12, pages 383–385; Adam Tickell (1995) 'Reflections on "activism and the academy"', *Environment and Planning D: Society and Space*, volume 13, pages 235–237; Noel Castree

(1999b) '"Out there"? "In here"? Domesticating critical geography', *Area*, volume 21, pages 81–86; Andrew Blowers (1974) 'Relevance, research and the political process', *Area*, 6, pages 32–36; Doreen Massey (2000) 'Practising political relevance', *Transactions of the Institute of British Geographers*, New Series, 24, page 131–134; Alison Blunt and Jane Wills (2000) *Dissident Geographies: An Introduction to Radical Ideas and Practice* (Pearson); Paul Chatterton, David Featherstone and Paul Routledge (2013) 'Articulating climate justice in Copenhagen: antagonism, the commons, and solidarity', *Antipode* 45, 602–620; and Paul Routledge (2012) 'Sensuous solidarities: emotion, politics and performance in Clandestine Insurgent Rebel Clown Army', *Antipode* 44, 428–452.

If you are interested in finding out more about the policy-process, see: Christopher Ham and Michael Hill (1993) *The Policy Process in the Modern Capitalist State* (Harvester Wheatshef); Martin Burch and Bruce Wood (1989) *Public Policy in Britain* (Blackwell); Claus Offe (1984) *The Contradictions of the Welfare State* (Hutchinson); and Charles Lindbolm (1968) *The Policy-Making Process* (Prentice-Hall); and Richard Rose (1993) *Lesson Drawing in Public Policy: A Guide to Learning Across Time and Space* (Chatham House).

For further on the strategic-relational approach to the state, see: Bob Jessop (1990a) *State Theory* (Polity); Bob Jessop (2001b) 'Institutional re(turns) and the strategic-relational approach', *Environment and Planning A*, volume 33, pages 1213–1235; Bob Jessop (2008) *State Power: A Strategic Relational Approach* (Polity); Bob Jessop (2014) *The State* (Polity); Gordon MacLeod and Mark Goodwin (1999) 'Space, scale and state strategy: Rethinking urban and regional governance', *Progress in Human Geography*, volume 23, pages 503–527; Martin Jones (1997) 'Spatial selectivity of the state? The regulationist enigma and local struggles over economic governance', *Environment and Planning A*, volume 29, pages 831–864; and Jamie Peck and Martin Jones (1995) 'Training and Enterprise Councils: Schumpeterian workfare state, or what', *Environment and Planning A*, volume 27, pages 1361–1396.

Glossary

Active citizenship   the idea that citizens are not the passive recipients of rights and state benefits, but that the citizen has a responsibility to be actively involved in the governing process. Commonly associated with the strategy of governmentality (qv) of 'governing through communities'.

Capitalism   a specific social and economic system that is divided into two classes: those owning the means of production (land, machinery and factories, etc.) and those selling labour power. Under the capitalist mode of production, labour power is exploited to provide surplus value (or profit) and capitalists compete for this profit through a system that necessitates the 'accumulation of capital' (see Box 3.1).

Citizenship   a mark of belonging to a political entity or collective that both guarantees rights for the individual and carries responsibilities towards the collective. Citizenship codifies the relationship between the individual and the state (see Box 4.1).

Civic nationalism   a type of nationalism that is based on the organisation of the state and which highlights the fact that nations are produced as a result of certain processes. It is often a more inclusive form of nationalism.

Colony   a political and spatial form, often based on ideas of domination and which is created by the colonisation of one territory and people by a state, organisation or group of people. Thus, the act of colonialism is characterised by unequal economic, political and cultural relationships.

Community   a collective of individuals who share a mutual sense of identity and solidarity. Communities are frequently defined in terms of a territorial association, but need not necessarily be so.

Critical geopolitics   a sub-field of political geography that critically analyses the production, circulation and consumption of geopolitical knowledge (see also geopolitics).

Cultural turn   the popularisation in human geography in the late 1980s and early 1990s of the study of cultural relations, processes and entities, including issues of identity, difference and representation. Associated with the use of qualitative research methods and the influence of cultural studies and of post-structuralist and post-modernist thought.

Democracy   a political system based on the principle of government by the people through majority decision-making.

Devolution   a process whereby political power is transferred from a national state to regions within the state.

Discourse   a body of knowledge that structures a particular way of understanding the world (see Box 1.4).

Electoral geography   a sub-field of political geography that is concerned with the analysis of the spatial aspects of elections, including the influence of geographical factors of voting behaviour and election outcomes, and the spatial patterns of election results.

Empire   a political form, which is based on a subservient relationship between a metropolitan state and other lands or people. The political form of empire is, thus, closely related to the process of imperialism.

Ethnic nationalism   a type of nationalism that emphasises the common cultural and historical links between a named human population. It is often associated with ideas concerning the 'naturalness' of nations. It can, under certain circumstances, be linked to extreme and exclusionary forms of nationalism.

Federalism   a political system, which emphasises the notion of subsidiarity. In this context, it is believed that decisions should be made at the smallest practical spatial scale.

Feminism   as a political movement, feminism advocates the rights of women to equality in society; as an intellectual movement, feminism challenges masculinist discourses and approaches an understanding of the world from a female perspective.

Gentrification   the renovation of property in relatively less favoured areas by and for affluent incomers displacing lower-income groups.

Geopolitics   a sub-field of political geography concerned with political relations between states, the external strategies of states and the global balance of power.

Geopolitik   literally the German translation of geopolitics (qv), but particularly associated with a partisan form of political geography practised in Germany in the 1920s and 1930s that was used to support the racist and expansionist policies of the Nazi party.

Gerrymander   The deliberate manipulation of the territory of electoral districts for partisan gain. Named after the nineteenth century governor of Massachusetts, Eldridge Gerry.

Globalisation   the advanced interconnection and interdependence of localities across the world, economically, socially, politically and culturally.

Glocalisation   a term coined by Erik Swyngedouw to emphasise the simultaneous erosion of power from the nation scale upwards to the global and downwards to the local.

Governance   the process of governing through networks of organisations that involve state institutions, private sector corporations and third-sector groups, often working in partnership arrangements. Governance can also involve an emphasis on active citizenship.

Governmentality   the techniques and strategies by which a society is rendered governable.

Heartland   a geopolitical term referring to an area of central Eurasia, similar to the territory of present-day Russia, whose control Mackinder argued was crucial to the global balance of power. Also known as the 'pivot area'.

Identity politics   the way in which peoples' politics, in recent years, are increasingly being shaped by aspects of their identity. This can be contrasted with the traditional domination of class conflict within politics.

Landscape   the assemblage of physical objects that comprise the visual surface appearance of an area of land.

Landscape of power   the symbolic representation of power relations through the landscape (qv), including both monumental landscapes with an explicit political meaning and more subtle signifiers in everyday landscapes, such as the juxtapositioning of skyscrapers and slums.

Lebensraum   literally 'living space', a term borrowed from biology by Ratzel to indicate the territory required for the comfortable existence of a state. The concept was used to justify expansionist policies.

Localism   has two meanings in political geography: first, as the devolution of governance and decision-making to the local level, as part of a new strategy of governmentality; and second, the mobilisation of citizens to defend perceived local interests.

Locality   a place defined at local scale with a territorial expression. The term implies a spatial unit that can be attributed with distinctive characteristics and differentiated from other localities (see Box 6.1).

Local state   a collective term for the apparatuses of the state (qv) that exist and operate at a local scale, usually with reference to a specific locality (qv). The local state includes not only 'local government' (local-scale elected or appointed public authorities) but also the local branches of the judiciary and security agencies.

Malapportionment   a term in electoral geography (qv) referring to the disproportionate distribution of seats in a legislature to a geographical district compared to its entitlement on an objective population-based allocation.

Nation   a named human population that is perceived as possessing a common culture, customs and territory.

Nationalism   an ideology that seeks to promote the existence of nations within the world. In addition, it can refer to nations' attempts to reach the political goal of being constituted as a nation-state.

Nation-state   a political form in which the boundaries of a state and nation coincide.

Neighbourhood effect   a theory in electoral geography (qv) that suggests that voting behaviour is influenced by the geographical situation of the voter. In other words, residents of a neighbourhood are more likely to vote the same way than would be anticipated on the basis of social or economic characteristics.

Neoliberalism   a political ideology that promotes economic freedom, including free trade and competition in a 'free market', and contends that the role of the state in the economy should be restricted to facilitating free market capitalism. Neoliberalism is commonly regarded as the dominant political ideology in the early twenty-first century and is associated with trade liberalisation, privatisation and welfare reforms.

Place   a point, or area, of space that can be identified through verbal, written, cartographic or visual representation.

Pluralism   a political theory that holds that power is widely dispersed within society and that a diverse range of groups have an equal opportunity to influence the political process.

Post-structuralism   a philosophical movement of the late twentieth century that rejects notions of essential truth and the rational subject and proposes instead that meanings are produced within language and subjectivity is constructed through discourse (qv) (see Box 1.3).

Power   the capacity to do something. Politics can be described as the pursuit and discharge of power, but there are many different conceptualisations of exactly what power is and how it works (see Box 1.1).

Public space   an area of space to which all people in theory have a right to access without restriction, selection or payment. In practice, public space is regulated both formally and informally and the freedom to use public space differs between groups. Many public spaces are privately owned but open for public use by convention or for commercial purposes.

Quantitative revolution   a period of transformation in human geography in the 1950s and 1960s associated with the introduction of statistical and mathematical methods for geographical research and analysis, replacing the previous concern with areal differentiation and regional studies.

Region   a more or less bounded area possessing some relative unity or organising principle that distinguishes it from other regions. Regions, however, are never closed and are actively produced and reproduced by different forms of agency.

Régulation approach   a set of neo-Marxist ideas on political economy, whereby economies and societies are seen to emerge through social, economic and institutional frameworks and supports, despite the instabilities and crisis-tendencies within capitalism (see Box 3.2).

Resistance   the act of opposing or withstanding the exercise of power. Geographies of resistance are concerned with studying the spatial aspects of political opposition to the state and other centres of power.

Scale   a level of representation that is differentiated from other scales by variations in magnitude. Geographical scale is differentiated by the spatial dimensions encompassed by each level of magnitude and is commonly referred to in terms of fixed (but also process-based and relational) increments including (with increasing magnitude): local, regional, national and global scale.

Social capital   the worth and potential that is invested in social networks and contacts between people (see Box 8.4).

Social construction   the ascribing of meaning to things by and through social interactions. A social construct has no fixed meaning outside the social context of its definition (see Box 6.3).

Social movement   a network of individuals, groups and/or organisations engaged in political or cultural activity based on a shared identity. The components of a social movement may be highly fragmented and diverse in nature with no single centre of leadership (see Box 8.7).

Spatial science   a form of human geography that applied scientific principles and models to the analysis of spatial processes and spatial variations.

Tactical voting   the practice by which an elector votes for their second preference candidate in order to prevent a third candidate from being elected.

Urban regime theory   a model that proposes that stability in urban politics is achieved through the construction of 'urban regimes' that draw together the resources of public and private actors to produce a 'capacity to act'.

War on terror   a term coined by President George W. Bush to describe the United States' geopolitical strategy following the al-Qaeda attack on New York and Washington on 11 September 2001. The strategy has encompassed military interventions in countries perceived to be supportive of terrorism or threatening to global security, notably Afghanistan and Iraq, as well as operations against bases of groups such as al-Qaeda, and a tightening of national security arrangements and increased surveillance of citizens.

Workfare   a model of welfare reform based on the movement from a universal rights and needs-based entitlement to income support and to a selective system combining welfare with work in order to enforce new social responsibilities (see Box 3.9).

World systems analysis   a conceptual approach that proposes that social change at any scale can only be understood in the context of a wider world system, and that change needs to be approached through a long-term historical perspective (see Box 1.2).

References

Abraham, D. (2007) Bush regime from elections to detentions: a moral economy of Carl Schmitt and human rights, *University of Miami Law Review,* 62: 249.

Abrahamsen, R. (2003) African studies and the postcolonial challenge, *African Affairs,* 102: 189–210.

Agamben, G. (2005) *State of Exception,* Chicago and London: University of Chicago Press.

Aglietta, M. (1978) Phases of US capital expansion, *New Left Review,* 110: 17–28.

Aglietta, M. (2000) *A Theory of Capitalist Regulation: The US Experience,* new edition, London: Verso.

Agnew, J. (1998) *Geopolitics,* London: Routledge.

Agnew, J. (2002) *Making Political Geography,* London: Arnold.

Agnew, J. (2003) Contemporary political geography: intellectual heterodoxy and its dilemmas, *Political Geography,* 22: 603–606.

Agnew, J. (2007) Know-where: geographies of knowledge of world politics, *International Political Sociology* 1, no. 2: 138–148.

Agnew, J., Mitchell, K. and Toal, G. (eds.) (2003) *A Companion to Political Geography,* Oxford: Blackwell.

Alexander, L. (1963) *World Political Patterns,* Chicago: Rand McNally.

Allen, J. (2003) *Lost Geographies of Power,* Oxford: Blackwell.

Allen, J. (2006) Ambient power: Berlin's Potsdamer Platz and the seductive logic of public spaces, *Urban Studies* 43: 441–455.

Allen, J. (2011) Topological twists: power's shifting geometries, *Dialogues in Human Geography,* 1: 283–298.

Allen, J., Massey, D. and Cochrane, A. (1998) *Rethinking the Region,* London: Routledge.

Amin, A. (ed.) (1994) *Post-Fordism: A Reader,* Oxford: Blackwell.

Amin, A. (2002) Spatialities of globalisation, *Environment and Planning A,* 34: 385–399.

Amin, A. (2005) Local community on trial, *Economy and Society,* 34: 612–633.

Amin, A. and Robins, K. (1990) The re-emergence of regional economies? The mythical geography of flexible accumulation, *Environment and Planning D, Society and Space,* 8: 7–34.

Anderson, B. (1983) *Imagined Communities: Reflections on the Origin and Spread of Nationalism.* London: Verso.

Anderson, J. (1996) The shifting stage of politics: new medieval and postmodern territorialities, *Environment and Planning D: Society and Space,* 14: 133–53.

Anderson, J. (1998) 'Nationalist ideology and territory' in Johnston, R.J., Knight, D.B. and Kaufman, E. (eds) *Nationalism Self-Determination and Political Geography,* London: Croom Helm, PP. 18–39.

Andresen, S., Skodvin, T., Underdal, A. and Wettestad, J. (2000) *Science and Politics in International Environmental Regimes,* Manchester: Manchester University Press

Antonsich, M. (2009) The 'revenge' of political geographers, *Political Geography,* 28: 211–212.

Antonsich, M. and Jones, P. (2010) Mapping the Swiss referendum on the minaret ban, *Political Geography,* 29: 57–62

Antonsich, M., Minghi, J., Johnston, R. and Berry, B. (2009) Interventions on the 'moribund backwater' forty years on, *Political Geography,* 28: 388–394.

Appleton, L. (2002) Distillations of something larger: the local scale and American national identity, *Cultural Geographies*, 9: 421–447.

Archer, J. C. (2002) The geography of an interminable election: Bush v. Gore, 2000, *Political Geography*, 21(1): 71–77.

Archer, K. (2012) Rescaling global governance: imagining the demise of the nation-state, *Globalizations*, 9: 241–256.

Area (1980) Observations: a future for cultural geography? *Area*, 12: 105–113.

Arnold, D. (1998) India's place in the tropical world, 1770–1930, *Journal of Imperial and Commonwealth History*, 26: 1–21.

Arts, B. (2004) The global-local nexus: NGOs and the articulation of scale, *Tidjschrift voor Economische en Sociale Geografie*, 95: 498–510.

Arzheimer, K. and Evans, J. (2012) Geolocation and voting: candidate-voter distance effects on party choice in the 2010 UK general election in England, *Political Geography*, 31, 301–310.

Ashcroft, B., Griffiths, G. and Tiffin, H. (1998) *Key Concepts in Post-Colonial Studies*, London: Routledge.

Atkinson, D. and Cosgrove, D. (1998) Urban rhetoric and embodied identities: city, nation and empire at the Vittorio Emanuel II monument in Rome, 1870–1945, *Annals of the Association of American Geographers*, 88(1): 28–49.

Azaryahu, M. (1997) German reunification and the politics of street names: the case of East Berlin, *Political Geography*, 16(6): 479–494.

Azaryahu, M. and Kellerman, A. (1999) Symbolic places of national history and revival: a study in Zionist mythical geography, *Transactions of the Institute of British Geographers*, 24: 109–23.

Baigent, E. (2003) Monarchs, ministers and maps: the geometric mapping of early modern Sweden in international perspective, in B. R. Hansen (ed.) *Nationalutgava av de Alder Geometriska Kartorna*, Stockholm: Kungl, pp. 23–60.

Bailey, N. and Pill, M. (2011) The continuing popularity of the neighbourhood and neighbourhood governance in the transition from the 'big state' to the 'big society' paradigm, *Environment and Planning C: Government and Policy*, 29: 927–942.

Baillie Smith, M. and Laurie, N. (2011) International volunteering and development: global citizenship and neoliberal professionalization today, *Transactions of the Institute of British Geographers*, 36: 545–559.

Baker, A. R. H. (1998) Military service and migration in nineteenth century France: some evidence from Loir-et-Cher, *Transactions of the Institute of British Geographers*, 23: 193–206.

Bakshi, P., Goodwin, M., Painter, J. and Southern, A. (1995) Gender, race, and class in the local welfare state: moving beyond regulation theory in analysing the transition from Fordism, *Environment and Planning A*, 27: 1539–1554.

Banks, M. and Mackian, S. (2000) Jump in! The water's warm: A comments on Peck's 'grey geography', *Transactions of the Institute of British Geographers*, 25: 249–254.

Baratz, M. S. and White, S. B. (1996) Childfare: a new direction for welfare reform, *Urban Geography*, 33: 1935–1944.

Barnes, T. J. and Minca, C. (2013) Nazi spatial theory: the dark geographies of Carl Schmitt and Walter Christaller, *Annals of the Association of American Geographers*, 103: 669–687.

Barnett, C. (2004) Media, democracy and representation: disembodying the public, in C. Barnett and M. Low (eds.), *Spaces of Democracy: Geographical Perspectives on Citizenship, Participation and Representation*, London: Sage, pp. 185–206.

Barnett, C. and Low, M. (eds.) (2004) *Spaces of Democracy: Geographical Perspectives on Citizenship, Participation and Representation*, London: Sage.

Barry, B. (1997) Sustainability and intergenerational justice, *Theoria*, 45: 43–65.

Bassett, K. and Short, K. (1980) *Housing and Residential Structure: Alternative Approaches*, London: Routledge & Kegan Paul.

Bassin, M. (1987) Imperialism and the nation state in Friedrich Ratzel's political geography, *Progress in Human Geography*, 11: 473–495.

Bassin, M. (2000) Studying ourselves: history and philosophy of geography, *Progress in Human Geography*, 24: 475–487.

Beaumont, J., Loopmans, M. and Uitermark, J. (2005) Politicization of research and the relevance of geography: some experiences and reflections for an ongoing debate, *Area*, 37: 118–126.

Bell, D. (2007) Fade to grey: some reflections on policy and mundanity, *Environment and Planning* A, 39: 541–554.

Bell, J. E. and Staeheli, L. A. (2001) Discourses of diffusion and democratization, *Political Geography*, 20: 175–195.

Bénit-Gbaffou, C. (2012) Party politics, civil society and local democracy – reflections from Johannesburg, *Geoforum*, 43: 178–189.

Benton-Short, L. (2007) Bollards, bunkers and barriers: securing the national mall in Washington, DC, *Environment and Planning D: Society and Space*, 25: 424–446.

Bernal, V. (2006) Diaspora, cyberspace and political imagination: the Eritrean diaspora online, *Global Networks*, 6: 161–179.

Bernstein, S. (2000) Ideas, social structure and the compromise of liberal environmentalism, *European Journal of International Relations*, 6: 464–512.

Bernstein, S. (2001) *The Compromise of Liberal Environmentalism*, New York: Columbia University Press.

Berry, B. (1969) Review of Russett, international regions and the international system, *Geographical Review*, 59: 450.

Berry B. (1972) More on relevance and policy analysis, *Area*, 4: 77–80.

Berry B. (1994) Editorial: Let's have more policy analysis, *Urban Geography*, 15: 315–317.

Bezmez, D. (2013) Urban citizenship, the right to the city, and politics of disability in Istanbul, *International Journal of Urban and Regional Research*, 37: 93–114.

Bhahba, H. (1990) *Nation and Narration*, London: Routledge.

Billig, M. (1995) *Banal Nationalism*, London: Sage.

Blaikie, P. (1985) *The Political Economy of Soil Erosion in Developing Countries*, London: Longman.

Blaikie, P. and Brookfield, H. (1987) *Land Degradation and Society*, London: Methuen.

Blaut, J. (1992) Fourteen ninety-two, *Political Geography Quarterly*, 11: 355–385.

Blaut, J. (1993) *The Colonizer's Model of the World*, New York: Guilford Press.

Blaut, J. (2000) *Eight Eurocentric Historians*, New York: Guilford Press.

Blomley, N. (1994) Activism and the academy, *Environment and Planning D: Society and Space*, 12: 383–385.

Blowers, A. T. (1974) Relevance, research and the political process, *Area*, 6: 32–36.

Blunt, A. (1994) *Travel, Gender and Imperialism: Mary Kingsley and West Africa*, New York; Guilford.

Blunt, A. and Willis, J. (2000) *Dissident Geographies: An Introduction to Radical Ideas and Practice*, London: Pearson.

Blunt, A. and McEwan, C. (eds.) (2002) *Postcolonial Geographies*, London: Continuum.

Bohuon, A. and Luciani, A. (2009) Biomedical discourse on women's physical education and sport in France (1880–1922), *The International Journal of the History of Sport*, 26: 573–593.

Bondi, L. and Domosh, M. (1998) On the contours of public space: a tale of three women, *Antipode*, 30: 270–289.

Bookchin, M. (2005) *The Ecology of Freedom*. Oakland: AK Press.

Bonefeld, W. and Holloway, J. (eds.) (1991) *Post-Fordism and Social Form: A Marxist Debate on the Post-Fordist State*, London: Capital and Class/Macmillan.

Bookchin, M. (2007) *Social Ecology and Communalism*, Oakland: AK Press.

Bosco, F. J. (2006) The Madres de Plaza de Mayo and three decades of human rights' activism: embeddedness, emotions and social movements, *Annals of the Association of American Geographers*, 96: 342–365.

Bosco, F., Aitken, S. and Herman, T. (2011) Women and children in a neighbourhood advocacy group: engaging community and refashioning citizenship at the United States-Mexico border, *Gender, Place and Culture*, 18: 155–178.

Bowman, I. (1921) *The New World: Problems in Political Geography*, New York: World Book Company.

Boyer, R. (1990) *The Regulation School: A Critical Introduction*, New York: Columbia University Press.

Boyer, R. and Saillard, Y. (eds.) (2002) *Régulation Theory: The State of the Art*, London: Routledge.

Bradford Landau, S. and Condit, C. (1999) *The Rise of the New York Skyscraper*, New Haven, CT: Yale University Press.

Brenner, N. (1998) Between fixity and motion: accumulation, territorial organization and the historical geography of spatial scales, *Environment and Planning D: Society and Space*, 16: 459–481.

Brenner, N., Jessop, B., Jones, M., and Macleod, G. (eds.) (2003) *State/Space: A Reader*, Oxford: Blackwell.

Brenner, N. (2004) *New State Spaces: Urban Governance and the Rescaling of Statehood*, Oxford: Oxford University Press.

Brenner, R. and Glick, M. (1991) The regulation approach: Theory and history, *New Left Review*, 188: 45–119.

Brenner, N. and Theodore, N. (eds.) (2002) *Spaces of Neoliberalism: Urban Restructuring in North America and Western Europe*, Blackwell: Oxford.

Brenner, N., Peck, J. and Theodore, N. (2010) After neoliberalization?, *Globalizations*, 7: 327–345.

Brenner, N., Jessop, B., Jones, M. and Macleod, G. (2003) 'Introduction: State Space in question' in N. Brenner, B. Jessop, M. Jones and G. Macleod (eds.) *State/Space: A Reader*, Oxford: Blackwell, pp 1–26.

Brooking, T. and Pawson, E. (2011) *Seeds of Empire: the Environmental Transformation of New Zealand*, London: I. B. Taurus.

Brown, G. (1998) Address at the launch of the Tayside Pathfinder, London: HM Treasury.

Browning, C. (1992), *The Path to Genocide: Essays in Launching the Final Solution*, Cambridge: Cambridge University Press.

Brownlow, A. (2011) Between rights and responsibilities: insurgent performance in an invisible landscape, *Environment and Planning A*, 43: 1268–1286.

Brubaker, R. (2004) *Ethnicity Without Groups*, Cambridge, MA: Harvard University Press.

Brusco, S. and Righi, E. (1989) Local government, industrial policy and social consensus: the case of Modena (Italy), *Economy and Society*, 18: 405–424.

Bryant, R. L. and Bailey, S. (1997) *Third World Political Ecology*, London: Routledge.

Brzezinski, Z. (1998) *The Grand Chessboard: American Primacy and its Geostrategic Imperatives*, New York: Basic Books.

Bullard, R. (1990) *Dumping in Dixie: Race, Class and Environmental Quality,* Boulder, CO: Westview.

Bullen, A. and Whitehead, M. (2005) Negotiating the networks of space, time and substance: a geographical perspective on the sustainable citizen, *Citizenship Studies*, 9: 499–516.

Burch, M. and Wood, B. (1989) *Public Policy in Britain*, Oxford: Blackwell.

Burgess, J. (2005) Follow the argument where it leads: some personal reflections on 'policy-relevant' research, *Transactions of the Institute of British Geographers*, 30: 273–281.

Burnett, A. D. and Taylor, P. J. (1981) *Political Studies from Spatial Perspectives*, Chichester and New York: John Wiley.

Busby, J. W. (2008) Who cares about the weather? Climate change and US national security, *Security Studies*, 17: 468–504.

Busteed, M. A. (1975) *Geography and Voting Behaviour*, London: Oxford University Press.

Cahill, C. (2007) Negotiating grit and glamour: young women of color and the gentrification of the Lower East Side, *City and Society*, 19: 202–31.

Calhoun, C. (1997) *Nationalism*, Minneapolis: University of Minnesota Press.

Calvocoressi, P. (1991) *World Politics Since 1945*, London: Longman.

Cammaert, B. and Van Audenhove, L. (2005) Online political debate, unbounded citizenship and the problematic nature of a transnational public sphere, *Political Communication*, 22: 179–196.

Campbell, D. (2007) Geopolitics and visuality: sighting the Darfur conflict, *Political Geography*, 26: 357–382.

Castells, M. (1983) *The City and the Grassroots*, Berkeley: University of California Press.

Castells, M. (2012), *Networks of Outrage and Hope: Social Movements in the Internet Age*, Cambridge: Polity Press.

Castree, N. (1999a) Envisioning capitalism: geography and the renewal of Marxian political economy,

Transactions of the Institute of British Geographers, 24: 137–158 .

Castree, N. (1999b) 'Out there'? 'In here'? Domesticating critical geography, *Area*, 21: 81–86.

Castree, N. (2007a) Neo-liberalising nature: processes, outcomes and effects, *Environment and Planning A,* 40: 153–173.

Castree, N. (2007b) Neo-liberalising nature: the logics of deregulation and reregulation, *Environment and Planning A,* 40: 131–152.

Castree, N. (2012) *Making Sense of Nature,* London: Routledge.

Cerny, C. (1997) Paradoxes of the competition state: the dynamics of political globalization, *Government and Opposition*, 32: 251–274.

Challies, E. and Murray, W. (2011) The interaction of global value chains and rural livelihoods: the case of smallholder raspberry growers, *Journal of Agrarian Change*, 11: 29–59.

Chatterjee, P. (1986) *Nationalist Thought and the Colonial World*, London: Zed Books.

Chatterton, P., Featherstone, D. and Routledge, P. (2013) Articulating climate justice in Copenhagen: antagonism, the commons, and solidarity, *Antipode*, 45: 602–620.

Che, D. (2005) Constructing a prison in the forest: conflicts over nature, paradise and identity, *Annals of the Association of American Geographers*, 95: 809–831.

Cheshire, L. (2006) *Governing Rural Development*, Aldershot: Ashgate.

Christoff, P. (2006) Post-Kyoto? Post-Bush? Towards an effective 'climate coalition of the willing', *International Affairs,* 82: 831–860.

Christophers, B. (1998) *Positioning the Missionary*, Vancouver: University of British Columbia Press.

Ciută, F. and Klinke, I. (2010) Lost in conceptualization: reading the 'new Cold War' with critical geopolitics, *Political Geography*, 29: 323–332.

Clark, N. (2005) Ex-orbitant globality, *Theory, Culture and Society*, 22: 165–185.

Clark, G. and Dear, M. (1984) *State Apparatus*, London: Allen and Unwin.

Clarke, S. and Gaile, G. (1998) *The Work of Cities*, Minneapolis: University of Minnesota Press.

Clarke, D., Doel, M., and McDonough, F. X. (1996) Holocaust topologies: singularity, politics, space, *Political Geography*, 15: 457–489.

Clayton, D. (2000a) Imperialism, in R. J. Johnston, D. Gregory, G. Pratt and M. Watts (eds.) *The Dictionary of Human Geography*, Oxford: Blackwell, pp. 375–378.

Clayton, D. (2000b) *Islands of Truth: The Imperial Fashioning of Vancouver Island*, Vancouver: UBC Press.

Clough, N. (2012) Emotion and the center of radical politics: on the affective structures of rebellion and control, *Antipode*, 44: 1667–1686.

Coaffee, J. and Johnston, L. (2005) The management of local government modernisation – area decentralisation and pragmatic localism, *International Journal of Public Sector Management*, 18: 164–177.

Cochrane, A. (2006) Making up meanings in a capital city: power, memory and monuments in Berlin, *European Urban and Regional Studies*, 13: 5–24.

Cochrane, A. and Pain, K. (2000) A globalizing society, in D. Held (ed.) *A Globalizing World?* New York: Routledge, pp. 5–46.

Cochrane, A. Peck, J. and Tickell, A. (1996) Manchester plays games: exploring the local politics of globalization, *Urban Studies*, 33: 1319–36.

Cockburn, C. (1977) *The Local State: Management of Cities and People*, London: Pluto Press.

Cohen, S. (2007) Winning while losing: the apprentice boys of Derry walk their beat, *Political Geography*, 26: 951–967.

Cohen, J. and Arato, A. (1992) *Civil Society and Political Theory*, Cambridge, MA: MIT Press.

Cohen, S. B. and Rosenthal, L. D. (1971) A geographical model for political systems analysis, *Geographical Review*, 61: 5–31.

Cole, J. P. (1959) *Geography of World Affairs*, Harmondsworth: Penguin.

Coleman, M. (2004) The naming of terrorism and evil outlaws: geopolitical place-making after 11 September, in S. Brunn (ed.) *11 September and its Aftermath: The Geopolitics of Terror*, London: Frank Cass, pp. 87–104.

Collins, D. and Kearns, R. (2001) Under curfew and under siege? Legal geographies of young people, *Geoforum*, 32: 389–403.

Conrad, J. (1926) Last Essays, London and Toronto: J.M. Dent & Sons.

Conversi, D. (1995) Reassessing current theories of nationalism: nationalism as boundary maintenance and creation, in J. Agnew (ed.) *Political Geography: a Reader*. London, Arnold. Reprinted from *Nationalism and Ethnic Politics*, 1: 73–85.

Cooke, P. (ed.) (1989) *Localities*, London: Unwin Hyman.

Coppock, J. T. (1974) Geography and public policy: challenges, opportunities and implications, *Transactions of the Institute of British Geographer*, 63: 1–16.

Corbridge, S., Williams, G., Srivastava, M. and Véron, R. (2005) *Seeing the State: Governance and Governmentality in India*, Cambridge: Cambridge University Press.

Cornog, E. W. (1988) To give character to our city: New York's City Hall, *New York History*, 69: 389–423.

Cosgrove, D. and Daniels, S. (1988) *The Iconography of Landscape,* Cambridge: Cambridge University Press.

Cowen, D. and Gilbert, E. (eds) (2008) *War, Citizenship, Territory*, New York: Routledge.

Cox, K. (1997) Spaces of dependence, spaces of engagement and the politics of scale, or: looking for local politics, *Political Geography*, 17(1): 1–24.

Cox, K. (2002) *Political Geography: Territory, State and Society*, Oxford and Maldan, MA: Blackwell.

Cox, K. (2003) Political geography and the territorial, *Political Geography*, 22: 607–610.

Cox, K. and Mair, A. (1988) Locality and community in the politics of local economic development, *Annals of the Association of American Geographers*, 78(2): 307–325.

Crampton, J. W. (2004) *The Political Mapping of Cyberspace*, Chicago: University of Chicago Press.

Crang, P. (1996) Displacement, consumption and identity, *Environment and Planning A*, 28: 47–67.

Crang, M. (1999) Nation, region and homeland: history and tradition in Dalarna, Sweden. *Ecumene* 6: 447–70.

Crosby, A. (1986) *Ecological Imperialism: The Biological Expansion of Europe, 900–1900*, Cambridge: Cambridge University Press.

Crutzen, P. (2002) Geology of Mankind, *Nature*, 125: 23.

Cupples, J. (2009) Rethinking electoral geography: spaces and practices of democracy in Nicaragua, *Transactions of the Institute of British Geographers*, 34: 110–124.

Dahl, R. (1961) *Who Governs? Democracy and Power in an American City*, New Haven: Yale University Press.

Dalby, S. (1990) Critical geopolitics: discourse, difference and dissent, *Environment and Planning D: Society and Space*, 9: 261–283.

Dalby, S. (2004) Calling 9/11: geopolitics, security and America's new war, in S. Brunn (ed.) *11 September and its Aftermath: The Geopolitics of Terror*, London: Frank Cass, pp. 61–86.

Dalby, S. (2007) Ecological interventions and anthropocene ethics, *Ethics and International Affairs,* 21.

Dalby, S. (2009) Geopolitics, the revolution in military affairs and the Bush doctrine, *International Politics*, 46: 234–252.

Dalby, S. (2010) Recontextualising violence, power and nature: the next twenty years of critical geopolitics? *Political Geography*, 29: 280–288.

Dalby, S. (2013) Realism and geopolitics, in K. Dodds, M. Kuus and J. Sharp (eds.) *The Ashgate Research Companion to Critical Geopolitics*, Aldershot: Ashgate, pp. 33–48.

Daniels, S. (1993) *Fields of Vision: Landscape Imagery and National Identity in England and the United States*, Cambridge: Polity Press.

Darling, J. (2010) A city of sanctuary: the relational re-imaging of Sheffield's asylum politics, *Transactions of the Institute of British Geographers*, 35: 125–140.

Darwin, C. (1860) *On the Origins of Species by Means of Natural Selection*, London: John Murray.

Darwin, C. (1882) *The Descent of Man, and Selection in Relation to Sex* (2nd edn), London: John Murray.

Dauvergne, P. and Neville, K. (2011) Mindbombs of right and wrong: cycles of contention in the activist campaign to stop Canada's seal hunt, *Environmental Politics*, 20: 192–209.

Davies, A. D. (2012) Assemblage and social movements: Tibet support groups and the spatialities of political organisation, *Transactions of the Institute of British Geographers*, 37: 273–286.

Davies, J. and Pill, M. (2012) Hollowing out neigh-bourhood governance? Rescaling revitalisation in Baltimore and Bristol, *Urban Studies*, 49: 2199–2217.

Dean, M. (1999) *Governmentality: Power and Rule in Modern Society*. London: Sage.

Dear, M. (1999) The relevance of postmodermism, *Scottish Geographical Journal*, 115: 143–150.

De Bilj, H. J. (1967) *Systematic Political Geography*. New York: John Wiley.

Della Porta, D. and Diani, M. (1999) *Social Movements: An Introduction*, Oxford: Blackwell.

Della Porta, D. and Piazza, G. (2007) Local contention, global framing: the protest campaigns against the TAV in Val di Susa and the bridge on the Messina Straits, *Environmental Politics*, 16: 864–882.

Desforges, L. (2004) The formation of global citizen-ship: international non-governmental organiza-tions in Britain, *Political Geography*, 23: 549–569.

Desforges, L., Jones, R. and Woods, M. (2005) New geographies of citizenship, *Citizenship Studies*, 9: 239–252.

Diani, M. (1992) The concept of social movement, *The Sociological Review*, 40: 1–25.

Dicken, P. (2011) *Global Shift* (6th edn), London: Sage.

Dickinson, R. E. (1976) *Regional Concept: The Anglo-American Leaders*, London: Routledge and Kegan Paul.

DiGiovanna, S. (1996) Industrial districts and regional economic development: a regulation approach, *Regional Studies*, 30: 373–386.

Dikshit, R. D. (1977) The retreat from political geography, *Area*, 9: 234–239.

Dinius, O. J. and Vergara, A. (eds) (2011) *Company Towns in the Americas: Landscape, Power and Working Class Communities*, Athens: University of Georgia Press.

Dittmer, J. (2005) Captain America's empire: reflections on identity, popular culture and post-9/11 geopolitics, *Annals of the Association of American Geographers*, 95: 626–643.

Dixon, D. and Marston, S. (2011) Introduction: feminist engagements with geopolitics, *Gender, Place and Culture,* 18: 445–453.

Dixon, D., Hawkins, H. and Straughan, E. (2012) Of human birds and living rocks: remaking aesthetics for post-human worlds, *Dialogues in Human Geography,* 2: 249–270.

Dobson, A. (1995) *Green Political Thought*, London: Routledge

Dobson, A. (2003) *Citizenship and the Environment,* Oxford: Oxford University Press.

Dodds, K. (1994) Geopolitics in the Foreign Office: British representations of Argentina 1945–1961, *Transactions of the Institute of British Geographers*, 19: 273–290.

Dodds, K. (1996) The 1982 Falklands War and critical geopolitical eye: Steve Bell and the If . . . cartoons, *Political Geography*, 15: 571–596.

Dodds, K. (2003) Licensed to stereotype: geopolitics, James Bond and the spectre of Balkanism, *Geopolitics,* 8: 125–156.

Dodds, K. (2007) Steve Bell's eye: cartoons, geopolitics and the visualization of the 'War on Terror', *Security Dialogue*, 38: 157–177.

Dodds, K. (2010) Classical geopolitics revisited, in R. Denemark (ed.) *The International Studies Encyclopaedia*, Oxford: Blackwell, pp. 302–322.

Donnan, H. and Wilson T.W. (1999) *Borders: Frontiers of Identity, Nation and State*, Oxford: Berg.

Dorling, D. and Shaw, M. (2002) Geographies of the agenda: public policy, the discipline and its (re)'turns', *Progress in Human Geography*, 26: 629–646.

Dowler, L. and Sharp, J. (2001) A feminist geopolitics? *Space and Polity*, 5: 165–176.

Drake, C. and Horton, J. (1983) Comment on editorial essay: sexist bias in political geography, *Political Geography Quarterly*, 2: 329–337.

Driver, F. (1992) Geography's empire: histories of geographical knowledge, *Environment and Planning D: Society and Space*, 10: 23–40.

Driver, F. (2001) *Geography Militant: Cultures of Exploration and Empire*, Oxford: Blackwell.

Driver, F. and Martins, L. (eds.) (2005) *Tropical Visions in an Age of Empire*, Chicago: University of Chicago Press.

Driver, F. and Yeoh, B. (2000) Constructing the tropics: introduction, *Singapore Journal of Tropical Geography*, 21: 1–5.

Duncan, S. (1974) 'Research directions in social geography: housing opportunities and constraints' *Transactions of the Institute of British Geographers*, New Series, 1, 10–19.

Duncan, J. S. (2007) *In the Shadows of the Tropics: Climate, Race and Biopower in 19th Century Ceylon*, Aldershot: Ashgate.

Duncan, S. and Goodwin, M. (1988) *The Local State and Uneven Development: Behind the Local Government Crisis.* Cambridge: Polity.

Dunford, M. and Greco, L. (2006) *After the Three Italies: Wealth, Inequality and Industrial Change*, Chichester: Wiley Blackwell.

Dunn, K. (2005) Repetitive and troubling discourses of nationalism in the local politics of mosque development in Sydney, Australia, *Environment and Planning D: Society and Space*, 23: 29–50.

East, W. G. (1937) The nature of political geography, *Politica*, 2: 259–286.

East, W. G. and Moodie, A. E. (1956) *The Changing World: Studies in Political Geography*, London: George Harrap.

Eckersley, R. (2007) Ecological interventions: prospects and limits, *Ethics and International Affairs*, 21: 293–316.

Edensor, T. (1997) National identity and the politics of memory: remembering Bruce and Wallace in symbolic space, *Environment and Planning D: Society and Space*, 29: 175–194.

Ekers, M., Hamel, P. and Keil, R. (2012) Governing suburbia: modalities and mechanisms of sub-urban governance, *Regional Studies*, 46: 405–422.

Elden, S. (2002) The war of races and the constitution of the state: Foucault's 'Il faut défendre la société' and the politics of calculation, *Boundary*, 2, 29: 125–151.

Elden, S. (2005) Missing the point: globalization, deterritorialization and the space of the world, *Transactions of the Institute of British Geographers*, 30: 8–19.

Elwood, S. (2004) Partnerships and participation: reconfiguring urban governance in different state contexts, *Urban Geography*, 25: 755–770.

England, K. (2003) Towards a feminist political geography? *Political Geography*, 22: 611–616.

England, M. (2008) When 'good neighbors' go bad: territorial geographies of neighbourhood associations, *Environment and Planning A*, 40: 2879–2894.

England, K. and Ward, K. (eds.) (2007) *Neoliberalization: States, Networks, Peoples*, Oxford: Blackwell.

Enloe, C. (1993) *The Morning After: Sexual Politics at the End of the Cold War*, Berkeley: University of California Press.

Esping-Andersen, G (1990) *The Three Worlds of Welfare Capitalism*, Cambridge: Polity.

Esser, J. and Hirsch, J. (1989) The crisis of Fordism and the dimensions of a 'postfordist' regional and urban structure, *International Journal of Urban and Regional Research*, 13: 417–437.

Etherington, D. and Jones, M. (2004a) Beyond contradictions of the workfare state? Denmark, welfare-*through*-work, and the promise of job-rotation, *Environment and Planning C: Government and Policy*, 22: 129–148.

Etherington, D. and Jones, M. (2004b) Welfare-*through*-work and the re-regulation of labour markets in Denmark, *Capital and Class*, 83: 19–45.

Evans, J. P. (2012) *Environmental Governance*, London: Routledge

Evans, D. C. and Peattie, M. R. (1997) *Kaigun: Strategy, Tactics, and Technology in the Imperial Japanese Navy, 1887–1941*, Washington DC: Naval Institute Press.

Fainstein, S. S. and Hirst, C. (1995) Urban social movements, in D. Judge, G. Stoker and H. Wolman (eds.) *Theories of Urban Politics*, London and Thousand Oaks: Sage, pp. 181–204.

Fanon, F. (2001) [1961] *The Wretched of the Earth*, London: Penguin.

Fara, P. (2004) *Sex, Botany & Empire: The Story of Carl Linnaeus and Joseph Banks,* New York: Columbia University Press

Featherstone, D. (2003) Spatialities of transnational resistance to globalization: the maps of grievance of the Inter-Continental Caravan, *Transactions of the Institute of British Geographers,* 28: 404–421.

Featherstone, D. (2008) *Resistance, Space and Political Identities: The Making of Counter-global Networks.* Chichester: Wiley-Blackwell.

Featherstone, D., Ince, A., Mackinnon, D., Strauss, K. and Cumbers, A. (2012) Progressive localism and the construction of political alternatives, *Transactions of the Institute of British Geographers,* 37: 177–182.

Ferguson, N. (2004) *Empire,* Harmondsworth: Penguin.

Ferro, M. (1997) *Colonization: A Global History,* London: Routledge.

Fevre, R., Borland, J. and Denney, D. (1999) Nation, community and conflict: housing policy and immigration in North Wales, in R. Fevre and A. Thompson (eds.) *Nation, Identity and Social Theory: Perspectives from Wales,* Cardiff: University of Wales Press, pp. 129–148.

Fisher, R. and Johnston, H. (1993) *From Maps to Metaphors: The Pacific World of George Vancouver,* Vancouver: UBC Press.

Flint, C. (2003) Dying for a 'P'? Some questions facing contemporary political geography, *Political Geography,* 22: 617–620.

Flint, C. (ed.) (2005) *The Geography of War and Peace,* New York: Oxford University Press.

Flint, C. (2006) *Introduction to Geopolitics,* Abingdon: Routledge.

Flint, C. and Taylor, P (2011) *Political Geography: World-economy, Nation-state and Locality* (6th edn), Harlow: Prentice Hall.

Florida, R. and Jonas, A. (1991) US urban policy: The postwar state and capitalist regulation, *Antipode,* 23: 349–384.

Fluri, J. (2009) Geopolitics of gender and violence 'from below', *Political Geography,* 28: 259–265.

Fluri, J. (2011) Armored peacocks and proxy bodies: gender geopolitics in aid/development spaces of Afghanistan, *Gender, Place & Culture,* 18: 519–536.

Forest, B., Johnson, J. and Till, K. (2004) Post-totalitarian national identity: public memory in Germany and Russia, *Social and Cultural Geography,* 5: 357–380.

Foucault, M. (1977) *Discipline and Punish: The Birth of the Prison,* London: Allen Lane.

Foucault, M. (1979) *The History of Sexuality, volume 1, An Introduction,* London: Allen Lane.

Foucault, M. (1991) Governmentality, in G. Burchell, C. Gordon and P. Miller (eds.) *The Foucault Effect: Studies in Governmentality,* London: Harvester Wheatsheaf, pp. 87–104.

Fougner, T. (2006) The state, international competitiveness and neoliberal globalisation: is there a future beyond 'the competition state'?, *Review of International Studies,* 32: 165–185.

Friedman, T. (2005) *The World is Flat,* New York: Farrar, Straus and Giroux.

Fukushima, Y. (1997) Japanese geopolitics and its background: what is the real legacy of the past? *Political Geography,* 16: 407–421.

Funke, P. (2012) The global social forum rhizome: a theoretical framework, *Globalizations,* 9: 351–364.

Garmany, J. (2008) The Spaces of Social Movements: O Movimento dos Trabalhadores Rurals Sem Terra from a socio-spatial perspective, *Space and Polity,* 12: 311–328.

Gellner, E. (1983) *Nations and Nationalism,* Oxford: Blackwell.

Gerth, H. and Mills, C. W. (1970) *From Max Weber: Essays in Sociology,* London: Routledge and Kegan Paul.

Giaccaria, P. and Minca, C. (2011) Topographies/topologies of the camp: Auschwitz as a spatial threshold, *Political Geography,* 30: 3–12.

Gibson-Graham, J. K. (1996) *The End of Capitalism (as we Knew it): A Feminist Critique of Political Economy.* Oxford: Blackwell.

Giddens, A. (1985) *The Nation-state and Violence,* Cambridge, UK: Polity Press.

Gilbert, E. and Helleiner, E. (eds.) (1999) *Nation-states and Money: The Past, Present and Future of National Currencies,* London: Routledge.

Gilmartin, M. (2009) Border thinking: Rossport, Shell and the political geographies of a gas pipeline, *Political Geography,* 28: 274–282.

Glassman, J. (2010) 'The provinces elect governments, Bangkok overthrows them': urbanity, class and post-democracy in Thailand, *Urban Studies*, 47: 1301–1323.

Goblet, Y. (1955) *Political Geography and the World Map*, New York: Praeger.

Goodwin, M. and Painter, J. (1996) Local governance, the crises of Fordism and the changing geographies of regulation, *Transactions of the Institute of British Geographers*, 21: 635–648.

Goodwin, M, Duncan, S. and Halford, S. (1993) Regulation theory, the local state and the transition in urban politics, *Environment and Planning D: Society and Space*, 11: 67–88.

Goodwin, M., Cloke, P. and Milbourne, P. (1995) Regulation theory and rural research: Theorising contemporary rural change, *Environment and Planning A*, 27: 1245–1260.

Gordon, C. (1991) Introduction, in G. Burchell, C. Gordon and P. Miller (eds.) *The Foucault Effect: Studies in Governmentality*, London: Harvester Wheatsheaf, pp. 1–51.

Gottlieb, R. (1990) *Forcing the Spring: The Transformation of the American Environmental Movement*, Washington DC: Island Press.

Gottlieb, R. (1993) *Forcing the Spring: The Transformation of the American Environmental Movement,* Washington, DC: Island Press.

Gould, K. (2006) Castles in context: power, symbolism and landscape, 1066 to 1500, *Landscape Research*, 31: 424–426.

Graham, J. (1992) Post-Fordism as politics: The political consequences of narratives on the left, *Environment and Planning D: Society and Space*, 10: 393–410.

Graham, S. (ed.) (2004) *Cities, War and Terrorism: Towards an Urban Geopolitics*, Oxford: Blackwell.

Graham, S. (2006) Cities and the 'War on Terror', *International Journal of Urban and Regional Research*, 30: 255–276.

Graham, S. (2010) *Cities Under Seige: The New Military Urbanism*, London: Verso.

Gregory, D. (1995) Between the book and the lamp: imaginative geographies of Egypt, 1849–50, *Transactions of the Institute of British Geographers*, 20: 29–57.

Gregory, D. (2004) *The Colonial Present*, Oxford: Wiley-Blackwell.

Gregory, D. (2006) The black flag: Guantánamo and the space of exception, *Geografiska Annaler*, 88B: 405–427.

Gruffudd, P. (1994) Back to the land: historiography, rurality and the nation in interwar Wales. *Transactions of the Institute of British Geographers*, 19: 61–77.

Habermas, J. (1991), *The Structural Transformation of the Public Sphere: an Inquiry into a Category of Bourgeois Society*, Cambridge, MA: MIT Press.

Hagen, J. (2008) Parades, public space and propaganda: the Nazi culture parades in Munich, *Geografiska Annaler*, 90B: 349–367.

Hall, P. (1974) The new political geography, *Transactions of the Institute of British Geographers*, 63: 48–52.

Ham, C. and Hill C. (1993) *The Policy Process in the Modern Capitalist State*, London: Harvester Wheatsheaf.

Hannah, M. (2000) *Governmentality and the Mastery of Territory in Nineteenth-Century America*, Cambridge: Cambridge University Press.

Hannah, M. (2006) Torture and the ticking bomb: the war on terrorism as a geographical imagination of power/knowledge, *Annals of the Association of American Geographers*, 96: 622–640.

Hannah, M. (2010) (Mis)adventures in Rumsfeld Space, *GeoJournal*, 75: 397–406.

Hardt, M. and Negri, A. (2012) *Declaration*, Argo-Navis.

Harley, J. B. (1992) Rereading the maps of the Columbian encounter, *Annals of the Association of American Geographers*, 82: 522–542.

Harteshorne, R. (1950) The functional approach in political geography, *Annals of the Association of American Geographers*, 40: 95–130.

Harteshorne, R. (1954) Political geography, in P. E. James and C. F. Jones (eds.) *American Geography: Inventory and Prospects*, Syracuse: Syracuse University Press, pp. 167–225.

Harvey, D. (1973) *Social Justice and the City*, Oxford: Blackwell.

Harvey, D. (1974) What kind of geography for what kind of public policy?, *Transactions of the Institute of British Geographers*, 63: 18–24.

Harvey, D. (1979) Monument and myth, *Annals of the Association of American Geographers*, 69: 362–381.

Harvey, D. (1989a) *The Condition of Postmodernity*, Oxford: Blackwell.

Harvey, D. (1989b) From managerialism to entrepreneurialism: the transformation in urban governance in late capitalism, *Geografiska Annaler*, 71B: 3–17.

Harvey, D. (2003) *The New Imperialism*, Oxford: Oxford University Press.

Harvey, D. (2005) *A Brief History of Neoliberalism*, Oxford: Oxford University Press.

Harvey, D. (2008) The right to the city, *New Left Review*, 53: 23–40.

Harvey, D. (2010) *The Enigma of Capital: And the Crises of Capitalism*, Oxford: Oxford University Press.

Harvey, D. (2011) *The Enigma of Capital and the Crises of Capitalism*, London: Profile Books.

Hawthorn, G. (1995) The crises of southern states, in J. Dunn (ed.) *Contemporary Crisis of the Nation State?* Oxford: Blackwell.

Hayden, D. (1995) *The Power of Place: Urban Landscape as Public History*, Cambridge, MA: MIT Press.

Headrick, D. (1981) *The Tools of Empire: Technology and European Imperialism in the Nineteenth Century*, New York: Oxford University Press.

Heffernan, M. (1994) A state scholarship: the political geography of French international science during the nineteenth century, *Transactions of the Institute of British Geographers*, 19: 21–45.

Heffernan, M. (1995) For ever England: the Western Front and the politics of remembrance in Britain, *Ecumene*, 2: 293–324.

Heffernan, M. (2000) *Fin de siecle, Fin du monde? On the origins of European geopolitics, 1890–1920*, in K. Dodds and D. Atkinson (eds.), *Geopolitical Traditions. A Century of Geopolitical Thought*, London: Routledge, pp. 27–51.

Held, D. (ed.) (1991) *Political Theory Today*. Cambridge: Polity Press.

Herb, G. H. (1997) *Under the Map of Germany: Nationalism and propaganda, 1918–1945*, London and New York: Routledge.

Herbert, S. (1996) The geopolitics of the police: Foucault, disciplinary power and the tactics of the Los Angeles Police Department, *Political Geography*, 15: 47–61.

Herbert-Cheshire, L. (2000) Contemporary strategies for rural community development in Australia: a governmentality perspective, *Journal of Rural Studies*, 16: 203–215.

Herod, A. (2010) *Scale*, London: Routledge.

Heske, H. (1987) Karl Haushofer: his role in German geopolitics in the Nazi period, *Political Geography Quarterly*, 5: 267–282.

Hinchliffe, S., Kearnes, M. B., Degan, M. and Whatmore, S. (2005) Urban wild things: a cosmopolitical experiment, *Environment and Planning D: Society and Space*, 23: 643–658.

Hird, M. J. (2010) Indifferent globality Gaia, symbiosis and 'other worldliness, *Theory, Culture & Society*, 27: 54–72.

Hobson, K. (2007) Political animals? On animals as subjects in an enlarged political geography, *Political Geography*, 26: 250–267.

Hogenstijn, M., Van Middelkoop, D., Terlouw K. (2008) The established, the outsiders and scale strategies: studying local power conflicts, *The Sociological Review*, 56: 144–161.

Holmes, L. and Ettinger, S. (1997) *Workfairness & the Struggle for Jobs, Justice & Equality*, New York: Workfairness.

Homer-Dixon, T. F. (1999) *Environment, Scarcity and Violence*, Princeton, NJ: Princeton University Press.

Hook, D. (2005) Monumental space and the uncanny, *Geoforum*, 36: 688–704.

Hooson, I. D. (ed.) (1994) *Geography and National Identity*, Oxford: Blackwell.

House, J. W. (1973) Geographers, decision takers and policy matters, in M. Chisholm and B. Rodgers (eds.), *Studies in Human Geography*, London: Heinemann/SSRC, pp. 272–301.

Howe, N. (2008) Thou shalt not misinterpret: landscape as legal performance, *Annals of the Association of American Geographers*, 98: 435–460.

Hubbard, P. (2005) Accommodating otherness: anti-asylum protest and the maintenance of white privilege, *Transactions of the Institute of British Geographers*, 30: 52–65.

Hubbard, P. (2006) NIMBY by another name? A reply to Wolsink, *Transactions of the Institute of British Geographers*, 31: 92–94.

Huber, E. and Stephens, J. D. (2001) *Development and Crisis of the Welfare State*, Chicago: The University of Chicago Press.

Hudson, R. and Williams, A. (1995) *Divided Britain*, Chichester: Wiley.

Hughes, A. and Reimer, S. (eds.) (2004) *Geographies of Commodity Chains,* London and New York: Routledge.

Hugill, P. (2006) The geostrategy of global business: Wal-Mart and its historical forbears, in S. Brunn (ed.), *Wal-Mart World*, New York and London: Routledge, pp. 3–14.

Hunter, F. (1953) *Community Power Structure*, Chapel Hill, NC: University of North Carolina Press.

Huntington, S. P. (1991) *Third Wave: Democratization in the Late Twentieth Century*, Norman, OK: University of Oklahoma Press.

Huntington, S. P. (2002) *The Clash of Civilizations and the Remaking of World Order*, New York: Simon & Schuster.

Hyndman, J. (2001) Towards a feminist geopolitics, *Canadian Geographer*, 45(2): 210–222.

Hyndman, J. (2003) Beyond either/or: a feminist analysis of September 11th, *ACME: an International E-journal of Critical Geographies*, 2: 1–13.

Hyndman, J. (2007) Feminist geopolitics revisited: body counts in Iraq, *Professional Geographer*, 59: 35–46.

Hyndman, J. and Mountz, A. (2008) Another brick in the wall? Neo-refoulement & the externalisation of asylum in Europe & Australia, *Government & Opposition*, 43: 249–269.

Ignatieff, M. (1993) *Blood and Belonging: Journeys into the New Nationalism*, New York: Random House.

Imrie, R. and Raco, M. (eds.) (2003) *Urban Renaissance? New Labour, Community and Urban Policy*, Bristol: Policy Press.

Ingram, A. and Dodds, K. (eds.) (2009) *Spaces of Security and Insecurity: Geographies of the War on Terror*, Aldershot: Ashgate.

Isin, E. (2002) *Being Political: Genealogies of Citizenship*, Minneapolis, MN: University of Minnesota Press.

Jackson, P. (1989) *Maps of Meaning: an Introduction to Cultural Geography*, London: Unwin Hyman.

James, B. (1999) Fencing in the past: Budapest's Statue Park Museum, *Media, Culture and Society*, 21: 291–311.

Jessop, B. (1990a) *State Theory: Putting the Capitalist State in its Place,* Cambridge: Polity Press.

Jessop, B. (1990b) Regulation theories in retrospect and prospect, *Economy and Society*, 19: 153–216.

Jessop, B. (1992) Fordism and post-Fordism: a critical reformulation, in M. Storper and A. Scott (eds.), *Pathways to Industrialization and Regional Development*, London: Routledge, pp. 43–65.

Jessop, B. (1994) Post-Fordism and the state, in A. Amin (eds.), *Post-Fordism: A Reader*, Oxford: Blackwell, pp. 251–279.

Jessop, B. (1995) 'Towards a Schumpeterian workfare regime in Britain? Reflection on Regulation, governance and welfare state, *Environment and Planning A*, 27, 1613–1626.

Jessop, B. (1997a) Survey article: The regulation approach, *The Journal of Political Philosophy*, 3: 287–326.

Jessop, B. (1997b) Twenty years of the (Parisian) regulation approach: the paradox of success and failure at home and abroad, *New Political Economy*, 2: 503–526.

Jessop, B. (2001a) *Regulation Theory and the Crisis of Capitalism: Five Volumes*, Cheltenham: Elgar.

Jessop, B. (2001b) Institutional re(turns) and the strategic-relational approach, *Environment and Planning A*, 33: 1213–1235.

Jessop, B. (2002) *The Future of the Capitalist State*, Cambridge: Polity.

Jessop, B. (2008) *State Power: A Strategic-Relational Approach*, Cambridge: Polity.

Jessop, B. (2013) Revising the regulation approach: critical reflections on the contradictions, dilemmas, fixes and crisis dynamics of growth regime, *Capital and Class*, 37: 5–24.

Jessop, B. (2014) *The State*, Cambridge: Polity.

Jessop, B. and Sum, N. (2001) Pre-disciplinary and post-disciplinary approaches, *New Political Economy*, 6: 89–101.

Jessop, B. and Sum, N. (2006) *Beyond the Regulation Approach: Putting Capitalist Economies in their Place,* Cheltenham: Elgar.

John, P. (2001) *Local Governance in Western Europe,* London: Sage.

Johnston, R. J. (1979) *Political, Electoral and Spatial Systems,* Oxford: Oxford University Press.

Johnston, R. J. (1980) Political geography without politics, *Progress in Human Geography,* 4: 439–446.

Johnston, R. J. (1981) British political geography since Mackinder: a critical review, in Burnett and Taylor, *op cit,* pp. 11–32.

Johnston, R. J. (1989) The state, political geography, geography, in R. Peet and N. Thrift (eds.), *New Models in Geography,* volume 1, London: Unwin Hyman, pp. 292–309.

Johnson, N. (1995) Cast in stone: monuments, geography and nationalism, *Society and Space,* 13: 51–66.

Johnston, R. J. (1996) *Nature, State and Economy: A Political Economy of the Environment,* Chichester: Wiley.

Johnson, N. (1997) Making space: Gaeltacht policy and the politics of identity, in B. Graham (ed.), *In Search of Ireland: A Cultural Geography,* London: Routledge, pp. 151–173.

Johnston, R. J. (2000) Public policy, geography and, in R. J. Johnston, D. Gregory, G. Pratt, and M. Watts (eds.), *The Dictionary of Human Geography,* Oxford: Blackwell.

Johnston, R. J. (2002a) Manipulating maps and winning elections: measuring the impact of malapportionment and gerrymandering, *Political Geography,* 21(1): 1–31.

Johnston, R. J. (2002b) If it isn't a gerrymander, what is it? *Political Geography,* 21(1): 55–65.

Johnston, R. J. and Pattie, C. (2006) *Putting Voters in their Place: Geography and Elections in Great Britain,* Oxford: Oxford University Press.

Johnston, R. J., Pattie, C. and Rossiter, D. (2001) He lost . . . but he won! Electoral bias and George W. Bush's victory in the US Presidential Election 2000, *Representation,* 38(2): 150–158.

Johnston, R. J., Propper, C., Burgess, S., Sarker, R., Bolster, A. and Jones, K. (2005a) Spatial scale and the neighbourhood effect: multinomial models of voting at two recent British General Elections, *British Journal of Political Science,* 35: 487–514.

Johnston, R. J., Propper, C., Sarker, R., Jones, K., Bolster, A. and Burgess, S. (2005b) Neighbourhood social capital and neighbourhood effects, *Environment and Planning A,* 37: 1443–1459.

Jones, C. (1996) Empire, in I. McLean (ed.), *Oxford Concise Dictionary of Politics,* Oxford: Oxford University Press.

Jones, M. (1996) 'Full steam ahead to a workforce state? Analysing the UK Employment Departments abolition' *Policy and Politics* 24, 137–157.

Jones, M. (1997) Spatial selectivity of the state? The regulationist enigma and local struggles over economic governance, *Environment and Planning A,* 29: 831–864.

Jones, M. (1999) *New Institutional Spaces: Training and Enterprise Councils and the remaking of economic governance,* London: Routledge.

Jones, R. (2007) *People/States/Territories: The Political Geographies of British State Transformation,* Oxford: Wiley Blackwell.

Jones, M (2008) Recovering a sense of political economy, *Political Geography,* 27: 377–399.

Jones, L. (2010) 'How do the American people know..?' Embodying post-9/11 conspiracy discourse, *GeoJournal,* 75: 359–371.

Jones, M. (2010) 'Impedimenta state': anatomies of neoliberal penality, *Criminology and Criminal Justice,* 10: 393–404.

Jones, R. and Desforges, L. (2003) Localities and the reproduction of Welsh nationalism, *Political Geography,* 22: 271–292.

Jones, R. and Phillips, R. (2005) Unsettling geographical horizons: Exploring pre-modern and non-European imperialism, *Annats of the Association of American Geographers,* 95, 141–61.

Jones, R. and Fowler, C. (2008) *Placing the Nation: Aberystwyth and the Reproduction of Welsh Nationalism,* Cardiff: University of Wales Press.

Jones, M. and Jones, R. (2004) Nation states, ideological power and globalization: can geographers catch the boat? *Geoforum,* 35: 409–424.

Jones, M. and Woods, M. (2013) New localities, *Regional Studies*, 47: 29–42.

Jupp, E. (2008) The feeling of participation: everyday spaces and urban change, *Geoforum*, 39: 331–343.

Jupp, E. (2012) Rethinking local activism: 'cultivating the capacities' of neighbourhood organising, *Urban Studies*, 49: 3027–3044.

Jurjevich, J. R. and Plane, D. A. (2012) Voters on the move: the political effectiveness of migration and its effects on state partisan composition, *Political Geography*, 31: 429–443.

Kaplan, R. (1994) The coming anarchy. *Atlantic Monthly,* available at http://www.theatlantic.com/magazine/archive/1994/02/the-coming-anarchy/304670/ [Accessed 27 June 2014]

Kaplan, R. (2012) *The Revenge of Geography*, New York: Random House.

Kasperson, R. and Minghi, J. (eds.) (1969) *The Structure of Political Geography*, Chicago: Aldine.

Kearns, A. (1992) Active citizenship and urban governance, *Transactions of the Institute of British Geographers*, 17: 20–34.

Kearns, A. (1995) Active citizenship and local governance: political and geographical dimensions, *Political Geography*, 14: 155–175.

Kearns, G. (2009) *Geopolitics and Empire: the Legacy of Halford Mackinder*, New York: Oxford University Press.

Kearns, G. (2010) Geography, geopolitics and empire, *Transactions of the Institute of British Geographers*, 35: 187–203.

Kearns, G. (2011) Geopolitics, in J. Agnew and D. Livingstone (eds.) *The Sage Handbook of Geographical Knowledge*, London: Sage, pp. 610–622.

Keay, J. (2001) *The Great Arc: The Dramatic Tale of How India was Mapped and Everest was Named*, London: Harper Collins.

Kebbel, T. E. (ed.) (1882) *Selected Speeches of the Late Right Honourable the Earl of Beaconsfield*, London: Longmans, Green and Co.

Kedourie, E. (1960) *Nationalism*, London: Hutchinson.

Kedourie, E. (ed.) (1971) *Nationalism in Asia and Africa*, London: Weidenfeld and Nicolson.

Khaldun, I. (1377) *Muqqadima: On the Cause Which Increases or Reduces the Revenues of Empire*, available at http://asadullahali.files.wordpress.com/2012/10/ibn_khaldun-al_muqaddimah.pdf [Accessed 27 June 2014] Kjellén, R. (1916) *Staten som Lifsform* [The State as a Living Organism], Stockholm: Hugo Gebers förlag.

Klare, M. (2001) *Resource Wars: The New Landscapes of Global Conflict*, New York: Henry Holt

Kobyashi, A. (2009) Geographies of peace and armed conflict: introduction, *Annals of the Association of American Geographers*, 99: 819–826.

Kofman, E. (1996) Gender relations, feminism and geopolitics: problematic closures and opening strategies, in E. Kofman and G. Youngs (eds.), *Globalization: Theory and Practice*, London: Pinter, pp. 209–24.

Kofman, E. (2003) Future directions in political geography, *Political Geography*, 22: 621–624.

Kofman, E. and Peake, L. (1990) Into the 1990s: a gendered agenda for political geography. *Political Geography Quarterly*, 9: 313–336.

Kohn, M. (2004) *Brave New Neighborhoods: The Privatization of Public Space*, New York: Routledge.

Koht, H. (1947) The dawn of nationalism in Europe, *American Historical Review* 52: 265–80.

Konefal, J., Mascarenhas, M., and Hatanaka, M. (2005) Governance in the global agro-food system: backlighting the role of transnational supermarket chains, *Agriculture and Human Values*, 22: 291–302.

Kong, L. (2005) Religious schools: for spirit, (f)or nation, *Environment and Planning D: Society and Space*, 23: 615–31.

Koopman, S. (2011) Alter-geopolitics: other securities are happening, *Geoforum*, 42: 274–284.

Krätke, S. (1999) A regulationist approach to regional studies, *Environment and Planning A*, 31: 683–704.

Ku, A. S. (2012) Remaking places and fashioning an opposition discourse: struggle over the Star Ferry pier and the Queen's pier in Hong Kong, *Environment and Planning D: Society and Space*, 30: 5–22.

Kula, W. (1986) *Measures and Men*, trans. R. Szreter, Princeton: Princeton University Press.

Kull, K. (1998) On semiosis, umwelt, and semiosphere, *Semiotica*, 120: 299–310.

Kurasawa, F. (2004) A Cosmopolitanism from below: alternative globalization and the creation of a solidarity without bounds, *Archives of European Sociology*, 45: 233–255.

Kurtz, C. (2007). Torture, necessity and existential politics, *California Law Review,* 95: 235–276.

Kurtz, H. and Hankins, K. (2005) Geographies of citizenship, *Space and Polity*, 9: 1–8.

Ladd, B. (2004) *The Companion Guide to Berlin,* Woodbridge, UK and Rochester, NY: Boydell and Brewer.

Larner, W. (1998) Hitching a ride on the tiger's back: globalisation and spatial imaginaries in New Zealand, *Environment and Planning D: Society and Space*, 16: 599–614.

Larsen, L. (2012) Re-placing imperial landscapes: colonial monuments and the transition to independence in Kenya, *Journal of Historical Geography,* 38: 45–56.

Latour, B. (1986) The powers of association, in J. Law (ed.), *Power, Action and Belief: a New Sociology of Knowledge,* London: Routledge, pp. 264–280.

Latour, B. (2004) *Politics of Nature: How to Bring the Sciences in Democracy,* Cambridge, MA: Harvard University Press.

Lauria, M. (ed.) (1997) *Reconstructing Urban Regime Theory*, Thousand Oaks, CA: Sage.

Laurie, N. and Calla, P. (2004) Development, postcolonialism and feminist political geography, in L. Peake, L. Staeheli, and E. Kofman (eds.), *Mapping Women, Making Politics: Feminist Perspectives on Political Geography*, London: Routledge, pp . 99–112.

Leach, B. (1974) Race, problems and geography, *Transactions of the Institute of British Geographers*, 63: 41–47.

Leadbeater, C. (2000) *Living on Thin Air: The New Economy*, Harmondsworth: Penguin Books.

Le Billon, P. (2001) The political ecology of war: natural resources and armed conflicts, *Political Geography,* 20: 561–584.

Lee, N. (2009) How is a political public space made? The birth of Tiananmen Square and the May Fourth movement, *Political Geography*, 28: 32–43.

Lee, R. (1995) Look after the pounds and the people will look after themselves: social reproduction,

regulation, and social exclusion in Western Europe, *Environment and Planning A*, 27: 1577–1594.

Lee, R. (2002) Geography, policy and geographical agendas: a short intervention in a continuing debate, *Progress in Human Geography*, 26: 627–628.

Lefebvre, H. (1968) *Le Droit à la ville*, Paris: Anthopos.

Legg, S. (2007) *Spaces of Colonialism: Delhi's Urban Governmentalities*, Oxford: Blackwell.

Leitner, H. (1990) Cities in pursuit of economic growth, *Political Geography Quarterly*, 9: 146–170.

Leitner, H., Sheppard, E. and Sziarto, K. M. (2008) The spatialities of contentious politics, *Transactions of the Institute of British Geographers*, 33: 157–172.

Lessard-Lachance, M. and Norcliffe, G. (2013) To storm the citadel: geographies of protest at the summit of the Americas in Quebec City, April 2001, *Annals of the Association of American Geographers*, 103: 180–194.

Ley, D. and Dobson, C. (2008) Are there limits to gentrification? The contexts of impeded gentrification in Vancouver, *Urban Studies*, 45: 2471–2498.

Lindbolm, C. E. (1968) *The Policy-Making Process*, Englewood Cliffs: Prentice-Hall.

Lipietz, A. (1988) Reflections on a tale: the Marxist foundations of the concepts of accumulation and regulation, *Studies in Political Economy*, 26: 7–36.

Lødemel, I. and Trickey, H. (eds.) (2000) *An Offer you Can't Refuse: Workfare in International Perspective*, Bristol: Policy Press.

Lorimer, J. (2010) International conservation volunteering and the geographies of global environmental citizenship, *Political Geography*, 29: 311–322.

Lovell, G. (2005) *Conquest and Survival in Colonial Guatemala: A Historical Geography of the Cuchumatán Highlands of Guatemala, 1500–1821*, third edition, London: McGill-Queen's University Press.

Lovering, J. (1990) Fordism's unknown successor: a comment on Scott's theory of flexible accumulation, *International Journal of Urban and Regional Research*, 14: 159–174.

Lovering, J. (1999) Theory led by policy: the inadequacies of the 'new regionalism' (illustrated from the case of Wales), *International Journal of Urban and Regional Research*, 23: 379–395.

Low, M. (2003) Political geography in question, *Political Geography*, 22: 625–631.

Low, M. (2004), 'Cities as spaces of democracy: complexity, scale and governance', in C. Barnett and M. Low (eds.) *Spaces of Democracy: Geographical Perspectives on Citizenship, Participation and Representation*, London: Sage, pp. 128–146.

Lukes, S. (1974) *Power: a Radical View*, London: Macmillan.

Martin, R. (2002) A geography for policy, or a policy for geography? Progress in Human Geography, 26, 642–644.

McCarthy, J. and Prudham, S. (2004) Neoliberal nature and the nature of neoliberalism, *Geoforum*, 35: 275–283.

McCann, E. and Ward, K. (eds) (2001) *Mobile Urbanism: City Policymaking in the global age*. Minnesota Press.

McCormick, J. (2008) *Understanding the European Union*, fourth edition. London: Palgrave Macmillan.

McCracken, D. (1997) *Gardens of Empire: Botanical Institutions of the Victorian British Empire*, Leicester: Leicester University Press.

MacDonald, S. (2006) Words in stone? Agency and identity in a Nazi landscape, *Journal of Material Culture*, 11: 105–126.

MacDonald, F. (2007) Anti-astropolitik: outer space and the orbit of geography, *Progress in Human Geography*, 31: 592–615.

McDowell L (2005) Love, money, and gender divisions of labour: some critical reflections on welfare-to-work policies in the UK, *Journal of Economic Geography*, 5: 365–379.

McEwan, C. and Bek, D. (2009) Placing ethical trade in context: WIETA and the South African wine industry, *Third World Quarterly*, 30: 723–742.

McFarlane, T. and Hay, I. (2003) The battle for Seattle: protest and popular geopolitics in the Australian newspaper, *Political Geography*, 22: 211–232.

McGuirk, P. and Dowling, R. (2009) Neoliberal privatization? Remapping the public and private in Sydney's masterplanned residential estates, *Political Geography*, 28: 174–185.

McKee, S. C. and Teigen, J. M. (2009) Probing the reds and blues: sectionalism and voter location in the 2000 and 2004 U.S. presidential elections, *Political Geography*, 28: 484–495.

MacKenzie, J. M. (1986) *Imperialism and Popular Culture*, Manchester: Manchester University Press.

McKibben, B. (1990) *The End of Nature*, London: Bloomsbury.

MacKinnon, D. (2011) Reconstructing scale: towards a new scalar politics, *Progress in Human Geography*, 35: 21–36.

MacLeod, G. (2001) New regionalism reconsidered: globalization and the remaking of political economic space, *International Journal of Urban and Regional Research*, 25: 804–829.

MacLeod, G. (2011) Urban politics reconsidered: growth machine to post-democratic city? *Urban Studies*, 48: 2629–2660.

MacLeod, G. and Goodwin, M. (1999) Reconstructing an urban and regional political economy: On the state, politics, scale, and explanation, *Political Geography*, 18: 697–730.

MacLeod, G. and Jones, M. (2011) Renewing urban politics, *Urban Studies*, 12: 2443–2472.

McPhail, I. R. (1971) Recent Trends in Electoral Geography, *Proceedings of the New Zealand Geography Conference*, 1: 7–12.

Machiavelli, N. (1532 [1998]) *The Prince*, translated and introduction by H. C. Mansfield, Chicago: University of Chicago Press.

Mackinder, H. J. (1902) *Britain and the British Seas*, Oxford: Clarendon Press.

Mackinder, H. J. (1904) The geographical pivot of history, *The Geographical Journal*, 170: 298–321.

Mackinder, H. J. (1919) *Democratic Ideals and Reality*, London: Constable and Co.

Macleavy, J. (2011) A 'new politics' of authority, workfare and gender? The UK coalition government's welfare reform proposals, *Cambridge Journal of Regions, Economy and Society*, 4: 355–367.

Madley, B. (2005) From Africa to Auschwitz: how German South West Africa incubated ideas and methods adopted and developed by the Nazis in Eastern Europe, *European History Quarterly* 35: 429–464.

Magnusson, W. (2003) Introduction: the puzzle of the political, in W. Magnusson and K. Shaw (eds) *A Political Space: Reading the Global through Clayoquot Sound*. Minneapolis: University of Minnesota Press, pp. 1–20.

Magnusson, W. and Shaw, K. (eds.) (2003) *A Political Space: Reading the global through Clayoquot Sound*, Minneapolis: University of Minnesota Press.

Mahan, A. (1890) *The Influence of Sea Power Upon History, 1660–1783*, Boston: Little, Brown, and Co.

Makiya, K. (2004) *The Monument: Art and Vulgarity in Saddam Hussein's Iraq*, London: I B Taurus.

Mamadouh, V. (2003) Some notes on the politics of political geography, *Political Geography*, 22: 663–675.

Mann, M. (1984) The autonomous power of the state: its origins, mechanisms and results. *European Journal of Sociology* 25, 185–213 (reprinted in J. Agnew (ed.) (1997) *Political Geography: a Reader*, London: Arnold).

Mann, M. (1986) *The Sources of Social Power vol 1. The Beginning to AD1760*, Cambridge: Cambridge University Press.

Mann, M. (1988) *States, War and Capitalism: Studies in Political Sociology*, Oxford: Basil Blackwell.

Markusen, A. (1999) Fuzzy concepts, scanty evidence and policy distance: the case for rigour and policy relevance in critical regional studies, *Regional Studies*, 33: 869–886.

Marshall, M. (1987) *Long Waves of Regional Development*, London: Macmillan.

Marston, S. A. (2003) Political geography in question, *Political Geography*, 22: 633–636.

Marston, S. and Mitchell, K. (2004), Citizens and the state: citizenship formations in space and time, in C. Barnett and M. Low (eds.) *Spaces of Democracy: Geographical Perspectives on Citizenship, Participation and Representation*, London: Sage, pp. 93–112.

Martin, D. G. (2003) 'Place-framing' as Place-making: constituting a neighborhood for organizing and activism, *Annals of the Association of American Geographers*, 93: 730–750.

Martin, L. (2007) Fighting for control: political displacement in Atlanta's gentrifying neighborhoods, *Urban Affairs Review*, 42: 603–628.

Martin, R. (2001) Geography and public policy: the case of the missing agenda, *Progress in Human Geography*, 25: 189–210.

Martin, L. (2011) The geopolitics of vulnerability: children's legal subjectivity, immigrant family detention and US immigration law and enforcement policy, *Gender, Place and Culture*, 18: 477–498.

Mason, K. (2013) Academics and social movements: knowing our place, making our space, *ACME: An International E-Journal for Critical Geographies*, 12: 23–43.

Mason, K. and Whitehead, M. (2012) Transition urbanism and the contested politics of ethical place making, *Antipode*, 44: 493–516.

Massey, D. (1994) *Space, Place and Gender*, Cambridge: Polity Press.

Massey, D. (2000) Practising political relevance, *Transactions of the Institute of British Geographers*, 24: 131–134.

Massey, D. (2001) Geography on the agenda, *Progress in Human Geography*, 25: 5–17.

Massey, D. (2002) Geography, policy and politics: a response to Dorling and Shaw, *Progress in Human Geography*, 26: 645–646.

Massey, D. (2004) Geographies of responsibility, *Geografiska Annaler*, 86B: 5–18.

Massey, D. (2005) *For Space*, London: Sage.

Mazer, K. M. and Rankin, K. N. (2011) The social space of gentrification: the politics of neighbourhood accessibility in Toronto's Downtown West, *Environment and Planning D: Society and Space*, 29: 822–839.

Mead, L. M. (1997) *The New Paternalism: Supervisory Approaches to Poverty*, Washington: Brookings Institution Press.

Megoran, N. (2010) Neoclassical geopolitics, *Political Geography*, 29: 187–189.

Megoran, N. (2011) War and peace? An agenda for peace research and practice in geography, *Political Geography*, 30: 178–189.

Meijering, L., Huigen, P. and van Hoven, B. (2007) Intentional communities in rural spaces, *Tijdschrift voor Economische en Sociale Geografie*, 98: 42–52.

Meinig, D. W. (1993) *The Shaping of America: A Geographical Perspective on 500 Years of History, Volume 2: Continental America, 1800–1867*, Yale: Yale University Press.

Mensagem, F. P., Sidaway J. D. and Power, M. (2005) 'The tears of Portugal': empire, identity, race, and destiny in Portuguese geopolitical narratives, *Environment and Planning D: Society and Space,* 23: 527–554.

Mercille, J. (2010) The radical geopolitics of the 2003 Iraq War, *GeoJournal*, 75: 327–337.

Mikesell, M. W. (1983) The myth of the nation state. *Journal of Geography*, 82: 257–60.

Mills, S. (2011) Scouting for girls? Gender and the scout movement in Britain, *Gender, Place and Culture,* 18: 537–556.

Minca, C. (2006) Giorgio Agamben and the new biopolitical *nomos*, *Geografiska Annaler*, 88B: 387–403.

Mirkovic, D. (1996) Ethnic conflict and genocide: reflections on ethnic cleansing in the former Yugoslavia, *Annals of the American Academy of Political and Social Science*, 548: 191–6.

Mitchell, D. (2000) *Cultural Geography: a Critical Introduction*. Oxford and Maldon, Mass: Blackwell.

Mohan, G. and Mohan, J. (2002) Placing social capital, *Progress in Human Geography*, 26: 191–210.

Moodie, A. E. (1949) *The Geography behind Politics*, London: Hutchinson.

Moore, T. (2002) Comments on Ron Johnston's 'Manipulating maps and winning elections: measuring the impact of malapportionment and gerrymandering', *Political Geography*, 21(1): 33–38.

Moore, K. and Lewis, D. (2009) *The Origins of Globalization*, Abingdon: Routledge.

Morrill, R., Knopp, L. and Brown, M. (2011) Anomalies in Red and Blue II: Towards an understanding of the roles of setting, values, and demography in the 2004 and 2008 US presidential elections, *Political Geography*, 30: 153–168.

Moulaert, F. (1996) Rediscovering spatial inequality in Europe: Building blocks for an appropriate 'regulationist' analytical framework, *Environment and Planning D: Society and Space*, 14: 155–179.

Moulaert, F. and Swyngedouw, E. (1989) Survey 15: a regulationist approach to the geography of flexible production systems, *Environment and Planning D: Society and Space*, 7: 327–345.

Mountz, A. (2003) Human smuggling, the transnational imaginary and the everyday geographies of the nation-state, *Antipode*, 35: 622–644.

Mountz, A. and Hyndman, J. (2006) Feminist approaches to the global intimate, *Women's Studies Quarterly*, 34: 446–463.

Muir, R. (1976) Political geography: dead duck or phoenix? *Area*, 8: 195–200.

Muir, R. (1981) *Modern Political Geography* (2nd edn), London: Macmillan.

Murphy, A. (1996) The sovereign state system as political-territorial ideal: historical and contemporary considerations, in T. Biersteker and C. Weber (eds.), *State Sovereignty as Social Construct*, Cambridge: Cambridge University Press, pp. 81–120.

Murray, W. (2006) *Geographies of Globalization*, Abingdon: Routledge.

Myers, S. L. (2011) Iraq restores monument symbolizing Hussein era, *New York Times*, 6 February 2011, p A6. Available online at: http://www.nytimes.com/2011/02/06/world/middleeast/06iraq.html?_r=0 [Accessed 2 July 2014].

Nairn, T. (1977) *The Break-up of Britain: Crisis and Neo-Nationalism*, London: New Left Books.

Narlikar, A. (2005) *The World Trade Organization: A Very Short Introduction*, Oxford: Oxford University Press.

Nash, C. (2002) Genealogical identities, *Environment and Planning D: Society and Space*, 20: 27–52.

Naylor, S. (2000) 'That very garden of South America': European surveyors of Paraguay, *Singapore Journal of Tropical Geography*, 21: 48–62.

Neilson, J. and Pritchard, B. (2009) *Value Chain Struggles: Institutions and Governance in the Plantation Districts of South India*, Oxford: Wiley-Blackwell.

Neumayer, E. (2006) Unequal access to foreign spaces: how states use visa restrictions to regulate mobility in a globalized world, *Transactions of the Institute of British Geographers*, 31: 72–84.

Newman, D. and Falah, G. (1997) Bridging the gap: Palestinian and Israeli discourses on autonomy and statehood, *Transactions of the Institute of British Geographers*, 22: 111–29.

Newman, D. and Paasi, A. (1998) Fences and neighbours in the postmodern world: boundary narratives in political geography, *Progress in Human Geography*, 22: 186–207.

Newman, K. and Wyly E. (2006) The right to stay put, revisited: gentrification and resistance to displacement in New York City, *Urban Studies*, 43: 23–57.

Nicholls, W. (2009) Place, networks, space: theorising the geographies of social movements, *Transactions of the Institute of British Geographers*, 34: 78–93.

Nicholls, W. (2011) Cities and the unevenness of social movement space: the case of France's immigrant rights movement, *Environment and Planning A*, 43: 1655–1673.

Nkrumah, K. (1965) *Neo-colonialism: The Last Stage of Imperialism*, London: Thomas Nelson & Sons.

Nogué, J. and Vicente, J. (2004) Landscape and national identity in Catalonia, *Political Geography*, 23: 113–132.

O'Loughlin, J. (2004), Global democratization: measuring and explaining the diffusion of democracy, in C. Barnett and M. Low (eds.), *Spaces of Democracy: Geographical Perspectives on Citizenship, Participation and Representation*, London: Sage, pp. 23–44.

O'Reilly, K. and Crutcher, M. E. (2006) Parallel politics: the spatial power of New Orleans' Labor Day parades, *Social and Cultural Geography*, 7: 245–265.

Ó'Tuathail, G. (1992) Putting Mackinder in his place, *Political Geography*, 11: 100–118.

Ó'Tuathail, G. (1996) *Critical Geopolitics*, London: Routledge.

Ó'Tuathail, G. (2004) 'Just looking out for a fight': American affect and the invasion of Iraq, *Antipode*, 35: 856–870.

Ó'Tuathail, G. (2013) Foreword: Arguing about Geopolitics, in K. Dodds, M. Kuus and J. Sharp (eds.) *The Ashgate Research Companion to Critical Geopolitics*, Aldershot: Ashgate, pp. ixx–xxi.

Ó'Tuathail, G. and Agnew, J. (1992) Geopolitics and discourse: practical geopolitical reasoning in American foreign policy, *Political Geography*, 11: 190–204.

OECD (1999) *The Local Dimension of Welfare-to-Work: An International Survey*, Paris: OECD.

Offe, C. (1984) *The Contradictions of the Welfare State*, Hutchinson, London

Ogborn, M. (1992) Local power and state regulation in nineteenth century Britain, *Transactions of the Institute of British Geographers*, 17: 215–226.

Ogborn, M. (1998) The capacities of the state: Charles Davenant and the management of the Excise, 1683–1698, *Journal of Historical Geography*, 24: 289–312.

Ogborn, M. (2002) Writing travels: power, knowledge and ritual on the English East India Company's early voyages, *Transactions of the Institute of British Geographers*, 27: 155–171.

Ogborn, M. (2007) *Indian Ink: Script and Print in the Making of the English East India Company*, Chicago: Chicago University Press.

Ophuls, W. (1997a) *Requiem for Modern Politics*, Boulder, CO: Westview Press.

Ophuls, W. (1997b) *Ecology and the Politics of Scarcity*, New York: W. H. Freeman & Co.

Ophuls, W. (2011) *Plato's Revenge: Politics in the Age of Ecology*, Cambridge, MA: MIT Press.

Orford, S., Rallings, C., Thrasher, M. and Borisyuk, G. (2009) Electoral salience and the costs of voting at national, sub-national and supra-national elections in the UK: a case study of Brent, UK, *Transactions of the Institute of British Geographers*, 34: 195–214.

Osborne, B. S. (1998) Constructing landscapes of power: the George Etienne Cartier Monument, Canada, *Journal of Historical Geography*, 24(4): 431–458.

Özkirimli, U. (2000) *Theories of Nationalism: A Critical Introduction*, London: Palgrave Macmillan.

Paasi, A. (1996) *Territories, Boundaries and Consciousness: The Changing Geographies of the Finnish-Russian Border*, Chichester: Wiley.

Pain, R. (2006) Social geography: seven deadly myths in policy research, *Progress in Human Geography*, 30: 250–259.

Pain, R. (2009) Globalized fear? Towards an emotional geopolitics, *Progress in Human Geography*, 33: 466–486.

Painter, J. (1995) *Politics, Geography and 'Political Geography'*, London: Arnold.

Painter, J. (2003) Towards a post-disciplinary political geography, *Political Geography*, 22: 637–639.

Painter, J. (2012) The politics of the neighbour, *Environment and Planning D: Space and Society*, 30: 515–533.

Painter, J. and Goodwin, M. (1995) Local governance and concrete research: Investigating the uneven development of regulation, *Economy and Society*, 24: 334–356.

Painter, J. and Goodwin, M. (2000) Local governance after Fordism: a regulationist perspective, in G. Stoker (ed.) *The New Politics of British Local Governance*, London: Macmillan, pp. 33–53.

Painter, J. and Jeffrey, A. (2009) *Politics, Geography and 'Political Geography'*, London: Arnold.

Panelli, R., Nairn, K. and McCormack, J. (2002) 'We make our own fun': reading the politics of youth with(in) community, *Sociologia Ruralis*, 42: 106–130.

Parlette, V. and Cowen, D. (2011) Dead malls: suburban activism, local spaces, global logistics, *International Journal of Urban and Regional Research*, 35: 794–811.

Parvin, P. (2000) Against localism: does decentralising power to communities fail minorities? *The Political Quarterly*, 80: 351–360.

Paterson, J. H. (1987) German geopolitics reassessed, *Political Geography Quarterly*, 6: 107–114.

Paul, D. E. (2005) The local politics of 'going global': making and unmaking Minneapolis-St Paul as a world city, *Urban Studies*, 42: 2103–2122.

Pawson, E. (1992) Two New Zealands: Maori and European, in K. Anderson and F. Gale (eds.), *Inventing Places: Studies in Cultural Geography*, London: Longman, pp. 15–33.

Peck, J. (1995) Moving and shaking: Business élites, state localism and urban privatism, *Progress in Human Geography*, 19: 16–46.

Peck, J. (1998) Workfare in the sun: politics, representation, and method in US welfare-to-work strategies, *Political Geography*, 17: 535–566.

Peck, J. (1999) Editorial: Grey geography?, *Transactions of the Institute of British Geographers*, 24: 131–135.

Peck, J. (2000) Jumping in, joining up and getting on, *Transactions of the Institute of British Geographers*, 25: 255–258.

Peck, J. (2001) *Workfare States*, Guilford, New York.

Peck, J. (2010) *Constructions of Neoliberal Reason*, Oxford: Oxford University Press.

Peck, J. and Jones, M. (1995) Training and Enterprise Councils: Schumpeterian workfare state, or what?, *Environment and Planning A*, 27: 1361–1396.

Peck, J. and Tickell, A. (1992) Local modes of social regulation? Regulation theory, Thatcherism and uneven development, *Geoforum*, 23: 347–363.

Peck, J. and Tickell A. (1995) The social regulation of uneven development: 'regulatory deficit', England's south east, and the collapse of Thatcherism, *Environment and Planning A*, 27: 15–40.

Peet, R. (1985) The social origins of environmental determinism, *Annals of the Association of American Geographers*, 75: 309–333.

Peet, R. (2003) *Unholy Trinity: The IMF, World Bank and WTO*, London and New York: Zed Books.

Peet, R. and Thrift, N. (1989) Political economy and human geography, in R. Peet and N. Thrift (eds.), *New Model in Geography: The Political-economy Perspective*, London: Unwin Hyman, pp. 3–29.

Pendle, G. (1954) *Paraguay: A Riverside Nation*, London: Royal Institute of International Affairs.

Pepper, D. (1984) *The Roots of Contemporary Environmentalism*, London: Croom Helm.

Pepper, D. (1996) *Modern Environmentalism: An Introduction*, London: Routledge.

Phadke, R. (2011) Resisting and reconciling big wind: middle landscape politics in the new American West, *Antipode*, 43: 754–776.

Phillips, R. (1997) *Mapping Men and Empire: A Geography of Adventure*, London: Routledge.

Phillips, R. (2009) Bridging east and west: Muslim-identified activists and organisations in the UK anti-war movements, *Transactions of the Institute of British Geographers*, 34: 506–520.

Phillips, R. and Jones, R. (2008) Imperial and anti-imperial constructions of civilization: engagements with pre-modern pasts, *Geopolitics*, 13: 730–735.

Pickerell, J. (2006) Radical politics on the net, *Parliamentary Affairs*, 59: 266–282.

Pickerell, J. and Krinsky, J. (2012) Why does Occupy matter? *Social Movement Studies*, 11: 279–287.

Pierce, J., Martin, D. and Murphy, J. (2011) Relational place-making: the networked politics of place, *Transactions of the Institute of British Geographers*, 36: 54–70.

Pile, S. and Keith, M. (eds.) (1997) *Geographies of Resistance*, London: Routledge.

Pine, A. (2010) The performativity of urban citizenship, *Environment and Planning A*, 42: 1103–1120.

Piore, M. and Sabel, C. (1984) *The Second Industrial Divide: Possibilities for Prosperity*, New York: Basic Books.

Plumbwood, V. (1993) *Feminism and the Mastery of Nature,* London: Routledge.

Pollard, J., Henry, N., Bryson, J. and Daniels, P. (2000) Shades of grey? Geographers and policy, *Transactions of the Institute of British Geographers*, 25: 243–248.

Pollard, J. S, McEwan, C., Laurie, N. D. and Stenning, A. C. (2009) Economic geography under postcolonial scrutiny, *Transactions of the Institute of British Geographers*, 34: 137–142.

Power, M. and Crampton, A. (2005) Reel geopolitics: cinemato-graphing political space, *Geopolitics*, 10: 193–203.

Prescott, J. R. V. (1965) *The Geography of Frontiers and Boundaries*, London: Hutchinson.

Prince, H. (1971) Questions of social relevance, *Area* 3: 150–153.

Prudham, S. (2007a) The fictions of autonomous invention: accumulation by dispossession, commodification, and life patents in Canada, *Antipode*, 39: 406–429.

Prudham, S. (2007b) Sustaining sustained yield: class, politics and post-war forest regulation in British Columbia, *Environment and Planning D: Society and Space*, 27: 258–283.

Prudham, S. (2008) Tall among the trees: organizing against globalist forestry in rural British Columbia, *Journal of Rural Studies*, 24: 182–196.

Purcell, M. (2001) Metropolitan political reorganization as a politics of urban growth: the case of San Fernando Valley secession, *Political Geography*, 20: 613–633.

Purcell, M. (2002a) The state, regulation, and global restructuring: reasserting the political in political economy, *Review of International Political Economy*, 9: 284–318.

Purcell, M. (2002b) Excavating Lefebvre: the right to the city and its urban politics of the inhabitant, *GeoJournal*, 58: 99–108.

Purcell, M. (2006) Urban democracy and the local trap, *Urban Studies*, 43: 1921–1941.

Purcell, M. (2008) *Recapturing Democracy: Neoliberalism and the Struggle for Alternative Urban Futures,* London: Routledge.

Purcell, M. (2013) *The Down-Deep Delight of Democracy*, Chichester: Wiley Blackwell.

Purcell, M. and Brown, J. (2005) Against the local trap: scale and the study of environment and development, *Progress in Development Studies*, 5: 279–297.

Putnam, R. (2000) *Bowling Alone: the Collapse and Revival of American Community*, New York: Simon and Schuster.

Quilley, S. (2013) Degrowth is not a liberal agenda: relocalization and the limits to low energy cosmopolitanism, *Environmental Values,* 22: 261–285.

Rabinow, P. (ed.) (1984) *The Foucault Reader*, Harmondsworth: Penguin.

Raco, M., Parker, G. and Doak, J. (2006) Reshaping spaces of local governance? Community strategies and the modernisation of local government in England, *Environment and Planning C: Government and Policy*, 24: 475–496.

Radcliffe, S. (1999) Embodying national identities: mestizo men and white women in Ecuadorian racial-national imaginaries, *Transactions of the Institute of British Geographers*, 24: 213–226.

Raento, P. and Brunn, S. (2005) Visualising Finland: postage stamps as political messengers, *Geografiska Annaler*, 87B: 145–164.

Raza, S. S. (2012) The North West Frontier of Pakistan, in S. O. Opondo and M. J. Shapiro (eds.), *The New Violent Cartography: Geo-Analysis After the Aesthetic Turn*, London and New York: Routledge, pp. 173–194.

Reynolds, D. (1993) Political geography: closer encounters with the state, contemporary political economy, and social theory, *Progress in Human Geography*, 17: 389–403.

Reynolds, S. (1984) *Kingdoms and Communities in Western Europe, 900–1300*, Oxford: Oxford University Press.

Robbins, P. (2003) Political ecology in political geography, *Political Geography*, 22: 641–645.

Robbins, P. (2004) *Political Ecology: a Critical Introduction*, Oxford: Blackwell.

Robinson, J. (2003) Political geography in a postcolonial context, *Political Geography*, 22: 647–651.

Robinson, T. and Noriega, S. (2010) Voter migration as a source of electoral change in the Rocky Mountain West, *Political Geography*, 29: 28–39.

Robson, B. (1972) Editorial comment: The corridors of geography, *Area*, 4: 213–214.

Robson, B. (1976) Houses and people in the city, *Transactions of the Institute of British Geographers* New Series, 1: 1.

Robson, B., Bradford, M., Deas, I., Hall, E., Harrison, E., Parkinson, M., Evans, R., Garside, P. and Harding, A. (1994) *Assessing the Impact of Urban Policy*, London: HMSO.

Rokkan, S. (1970) *Citizens, Elections, Parties*, New York: McKay.

Rokkan, S. (1980) Territories, centres and peripheries: toward a geoethnic-geoeconomic-geopolitical model of differentiation within Western Europe, in J. Gottmann (ed.), *Centre and Periphery: Spatial Variation in Politics*, London: Sage, pp. 163–204.

Rose, N. (1993) Government, authority and expertise in advanced liberalism, *Economy and Society*, 22: 283–299.

Rose, N. (1996) Governing 'advanced liberal democracies', in A. Barry, T. Osborne and N. Rose (eds.) *Foucault and Political Reason*, London: UCL Press, pp. 37–64.

Rose, R. (1997) *Lesson Drawing in Public Policy: A Guide to Learning Across Time and Space*, Chatham: Chatham House.

Routledge, P. (1997) The imagineering of resistance: Pollock free state and the practice of postmodern politics, *Transactions of the Institute of British Geographers*, 22: 359–376.

Routledge, P. (2000) Geopoetics of resistance: India's Baliapal movement, *Alternatives*, 25: 375–389.

Routledge, P. (2003) Convergence space: process geographies of grassroots globalization networks, *Transactions of the Institute of British Geographers*, 28: 333–349.

Routledge, P. (2012) Sensuous solidarities: emotion, politics and performance in clandestine insurgent Rebel Clown Army, *Antipode*, 44, 428–452.

Routledge, P. and Cumbers, A. (2009) *Global Justice Networks: Geographies of transnational solidarity*, Manchester: Manchester University Press.

Routledge, P. and Nativel, C. (2008) The entangled geographies of global justice networks, *Progress in Human Geography*, 32: 182–201.

Ruddiman, W. F. (2001) *Earth's Climate: Past and Future*, New York: W. H. Freeman and Co.

Ruddiman, W. F. (2005) *Plows, Plagues, and Petroleum*, Princeton, NJ: Princeton University Press.

Sack, R. (1983) Human territoriality: a theory, *Annals of the Association of American Geographers*, 73: 55–74.

Sack, R. (1986) *Human Territoriality: Its Theory and History*, Cambridge: Cambridge University Press.

Said, E. (1978) *Orientalism*, London: Penguin.

Said, E. (1993) *Culture and Imperialism*, London: Penguin.

Salmenkari, T. (2009) Geography of protest: places of demonstration in Buenos Aires and Seoul, *Urban Geography*, 30: 239–260.

Schmelzkopf, K. (2002) Incommensurability, Land Use and the Right to Space: Community Gardens in New York City, *Urban Geography*, 23: 323–343.

Schmitt, C. (1996) *The Concept of the Political*, Chicago: University of Chicago Press.

Schuermans, N. and De Maesschalck, N. (2010) Fear of Crime as a Political Weapon: Explaining the Rise of Extreme Right Politics in the Flemish countryside, *Social and Cultural Geography*, 11: 247–262.

Scott, A. (1988a) Flexible production systems and regional development: the rise of new industrial spaces in North America and Western Europe, *International Journal of Urban and Regional Research*, 12: 171–185.

Scott, A. (1988b) *New Industrial Spaces: Flexible Production Organisation and Regional Development in North America and Western Europe*, London: Pion.

Scott, J. C. (1998) *Seeing Like a State: How Certain Schemes to Improve the Human Condition Have Failed*, New Haven: Yale University Press.

Scott, H. (2009) *Contested Territory: Mapping Peru in the Sixteenth and Seventeenth Centuries*, Notre Dame, IN: University of Notre Dame Press.

Scott Cato, M. (2012) *The Bioregional Economy*, London: Earthscan.

Scottish Geographical Journal (1999) Relevance in human geography, *Scottish Geographical Journal* (special issue), 115: 91–165.

Secor, A. (2001) Toward a feminist counter-geopolitics: gender, space and Islamist politics in Istanbul, *Space and Polity*, 5: 191–211.

Secor, A. (2004) 'There is an Istanbul that belongs to me': citizenship, space and identity in the city, *Annals of the Association of American Geographers*, 94: 352–368.

Semple, E. C. (1901) The Anglo-Saxons of the Kentucky mountains: a study in anthropogeography, *The Geographical Journal*, 17: 588–623.

Semple, E. C. (1911) *Influences of Geographic Environment, on the Basis of Ratzel's System of Anthropo-geography*, London: H. Holt.

Sennett, R. (1977), *The Fall of Public Man*, New York: Alfred A. Knopf

Sharp, J. P. (2011) Subaltern geopolitics: introduction, *Geoforum*, 42: 271–273.

Sharp, J. P. (2013) Geopolitics at the margins? Reconsidering genealogies of critical geopolitics, *Political Geography*, 37: 20–29.

Sharp, J. P., Routledge, P., Philo, C. and Paddison, R. (eds.) (2000) *Entanglements of Power: Geographies of domination/resistance*, London: Routledge.

Shaw, I., Jones, III, J. P. and Butterworth, M. K. (2013) The mosquito's *umwelt*, or one monster's standpoint ontology, *Geoforum*, 48: 260–267.

Shields, R. (1991) *Places on the Margin: Alternative Geographies of Modernity*, London: Routledge.

Shin, M. (2009) Democratizing electoral geography: visualizing votes and political neogeography, *Political Geography*, 28: 149–152.

Short, J. R. (1991) *Imagined Country: Society, Culture and Environment*, London: Routledge.

Sidaway, J. (2009) Shadows on the path: negotiating geopolitics on an urban section of Britain's South West Coast Path, *Environment and Planning D: Society and Space*, 27: 1091–1116.

Sidaway, J., Mamadouh, V. and Power, M. (2013) Reappraising geopolitical traditions, in K. Dodds, M. Kuus and J. Sharp (eds.) *The Ashgate Research Companion to Critical Geopolitics*, Aldershot: Ashgate, pp. 165–188.

Sidorov, D. (2000) National monumentalization and the politics of scale: the resurrections of the Cathedral of Christ the Savior in Moscow, *Annals of the Association of American Geographers*, 90: 548–572.

Sites, W. (2003) *Remaking New York,* Minneapolis, MN: University of Minnesota Press.

Sluyter, A. (2003), Neo-environmental determinism, intellectual damage control, and nature/society science, *Antipode,* 4: 813–817.

Smith, D. (1971) Radical geography – the next revolution? *Area*, 3: 153–157.

Smith, N. (1984) *Uneven Development*, Oxford: Blackwell.

Smith, A. D. (1986) *The Ethnic Origins of Nations*, Oxford: Blackwell.

Smith, N. (1986) Bowman's new world and the Council on Foreign Relations, *Geographical Review*, 76: 438–460.

Smith, G. (1989) Privilege and place in Soviet society, in D. Gregory and R. Walford (eds.) *Horizons in Human Geography*, Basingstoke: Macmillan, pp. 320–340.

Smith, A. D. (1991) *National Identity*, London: Penguin.

Smith, S. J. (1993) Bounding the borders: claiming space and making place in rural Scotland, *Transactions of the Institute of British Geographers*, 18: 291–308.

Smith, A. D. (1996) Memory and modernity: reflections on Ernest Gellner's theory of nationalism, *Nations and Nationalism*, 2: 371–388.

Smith, N. (1996) *The New Urban Frontier: Gentrification and the Revanchist City*, London and New York: Routledge.

Smith, M. P. (2001) *Transnational Urbanism: Locating Globalization*, Malden: Blackwell.

Smith, A. D. (1998) *Nationalism and Modernism: A Critical Survey of Recent Theories of Nations and Nationalism*, London: Routledge.

Smith, N. (2003) Remaking scale: competition and co-operation in pre-national and post-national Europe, in N. Brenner, B. Jessop, M. Jones and G. MacLeod (eds.), *State/Space: A Reader*, Oxford: Blackwell, pp. 227–238.

Smith, N. (2005) *The Endgame of Globalization*, Abingdon: Routledge.

Smith, R. J. (2011) Graduated incarceration: the Israeli occupation in subaltern geopolitical perspective, *Geoforum, 42*: 316–328.

Smith, S. (2011) 'She says herself, "I have no future"': love, fate and territory in Leh District, India, *Gender, Place and Culture, 18*: 455–476.

Smith. C. and Kurtz, H. (2003) Community Gardens and Politics of Scale in New York City, *Geographical Review, 93*: 193–212

Soja, E. (1968) Communications and territorial integration in East Africa, *East Lakes Geographer, 4*: 39–57.

Spencer, H. (1857) Progress: its laws and cause, *Westminster Review, 67*: 445–485.

Springer, S. (2011) Public space as emancipation: meditations on anarchism, radical democracy, neoliberalism and violence, *Antipode, 43*: 525–562.

Spruyt, H. (1994) *The Sovereign State and its Competitors: an Analysis of Systems Change*, Princeton: Princeton University Press.

Spykman, N. J. (1942) *America's Strategy in World Politics*, New York: Harcourt, Brace & Co.

Spykman, N. J. (1944) *The Geography of the Peace*, New York: Harcourt, Brace & Co.

Staeheli, L. (2008) Citizenship and the problem of community, *Political Geography, 27*: 5–21.

Staeheli, L. (2010) Political geography: democracy and the disorderly public, *Progress in Human Geography, 34*: 67–78.

Staeheli, L. (2011) Political geography: where's citizenship? *Progress in Human Geography, 35*: 393–400.

Staeheli, L. and Mitchell, D. (2008) *The People's Property? Power, Politics and the Public*, New York: Routledge.

Staeheli, L., Ehrkamp, P., Leitner, H. and Nagel, C. (2012) Dreaming the ordinary: daily life and the complex geographies of citizenship, *Progress in Human Geography, 36*: 628–644.

Staiger, U. (2009) Cities, citizenship, contested cultures: Berlin's palace of the republic and the politics of the public sphere, *Cultural Geographies, 16*: 309–327.

Stangl, P. (2006) Restoring Berlin's unter den linden: ideology, word view, place and space, *Journal of Historical Geography, 32*: 352–376.

Steger, M. (2003) *Globalization: A Very Short Introduction*, Oxford: Oxford University Press.

Steil, J. and Ridgley, J. (2012) 'Small-town defenders': the production of citizenship and belonging in Hazleton, Pennsylvania, *Environment and Planning D: Space and Society, 30*: 1028–1045.

Stiglitz, G. (2010) *Freefall: Free Markets and the Sinking of the Global Economy*, London: Allen Lane.

Stoker, G. (2011) Was local governance such a good idea? A global comparative perspective, *Public Administration, 89*: 15–31.

Stone, C. (1989) *Regime Politics: Governing Atlanta 1946–1988*, Lawrence, KS: University of Kansas Press.

Storper, M. and Scott, A. (eds) (1992) *Pathways to Industrialization and Regional Development*, London: Routledge.

Storper, M. and Salais, R. (1997) *Worlds of Production: The Action Frameworks of the Economy*, Harvard: Harvard University Press.

Striffler, S. (2002) *In the Shadows of State and Capital: The United Fruit Company, Popular Struggle and Agrarian Restructuring in Ecuador, 1900–1995*, Durham, NC: Duke University Press.

Sum, N. and Jessop, B. (2013) *Towards a Cultural Political Economy: Putting Culture in its Place in Political Economy*, Cheltenham: Elgar.

Sumida, J. (1999) Alfred Thayer Mahan, Geopolitician, in C. S. Gray and G. Sloan (eds.), *Geopolitics: Geography and Strategy*, London: Frank Cass, pp. 399–413.

Sunley, P., Martin, R. and Nativel, C. (2006) *Putting Workfare in Place: Local Labour Markets and the New Deal*, Oxford: Blackwell.

Swyngedouw, E. (2005) Governance innovation and the citizen: the Janus face of governance-beyond-the-state, *Urban Studies*, 42: 1991–2006.

Swyngedouw, E. (2011a) *Designing the Post-political City and the Insurgent Polis*, London: Bedford Press.

Swyngedouw, E. (2011b) Interrogating post-democratization: reclaiming egalitarian political spaces, *Political Geography*, 30: 370–380.

Taylor, P. J. (1985) *Political Geography: World-economy, Nation-state and Locality,* first edition, London: Longman.

Taylor, P. J. (1988) World systems analysis and regional geography, *Professional Geographer*, 40: 259–265.

Taylor, P. J. (1994) From heartland to hegemony: changing the world in political geography, *Geoforum,* 25: 403–411.

Taylor, P. J. (1995) Beyond containers: internationality, interstateness, interterritoriality, *Progress in Human Geography*, 19: 1–15.

Taylor, P. (2004) *World City Network: A Global Urban Analysis*, London and New York: Routledge.

Taylor, P. J. (2005) New political geographies: global civil society and global governance through world city networks, *Political Geography*, 24: 703–730.

Taylor, P. J. and Flint, C. (2000) *Political Geography: World-economy, Nation-state and Locality*, fourth edition, London: Prentice Hall.

Taylor, P. J. and Johnston, R. J. (1979) *Geography of Elections*, London: Penguin.

Thomas, J. (2000) *The Battle in Seattle: The Story Behind and Beyond the WTO Demonstrations*, Golden, CO: Fulcrum.

Thornton, P. M. (2010) From liberating production to unleashing consumption: mapping landscapes of power in Beijing, *Political Geography*, 29: 302–310.

Thrift, N. (1983) On the determination of social action in time and space, *Environment and Planning D: Society and Space*, 1: 23–57.

Tickell, A. (1995) Reflections on 'activism and the academy', *Environment and Planning D: Society and Space*, 13: 235–237.

Tickell, A. and Peck, J. (1992) Accumulation, regulation and the geographies of post-Fordism: Missing links in regulationist research, *Progress in Human Geography*, 16: 190–218.

Tickell, A. and Peck, J. (1995) Social regulation after Fordism: regulation theory, neo-liberalism and the global-local nexus, *Economy and Society*, 24: 357–386.

Tickell, A. and Peck, J. (1996) The return of the Manchester Men: men's words and men's deeds in the remaking of the local state, *Transactions of the Institute of British Geographers*, 21(4): 595–616.

Till, K. (2005) *The New Berlin: Memory, Politics, Place*, Minneapolis: University of Minnesota Press.

Tilly, C. (1975) Reflections on the history of European state-making, in C. Tilly (ed.), *The Formation of Nation States in Western Europe*, Princeton: Princeton University Press, pp. 3–83.

Tilly, C. (1990) *Coercion, Capital and European states AD990–1992*, Oxford: Blackwell.

Torfing, J. (1999) Workfare with welfare: recent reforms of the welfare state, *Journal of European Social Policy,* 9: 5–28.

Trask, D. (1996) *The War with Spain in 1898*, Lincoln: University of Nebraska Press.

Trapese Collective (2008) *The Rocky Road to Transition*, London, Trapese Collective

Tunander, O. (2001) Swedish-German geopolitics for a new century: Rudolf Kjellen's 'The State as a Living Organism', *Review of International Studies,* 27: 451–463.

Uitermark, J. and Nicholls, W. (2012) How local networks shape a global movement: comparing Occupy in Amsterdam and Los Angeles, *Social Movement Studies*, 11: 295–301.

Unwin, T. (1992) *The Place of Geography*, Longman: Harlow.

Unwin, T. and Hewitt, V. (2001) Banknotes and national identity in central and eastern Europe, *Political Geography*, 20: 1005–1028.

Van der Horst, D. and Vermeylen, S. (2011) Local rights to landscape in the global moral economy of carbon, *Landscape Research*, 36: 455–470.

Valler, D., Wood, A. and North, P. (2000) Local governance and local business interests: a critical review, *Progress in Human Geography*, 24: 409–428.

Vandergeest, P. and Peluso, N. (1995) Territorialization and state power in Thailand, *Theory and Society*, 24: 385–426.

Vasudevan, A. (2011) Dramaturgies of dissent: the spatial politics of squatting in Berlin, 1968–, *Social and Cultural Geography*, 12: 283–303.

Virilio, P. (1999) *Polar Inertia*, London: Sage.

Walks, R. A. (2006) The causes of city-suburban political polarization? A Canadian case study, *Annals of the Association of American Geographers*, 96: 390–414.

Walks, R. A. (2010) Electoral behaviour behind the gates: partisanship and political participation among Canadian gated community residents, *Area*, 42: 7–24.

Wallerstein, I. (1974) *The Modern World System I. Capitalist Agriculture and the Origins of the European World Economy in the Sixteenth Century*, New York: Academic Press.

Wallerstein, I. (1979) *The Capitalist World Economy*, Cambridge, UK: Cambridge University Press.

Wallerstein, I. (1980) *The Modern World System II. Mercantilism and the Consolidation of the European World Economy, 1600–1750*, New York: Academic Press.

Wallerstein, I. (1989) *The Modern World System III. The Second Era of Great Expansion of the Capitalist World Economy, 1730–1840*, New York: Academic Press.

Wallerstein, I. (1991) *Unthinking Social Science*, Cambridge, UK: Polity Press.

Walsh, E. A. (1949) *Total Power: A Footnote to History*, Garden City NY: Doubleday & Company, Inc.

Walsh, F. (1979) Time-lag in political geography, *Area*, 11: 91–92.

Walton, S. (2007) Site the mine in our backyard! Discursive strategies of community stakeholders in an environmental conflict in New Zealand, *Organization and Environment*, 20: 177–203.

Walzer, N. and Jacobs, B. (eds.) (1998) *Public-Private Partnerships for Local Economic Development*, Westport: Praeger.

Ward, K. (2006) Geography and public policy: towards public geographies, *Progress in Human Geography*, 30: 495–503.

Warf, B. (2009) The US Electoral College and spatial biases in voter power, *Annals of the Association of American Geographers*, 99: 184–204.

Warf, B. and Waddell, C. (2002) Florida in the 2000 presidential election: historical precedents and contemporary landscapes, *Political Geography*, 21: 85–90.

Waters, M. (2001) *Globalization*, second edition, London: Routledge.

Watt, P. (2013) 'It's not for us': Regeneration, the 2012 Olympics and the gentrification of East London, *City*, 17: 99–118.

Watts, M. (2001) 1968 and all that . . . *Progress in Human Geography*, 25: 157–188.

Watts, M. Okonta, I. and Von Kemedi, D. (2004) *Economies of Violence in the Niger Delta, Nigeria*, University of California Berkeley Working Paper, available at http://oldweb.geog.berkeley.edu/ProjectsResources/ND%20Website/NigerDelta/WP/1-WattsOkantaVon.pdf [Accessed 7 July 2014]

Webber, M. (1991) The contemporary transition, *Environment and Planning D: Society and Space*, 9: 165–182.

Weber, E. (1977) *Peasants into Frenchmen: the Modernization of Rural France, 1870–1914*, London: Chatto and Windus.

Weigert, H. (ed.) (1949) *New Compass of the World: A Symposium in Political Geography*. London: Harrap.

Western, J. (1996) *Outcast Cape Town*, Berkeley, CA. and London: University of California Press.

Whatmore, S. (2002) *Hybrid Geogrpahies: Nature, Culture, Spaces*, London: Sage.

Whelan, Y. (2002) The construction and destruction of a colonial landscape: monuments to British monarchs in Dublin before and after independence, *Journal of Historical Geography*, 28: 508–33.

Whitehead, M. (2005) Between the marvellous and the mundane: everyday life in socialist city and the politics of the environment, *Environment and Planning D: Society and Space,* 23: 273–294.

Whitehead, M. (2006) *Spaces of Sustainability: Geographical Perspectives on the Sustainable Society.* Abingdon: Routledge.

Whitehead, M. (2008) Cold monsters and ecological leviathan: on the relationships between states and the environment, *Geography Compass,* 2: 414–432.

Whitehead, M. (2009) *State, Science and the Skies: Environmental Governmentality and the British Atmosphere,* Oxford: Blackwell.

Whitehead, M., Jones, R., and Jones, M. (2007) *The Nature of the State: Excavating the Political Ecologies of the Modern State,* Oxford: Oxford University Press.

Williams, C. H. and Smith, A. D. (1983) The national construction of social space, *Progress in Human Geography,* 7: 502–518.

Williams, G., Thampi, B. V., Narayana, D., Nandigama, S. and Bhattacharyya, D. (2011) Performing participatory citizenship: power and politics in Kerala's Kudumbashree programme, *Journal of Development Studies,* 47: 1261–1280.

Willis, C. (1995) *Form Follows Finance: Skyscrapers and Skylines in New York and Chicago,* New York: Princeton Architectural Press.

Wittmann, E. (2004). To what extent were ideas and beliefs about eugenics held in Nazi Germany shared in Britain and the United States prior to the Second World War? *Pour une histoire de la Société Internationale d'Histoire de la Médecine:* 16–19.

Wolsink, M. (2006) Invalid theory impedes our understanding: a critique on the persistence of the language of NIMBY, *Transactions of the Institute of British Geographers,* 31: 85–91.

Wong, T. and Wainwright, J. (2009) Offshoring dissent: spaces of resistance at the 2006 IMF/World Bank meetings, *Critical Asian Studies* 41: 403–428.

Wood, A. (1998) Making sense of entrepreneurialism, *Scottish Geographical Magazine,* 114: 120–123.

Woodward, R. (2004) *Military Geographies,* Oxford: Blackwell.

Woods, M. (1999) Performing power: local politics and the Taunton pageant of 1928, *Journal of Historical Geography,* 25: 57–74.

Woods, M. (2003) Deconstructing rural protest: the emergence of a new social movement, *Journal of Rural Studies,* 19: 309–325.

Woods, M. (2005) *Contesting Rurality: Politics in the British Countryside,* Aldershot: Ashgate.

Woods, M. (2007) Engaging the global countryside: globalization, hybridity and the reconstitution of rural place, *Progress in Human Geography,* 31: 485–507.

Woods, N. (2010) Global governance after the financial crisis: a new multilateralism or the last gasp of the great powers? *Global Policy,* 1: 51–63.

Woods, M. (2011) The local politics of the global countryside: boosterism, aspirational ruralism and the contested reconstitution of Queenstown, New Zealand, *Geojournal,* 76: 365–381.

Woods, M. and Gardner, G. (2011) Applied policy research and critical human geography: some reflections on swimming in murky waters, *Dialogues in Human Geography,* 1: 198–214.

Woods, M., Edwards, B., Anderson, J. and Gardner, G. (2006) Leadership in place: elites, institutions and agency in British rural community governance, in L. Cheshire, V. Higgins and G. Lawrence (eds.), *Rural Governance: International Perspectives,* Abingdon: Routledge, pp. 211–226.

Woods, M., Anderson, J., Guilbert, S. and Watkin, S. (2012) 'The country(side) is angry': emotion and explanation in protest mobilization, *Social and Cultural Geography,* 13: 567–585.

Woods, M., Anderson, J., Guilbert, S. and Watkin, S. (2013) Rhizomic radicalism and arborescent advocacy: a Deleuzo-Guattarian reading of rural protest, *Environment and Planning D: Society and Space,* 31: 434–450.

Woon, C. Y. (2011). Undoing violence, unbounding precarity: beyond the frames of terror in the Philippines, *Geoforum,* 42: 285–296.

World Commission on Environment and Development (1987) *Our Common Future,* Oxford: Oxford University Press.

Yanow, D. (1995) Built space as story: the policy stories that buildings tell, *Policy Studies Journal,* 23(3): 507–422.

Yoshinaka, A. and Murphy, C. (2009) Partisan gerrymandering and population instability:

completing the redistricting puzzle, *Political Geography*, 28: 451–462.

Young, L. (1998) *Japan's Total Empire: Manchuria and the Culture of War-time Imperialism*, Berkeley: University of California Press.

Yusoff, K. (2012) Aesthetics of loss: biodiversity, banal violence and biotic subjects, *Transactions of the Institute of British Geographers*, 37: 578–592.

Yuval-Davis, N. (1997) *Gender and Nation*, London: Sage.

Zajko, M. and Béland, D. (2008) Space and protest policing at international summits, *Environment and Planning D: Society and Space*, 26: 719–735.

Zografos, C. and Martinez-Alier, J. (2009) The politics of landscape value: a case study of wind farm conflict in rural Catalonia, *Environment and Planning A*, 41: 1726–1744.

Zukin, S. (1991) *Landscapes of Power: from Detroit to Disneyworld*, Berkeley: University of California Press.

Index

Page numbers in italics refer to figures and plates; those in bold to tables and those followed by 'g' refer to glossary entries.